Advances in
MARINE BIOLOGY

VOLUME 31

Advances in
MARINE BIOLOGY

Edited by

J. H. S. BLAXTER

Dunstaffnage Marine Research Laboratory, Oban, Scotland

and

A. J. SOUTHWARD

Marine Biological Association, The Laboratory, Citadel Hill, Plymouth,
England

ACADEMIC PRESS

San Diego London New York Boston
Sydney Tokyo Toronto

Academic Press, Inc.
525 B Street, Suite 1900, San Diego, California 92101-4495, USA

Academic Press Limited
24–28 Oval Road, London NW1 7DX, UK

ISBN 0-12-026131-6

A catalogue record for this book is available from the British Library

Typeset by Keyset Composition, Colchester, Essex
Printed in Great Britain by Hartnolls, Bodmin, Cornwall
97 98 99 00 01 02 EB 9 8 7 6 5 4 3 2 1

CONTRIBUTORS TO VOLUME 31

B. E. Brown, *Department of Marine Sciences and Coastal Management, University of Newcastle-upon-Tyne, Newcastle-upon-Tyne, NE1 7RU, UK*

J. F. Dower, *Department of Biology, Queen's University, Kingston, Ontario, Canada, K7L 3N6*

D. A. Egloff, *Department of Biology, Oberlin College, Oberlin, OH 44074-1082, USA*

P. W. Fofonoff, *Smithsonian Environmental Research Center, PO Box 28, Edgewater, Maryland 21037-0028, USA*

J. P. A. Gardner, *Island Bay Marine Laboratory, School of Biological Sciences, Victoria University of Wellington, PO Box 600, Wellington, New Zealand*

W. C. Leggett, *Department of Biology, Queen's University, Kingston, Ontario, Canada, K7L 3N6*

T. J. Miller, *Chesapeake Biological Laboratory, Center for Environmental and Estuarine Studies, University of Maryland, Solomons, Maryland 20688-0038, USA*

T. Onbé, *Faculty of Applied Biological Science, Hiroshima University, Higashi-Hiroshima 739, Japan*

K. Richardson, *Danish Institute for Fisheries Research, Charlottenlund Castle, DK-2920, Charlottenlund, Denmark*

CONTENTS

Hybridization in the Sea

J. P. A. Gardner

Reproductive Biology of Marine Cladocerans

D. A. Egloff, P. W. Fofonoff and T. Onbé

The Role of Microscale Turbulence in the Feeding Ecology of Larval Fish

J. F. Dower, T. J. Miller and W. C. Leggett

Adaptations of Reef Corals to Physical Environmental Stress

B. E. Brown

Harmful or Exceptional Phytoplankton Blooms in the Marine Ecosystem

K. Richardson

Hybridization in the Sea

J. P. A. Gardner

Island Bay Marine Laboratory, School of Biological Sciences, Victoria University of Wellington, PO Box 600, Wellington, New Zealand

ADVANCES IN MARINE BIOLOGY VOL. 31
ISBN 0-12-026131-6

1. INTRODUCTION

1.1. Background and Definitions

Hybridization and hybrid zones have long been of interest to evolutionary biologists because they offer excellent opportunities to study speciation and processes which contribute to reproductive isolation by either pre- or post-zygotic mechanisms. Such processes include assortative mating, hybrid unfitness and gamete incompatibility, all of which are believed to contribute to speciation.

Perhaps surprisingly, the term "hybrid zone" has proved difficult to define because definitions often include or imply an underlying mechanism concerning the origin or maintenance of the zone which might not be appropriate in all cases (for example, inferences about the width of the zone or the fitness of the individuals within the zone). The most workable definition is that proposed by Harrison (1990, p. 72): hybrid zones are "interactions between genetically distinct groups of individuals resulting in at least some offspring of mixed ancestry. Pure populations of the two genetically distinct groups are found outside the zone of interaction." This is the definition of a hybrid zone that is employed in this review. Similarly, isolated instances of hybridization may be defined as the production of one or more individuals resulting from the interbreeding of two genetically distinct parental individuals. As will be seen, the frequency of isolated cases of hybridization in the sea is considerably higher than that for the existence of hybrid zones or swarms (the latter may be defined as a collection of individuals of mixed ancestry resulting from high levels of interbreeding of genetically distinct parental types accompanied by introgression).

In the last decade, hybrid zones have received much attention (e.g. reviews by Endler, 1977; Moore, 1977; Barton and Hewitt, 1985, 1989; Harrison, 1990; Arnold, 1992; Arnold, 1993). The various types of hybrid zones (clines, tension zones, mosaics, reticulates, etc.) and the factors responsible for the establishment and maintenance of such zones have been well documented and the reader is referred to these excellent reviews. It is not the intention of this paper to re-examine these ideas and concepts, but to provide new information about natural hybridization in the sea, a subject which has received very little attention.

It is both interesting and surprising that in the above-cited reviews only one author mentions natural hybridization in the sea, where it is stated that very few well-documented cases exist (Harrison, 1993, p. 111). Whilst it is true that more is known about natural terrestrial hybridization, and that a greater number of better-documented examples exist from the terrestrial environment, it is also true that natural hybridization in the sea is not an uncommon phenomenon. Clearly, data from studies which have looked at hybridization in the sea, which has an estimated volume of $137 \times 10^6 \, \text{km}^3$ and is the largest ecosystem on earth (Gage and Tyler, 1991), which covers over 71% of the planet (Briggs, 1991), and which contributes many unique phyla (May, 1988; Winston, 1992), must aid substantially our understanding of natural hybridization, its causes and effects.

Hybridization is often viewed as one of the most important steps in speciation because if interbreeding taxa do not fuse then they are thought to evolve mechanisms such as differential mate choice and/or gamete incompatibility to prevent hybridization and thereby promote speciation (e.g. Mayr, 1942). In the context of the marine environment, relatively little is known about speciation (Knox, 1963; Palumbi, 1992, 1994). The marine environment is continuous (Battaglia, 1957; Sara, 1985) and less ephemeral than other environments (Hubbs and Kuronuma, 1941). Furthermore, the kinds of relatively pronounced ecotones found in terrestrial systems are much less frequent in the marine environment. In conjunction with the often widespread dispersal potential of many marine species, genetic differences between spatially segregated populations are often slight (e.g. Skibinski et al., 1983). It has, however, been noted that ecological heterogeneity (e.g. the intertidal zone, or brackish water regions), combined with (micro)geographic isolation, should promote adaptive differentiation and ultimately evolutionary divergence in the marine environment (Battaglia, 1957). This in turn can promote hybridization because, according to Harrison's (1990) definition, genetic differentiation is a prerequisite for hybridization and the formation of hybrid zones.

Examination of the disparate and sometimes obscure data pertaining to natural hybridization in the sea indicates that many patterns exist which must be explained for a better understanding of the process of hybridization itself, as well as to provide greater insight into speciation in the sea. The major intentions of this review are: (1) to point out that hybridization in the sea is not as rare as previously thought; (2) to draw attention to hybridization bias amongst certain groups of marine organisms; (3) to illustrate biogeographic and environmental patterns in the location of hybridization; (4) to compare marine hybridization with hybridization in other environments; and (5) to discuss marine hybridization in the context of fisheries science and conservation.

1.2. Scope of the Review

A relatively broad approach is adopted in this review to permit inclusion of as many examples of marine hybridization as possible. Examples are drawn from environmental conditions ranging from fully marine to estuarine. The inclusion of hybridization in estuaries is justified on the grounds that: (1) estuaries are ecologically important to many truly marine species for moulting, reproduction, feeding, as larval nursery grounds, etc.; and (2) these areas play an important role in promoting marine speciation. Instances of hybridization involving both anadromous (e.g. salmonids) and catadromous (e.g. eels) species are excluded because such animals either breed in fresh water or spend most of their lives in fresh water, and are therefore not subject to the same evolutionary pressures as truly marine organisms. The scope of the review ranges from isolated examples of natural hybridization involving the production of a single (usually infertile) individual, through cases of marine hybrid swarms with associated high levels of introgression, to cases of speciation involving chromosome doubling. Although I have included as many examples as are known to me, this review is not exhaustive. Because of the subject matter (isolated instances of possible hybridization do not rank very highly on the scale of "publishability"), much of this information is obscure, and the reader may be familiar with examples which are not included here. My intention has been to draw from as many examples as possible to gain an overview of marine hybridization, its patterns, biases and effects.

   In a limited number of instances it is difficult to determine if natural hybridization does indeed occur because of conflicting reports in the literature. For example, the cockles *Cerastoderma edule* and *C. glaucum* occur sympatrically in north-west Europe (Brock, 1979; Gosling, 1980), and it has been suggested that individuals which are morphologically inter-mediate between these two good species are natural hybrids (Kingston, 1973). Elsewhere, evidence suggests that hybridization is not the explana-tion for this morphological intermediacy (Brock, 1978; Gosling, 1980). In a case such as this, despite some evidence of hybridization, the example is not included in the review because the majority of evidence indicates that natural hybridization does not occur between these two species.

A more recent example of differences in opinion concerning the possible occurrence of hybridization involves the seerfishes *Scomberomorus commerson* and *S. guttatus*. According to Srinivasa Rao and Lakshmi (1993) these two species interbreed off the south-east coast of India to produce a hybrid which was previously recognized as *S. lineolatus*. Collette (1994) objects strongly to this interpretation, but because the data presented by both sides are osteological as opposed to biochemical (e.g. allozyme variation) or molecular (DNA sequence data) it is difficult to judge the validity of either argument. This example is included in the present review for the sake of completeness without lending weight to either opinion.

Finally, it is apparent that many of the records of marine hybridization involve isolated instances of hybridization without any evidence of the existence of a marine hybrid zone. These records are included in the review along with the better-documented cases of hybrid zones and hybrid swarms, because they provide a valuable supplement to these data.

2. RECOGNITION OF HYBRIDIZATION

2.1. Theoretical Considerations

Mayr (1979, p. 113) recognized five levels of hybridization (the first four of which grade into one another) regardless of the environment in which the hybridization occurs:

1. Infrequent interbreeding of sympatric species resulting in the production of hybrid offspring that are unable, for ecological or behavioural reasons, to backcross to one or both of the parental species.
2. Frequent or occasional interbreeding of sympatric species resulting in the production of at least partially fertile offspring, some of which backcross to one or both parental species.
3. Partial interbreeding between two formerly geographically isolated populations which failed to acquire complete reproductive isolation resulting in the formation of a secondary zone of contact.
4. Extensive interbreeding, resulting from the complete breakdown of reproductive isolation between sympatric species which produces a hybrid swarm in which the complete range of parental viabilities may be observed.
5. Interspecific hybridization in which a new species is produced as a result of chromosome doubling (allopolyploidy). This case is virtually restricted to plants (this might reflect a bias in our knowledge rather than a genuine phenomenon).

A review of the literature indicates that examples of all five of the cases proposed by Mayr (1979) have been reported from the marine environment. It is impossible, however, to estimate the frequency with which each category occurs because many reports of marine hybridization provide too few details for a judgement to be made. The important point here is that the full range of hybridization, from instances which result in the production of a single infertile offspring, to the production of new species, have been reported from the sea. A summary of cases detailing natural hybridization in the sea is provided in Table 1 (page 56) along with information concerning the geographic location and biogeographic region (according to Briggs, 1974) of the hybridization event, as well as details pertaining to the causes and outcomes (if any) of the event. These data are discussed in greater detail in subsequent sections.

2.2. Practical Considerations

Most studies of hybridization in the sea have looked at morphological variation: this is often the primary, and sometimes the only criterion used to identify hybrids (see Table 1). Due to the blending nature of hybridization, hybrids are usually morphologically intermediate between the two parental taxa in most characters examined (Mayr, 1979; Campton, 1987, 1990). This can be expressed in two ways: the hybrid can be morphologically intermediate between the parental types for each character (e.g. possesses 46 vertebrae when the parental types possess 48 and 44), or can possess one suite of characters which are typical of one parent and at the same time possess a suite of different characters which are typical of the other parent. Both types of intermediacy are reported in cases of marine hybridization.

Protein electrophoresis has proved to be a particularly valuable tool by which hybrids and individuals of mixed ancestry can be identified. This technique permits an objective evaluation of the hybrid status of individuals when the two hybridizing taxa are fixed for electrophoretic differences at one or more allozyme loci. However, it is sometimes the case that no fixed allelic differences exist between hybridizing taxa (all differences are a matter of allele frequency differences), which makes it that much harder to quantify the hybridization because all genotype combinations are theoretically possible within "pure" populations of the taxa. Arguably the best approach to this problem is to determine the conditional probability of any given genotype and then to determine whether or not this genotype occurs at the expected frequency (Campton and Utter, 1985; Gardner, 1996; Miller and Benzie, submitted).

Other techniques which have been utilized to identify natural hybrids or to investigate the dynamics of natural hybridization in the sea include investigation of haemoglobin patterns (Sick *et al.*, 1963), sperm agglutination tests (Vasseur, 1952), cytogenetics (Menzel and Menzel, 1965; Gray *et al.*, 1991 and references therein), mitochondrial DNA variation (Skibinski, 1985; Edwards and Skibinski, 1987; Gardner and Skibinski, 1991b), immunology (Owen *et al.*, 1971) and behavioural differences (Feddern, 1968; Dahlberg, 1970; Ayling, 1980).

Hybrid indices are often employed in the identification and characterization of hybridization (e.g. Gosline, 1948; Feddern, 1968). The benefit of such indices is that variation in a number of different traits (e.g. morphometric, behavioural, electrophoretic, coloration) can be summed to give an overall score for each individual, and thereby permit an evaluation of the total amount of variability between individuals. Individuals with the highest and lowest index scores are parental types (this is confirmed by the application of this technique to pure individuals of the taxa under investigation), and individuals with intermediate scores are hybrids.

The application of hybrid indices has been largely superseded by the application of multivariate statistical techniques such as principal components analysis and canonical discriminant analysis to the description of morphometric and genetic (electrophoretic) variation (e.g. Boitard *et al.*, 1980, 1982; Skibinski, 1983; McDonald *et al.*, 1991; McClure and McEachran, 1992; Srinivasa Rao and Lakshmi, 1993; Gardner, 1996). When the first two components of the analysis explain a large proportion of the variability in the data set, then plots of principal component 1 (PC1) versus PC2, or canonical variate 1 (CV1) versus CV2, can identify parental types as relatively discrete clusters of points, the distance between which is maximized by the analytical technique. Falling somewhere between these two extremes would be the hybrids and other individuals of mixed ancestry. Sometimes this results in a continuum from one parental type to another through many intermediate stages (e.g. hybrid *Mytilus edulis* and *M. galloprovincialis* mussel populations from south-west England (Gardner, 1996)) or can be represented by three relatively discrete clusters in which the two parental-type clusters overlap only slightly with the hybrid cluster (e.g. seerfish populations in the Bay of Bengal, India (Srinivasa Rao and Lakshmi, 1993; but see Collette, 1994)).

3. THE IMPORTANCE OF TAXONOMY

3.1. Systematics, Taxonomy and Hybridization

The systematic relationships (the study of the relatedness of organisms reflecting their respective evolutionary histories) and taxonomy (the names given to organisms reflecting their systematic relationships) of species are constantly being revised in light of new data. Clearly, taxonomy plays an important role in the defining of hybrid zones. This point was emphasized by Mayr (1979, p. 118) for hybrid swarms: "A thorough knowledge of the taxonomy of the respective groups is a prerequisite for a sound analysis of such situations. This is presumably the reason why so few such cases have so far been described . . .". The recent discovery of numerous marine sibling species has emphasized still further the importance of correct taxonomy based upon sound systematics. In their review of coral reef niche partitioning and the taxonomy of reef-dwelling organisms, Knowlton and Jackson (1994) point out that for many animal groups the number of sibling species is far higher than previously suspected. Despite only small-scale differences in behaviour, morphology, colour patterns and genetics, concordant patterns of variation from distinct geographic locations support the erection of specific status for many of these organisms. Similar findings are reported by Knowlton (1993) who extended the discussion of sibling species from the reef ecosystem to the entire marine environment. One of the main conclusions from these works is the necessity for the rigorous application of taxonomy to all marine systems. The study of marine hybrid zones is hindered by poor taxonomy: it is important that marine hybrid zones receive due attention because this will substantially improve our understanding of evolutionary processes and systematics in the marine environment.

Laboratory-based hybridization studies can provide a valuable means by which conspecifics can be recognized when it is considered that their present systematic status is incorrect. For example, Warwick *et al.* (1990) used laboratory interbreeding experiments to test the specific status of *Littorina rudis* and *L. saxatilis* (Prosobranchia: Mollusca) collected from Venice lagoon and south-west England. Because these two previously recognized species were able to produce viable offspring from reciprocal crosses between them, the authors concluded that *L. rudis* is a junior synonym of *L. saxatilis*.

In some instances, a previously recognized species has been identified as the product of natural hybridization. For example, off the Pacific coast of the USA and Canada, the flounder *Inopsetta ischyra* has been recognized as a hybrid between the starry flounder *Platichthys stellatus* and the English sole *Parophrys vetulus* (Schultz and Smith, 1936; Herald, 1941; Hart, 1973), and in the Bay of Bengal, India, the seerfish or Spanish mackerel *Scomberomorus lineolatus* has been identified as the product of the *S. commerson* × *S. guttatus* cross (Srinivasa Rao and Lakshmi, 1993; but see Collette, 1994). In these two cases it is probable that the recognition of both "species" as hybrids was only possible because of the large catch sizes of these commercially important species. An example of a non-commercial fish "species" now being recognized as a hybrid is provided by Randall (1956). The surgeon fish *Acanthurus rackliffei* from Hull Island, an atoll in the Phoenix Islands, Oceania, is now thought to be a hybrid between the closely related *Acanthurus achilles* and *A. glaucopareius*. In cases such as these, the recognition of interspecific hybrids which had been afforded specific status has assisted considerably in the understanding of evolutionary relationships which had been clouded by incorrect taxonomy between these closely related organisms.

3.2. Species Concepts and Hybridization

Harrison's (1990) definition of a hybrid zone, as provided earlier, makes no assumptions about the origin or structure of the hybrid zone. This definition also makes no assumptions about the specific status of the taxa involved in the interaction, other than to stipulate that some degree of genetic differentiation must exist between them. Certain species concepts are not so liberal, although many of the newer species concepts promote the idea that species are evolutionary dynamic entities and not static elements. In the following section, hybridization is assessed briefly in the light of different species concepts. This topic might be considered as worthy of review in its own right. Excellent treatments are available in Mayr (1970), Grant (1971), Carson (1975), Wiley (1981), Cracraft (1983, 1989), Templeton (1989), Avise and Ball (1990), Wallace and Willis (1994) and Mallet (1995), amongst others.

The Biological Species Concept (BSC), as championed by proponents such as Mayr (1970), is founded on the idea that taxa are reproductively isolated from each other, primarily by intrinsic (genetic) rather than extrinsic (geographical separation) factors. This is a relatively strict interpretation of the species since it views a species as a closed system. Based upon the BSC, no two interbreeding taxa which produce fertile offspring can be viewed as discrete species, because they are not, according to the definition, reproductively isolated (there are 32 examples of hybrid crosses which result in fertile offspring presented in Table 1). Yet the specific status of most of the taxa listed in Table 1 would be likely to satisfy the majority of biologists. Other major criticisms of the BSC are that: (1) it lacks a phylogenetic perspective sufficient to aid in the understanding of the origins and evolutionary histories of the species; (2) the concept does not apply well to plants; and (3) it is not heuristic (Sokal and Crovello, 1970; Grant, 1971; Price, 1996).

The Evolutionary Species Concept (ESC) views an evolutionary species as "a lineage of ancestor-descendent populations which maintains its identity from other such lineages and which has its own evolutionary and historical fate" (Wiley, 1981, p. 25). As far as the ESC is concerned, hybridization is unimportant: the important consideration is whether or not the two species "maintain their separate identities, tendencies and fates" (Wiley, 1981, p. 27). The ESC has been criticized on the grounds that the very traits which are important in defining species are themselves undefined (Templeton, 1989). This means that phenotypic variation within an organismal group can be interpreted in a variety of different ways, i.e. it is subjective. Gardner (1992) noted that where the blue mussels *M. galloprovincialis* and *M. edulis* hybridize extensively in western Europe, it is often impossible to distinguish between the two taxa because there is so much character overlap. The ESC thus fails to identify parental, hybrid and backcross mussels. The individual identities, tendencies and fates of these lineages is therefore called into question. However, this shortcoming is unlikely to apply in the majority of hybridization cases of Table 1.

The Phylogenetic Species Concept (PSC) defines a species as "the smallest diagnosable cluster of individual organisms within which there is a parental pattern of ancestry and descent" (Cracraft, 1983, p. 170). Cracraft (1983, p. 172) notes that ". . . even if two sister-taxa hybridize, both can still be considered to be species if each is diagnosable as a discrete taxon . . . The critical point is that both species have had a distinct phylogenetic and biogeographic history prior to hybridization . . .". McKitrick and Zink (1988) suggest that hybridization between taxa simply indicates that reproductive isolation is not yet complete. The PSC places very little emphasis upon reproductive isolation but considerable emphasis on diagnostic characters. Avise and Ball (1990, p. 48) have criticized the PSC on the grounds that the number of species recognized depends on the resolving power of the tools available: "If each individual organism is genetically unique at a high level of resolution, then the grouping of individuals requires that we ignore distinctions that occur below some arbitrary threshold. The evolutionary significance of any such threshold must surely be questionable." With regard to hybridization, the PSC can only be judged to be useful in cases where truly diagnostic characters exist between species. This is not the case, for example, with smooth shelled blue mussels which exhibit frequency differences, but not absolute differences, for all traits so far investigated (Gardner, 1992, 1994).

The Recognition Species Concept (RSC) emphasizes mate-specific recognition systems of sexually biparental, eukaryotic reproduction. The species is defined as "that most inclusive population of individual biparental organisms which share a common fertilization system" (Paterson, 1985, p. 25). Thus, individuals of a species share common courtship behaviour, timing of reproduction, coloration, gamete compatibility, etc., in fact any and all components which contribute to reproductive success.

The major limitations of the RSC are its sole application to bisexual eukaryotes but to no other biotic groups, and the poor fashion in which it deals with the syngameon, the most inclusive unit of interbreeding in a hybridizing species group (Templeton, 1989).

The Cohesion Species Concept (CSC) defines a species as "the most inclusive population of individuals having the potential for phenotypic cohesion through intrinsic cohesion mechanisms" (Templeton, 1989, p. 12). Cohesion mechanisms include processes such as gene flow, reproductive isolation, developmental constraints, and stabilizing selection. The CSC builds upon the foundation of other species concepts by emphasizing the mechanisms that produce and maintain species cohesion. The CSC is a wide-ranging species concept which is equally well applied to asexual species as it is to sexual species. Perhaps the greatest limitation with the CSC is that the cohesive mechanisms themselves are often poorly understood and require much more attention. Because the CSC ranges in its applicability from asexual taxa at one extreme to the syngameon at the other, hybridization can be easily handled by the concept.

The general historical trend in species concepts appears to have involved a move away from a rigid view of species as closed systems (for example, the BSC), towards a more flexible view of species as evolutionary dynamic entities (in particular, the CSC). With this change, though, has come increasing recognition that the concept of hybrids and hybrid zones is less clear-cut than had previously been thought. A universally applicable species concept must be able to include and explain hybridization, just as it must be able to cope with asexual reproduction. On this score, several of the species concepts fall down. Those that come closest appear to be the Evolutionary Species Concept and, in particular, the Cohesion Species Concept. This review of marine hybridization is not based solely on the species definition of one species concept, but does tend to favour the CSC definition.

4. FACTORS PREVENTING OR PROMOTING MARINE HYBRIDIZATION

4.1. Factors Preventing Hybridization

Harrison's (1990) definition of a hybrid zone states that some degree of genetic differentiation must exist between the two interbreeding taxa before hybridization can be considered to have taken place. Yet many marine populations are characterized by high levels of genetic similarity over very large distances. This genetic homogeneity is usually attributed to high levels of gene flow between populations which results from the considerable dispersal potential of the juvenile (pre-metamorphosis and/or pre-settlement) phase of the organisms concerned. Such high levels of gene flow, and the associated relative absence of genetic heterogeneity, will tend to reduce the frequency with which hybridization can occur. Furthermore, many marine populations are very large (millions of individuals), which will slow genetic divergence between these populations (Palumbi, 1994). These factors will tend to reduce the possibility of hybridization in the sea.

4.2. Factors Promoting Hybridization

Physical factors which promote hybridization in the sea usually involve environmental heterogeneity. Battaglia (1957) points out that ecological heterogeneity in the sea (preferably associated with isolation since habitat segregation is an effective reproductive isolating mechanism (Mayr, 1979)) is the most effective source of adaptive differentiation, and subsequent genetic divergence. Freshwater input, for example, interrupts the continuity of the sea and introduces isolating factors and selective forces which may be sufficiently intense to form locally adapted genetic variants (Battaglia, 1957). The extent of this genetic heterogeneity can range from local races, through sub- and semi-species, to full species, depending upon the strength of the effect of the freshwater input and the time-scale involved. Ultimately, differences in salinity between geographically neighbouring regions can be sufficient to promote reproductive isolation. For example, amphipod species of the genus *Gammarus* which occur on either side of the Baltic Sea and North Sea salinity discontinuity are now reproductively isolated, not because of the distance between them, but because of the salinity tolerance differences that exist between them (Kolding, 1985). Thus, genetic heterogeneity can result from environmental heterogeneity, which in turn can lead to hybridization at the boundaries of the environmental discontinuity before reproductive isolation is established (if it ever is established).

The intertidal zone is a region which is typified for its occupants by high levels of physiological stress associated with extremes of environmental variability, and as such is a region which can exert a severe selective pressure (Battaglia, 1957). In many places, the vertical distance between the top and the bottom of the intertidal region is no more than 2 m, but this short distance encompasses a wide range of habitat types. The very narrowness of the intertidal zone will tend to bring organisms into contact. For example, isopods of the genus *Jaera* are limited to the intertidal region where it is possible to find up to four species co-occurring, each separated from the others on a micro-geographic scale by ecological preferences (Solignac, 1976). However, because these species are mobile, encounters between them inevitably occur in such a small area, and such encounters can lead to hybridization, for example, in regions of north-west Europe (see Table 1, p. 70).

The majority of cases of hybridization of plants and invertebrates listed in Table 1 (p. 70) are reported from intertidal populations (very few fish species are strictly intertidal, and therefore examples of marine hybridization involving fishes are entirely restricted to regions other than the intertidal zone). Of the 57 non-vertebrate examples given, 31 occur in the intertidal zone and a further 17 occur in the shallow subtidal region ($<$ 30 m below chart datum). To some extent these figures are a reflection of our familiarity with intertidal and shallow subtidal biota, but at the same time it is significant that the intertidal and shallow subtidal zones constitute only a minute percentage of the area of the marine ecosystem, and are therefore disproportionately highly represented. These data indicate a real propensity for hybridization amongst intertidal and shallow subtidal organisms, a phenomenon which is apparently related to the high degree of environmental variability of this region. Whether or not environmental heterogeneity has a similar effect upon fishes and other vertebrates is difficult to ascertain. Hybridization amongst both pelagic and benthic species is common, and since many fishes and marine mammals have wide ranges in their depth distributions it is difficult to draw any definitive conclusions concerning the influence of environmental heterogeneity upon the frequency of hybridization in these organisms.

Biological as well as physical factors will help to promote hybridization in the sea. A large number of sibling species (Knowlton, 1993) and species complexes are observed in the sea, with the result that individuals of these species have a high degree of genetic similarity to other such closely related species. This will help to promote hybridization (or at least will not assist the prevention of hybridization) because high levels of genetic similarity aid rather than prevent successful interbreeding.

In conclusion, the available data indicate that environmental heterogeneity plays an important role in promoting hybridization amongst plant and invertebrate species in the sea. Heterogeneous environments apparently promote and accelerate evolutionary processes which occur more slowly in other more uniform environments (Battaglia, 1957). No such conclusion can be reached for fish and mammalian species because too few data are currently available, and possibly because such a comparison cannot reasonably be made.

5. HOW FREQUENT IS HYBRIDIZATION IN THE SEA?

It is apparent that for two taxa to hybridize there must be, as a minimum, sufficient genetic compatibility between them to permit successful interbreeding, that is, to overcome the barriers of pre-zygotic reproductive isolation and post-zygotic genetic incompatibility.

High levels of genetic similarity characterize incipient species (which are undergoing speciation) and sibling species (which have recently diverged). Thus, regions with a high number of incipient or sibling species (such as coral reefs, represent "hot spots" for speciation, possibly resulting from increased niche partitioning amongst the coexisting species (Knowlton and Jackson, 1994)) would appear to be good potential locations in which to find marine hybrid zones. Knowlton (1993) has argued that sibling species are ubiquitous throughout the marine realm and, as a consequence, the possibility of observing natural hybridization in the sea should be good. However, it is often stated or indicated that natural hybridization in the sea is a rare phenomenon (Hubbs, 1955; Schwartz, 1972; Randall *et al.*, 1977; Mayr, 1979; Campton, 1987; Bert and Harrison, 1988). Campton (1990), in his review of the use of genetic markers in the analysis of hybridization in fisheries science, lists only two teleost and six invertebrate examples of marine hybridization, while Harrison (1990), in an excellent and extensive review of hybrid zones, mentions only one marine example.

One of the main reasons for considering hybridization in the sea to be such a relatively infrequent event probably stems from the fact that many examples involve one or only a few (< 10) hybrid individuals which have usually been determined by morphological intermediacy. Such descriptions often appear in the older literature or in non-mainstream publications, and for this reason can be easily overlooked. Despite the absence of data to confirm the genetic basis of the hybrid status of such individuals, it is clear that in most cases a long familiarity with a given species or group of organisms had been acquired by the authors before the hybrids were recognized (e.g. Owen *et al.*, 1971; Randall *et al.*, 1977).

Not surprisingly, many documented cases of natural marine hybridization involve commercially important species. Examples include hybrids between taxa of gadoids (Dymond, 1939), many flatfish (Norman, 1934 and references therein; Hubbs and Kuronuma, 1941; Fujio, 1977), clupeids (Dahlberg, 1969a and b, 1970; Turner, 1969), hard shell clams (Menzel, 1985; Hesselman *et al.*, 1988; Dillon, 1992; Bert *et al.*, 1993), blue mussels (Koehn *et al.*, 1984; Varvio *et al.*, 1988; Coustau *et al.*, 1991a; McDonald *et al.*, 1991; Freeman *et al.*, 1992; Sarver and Foltz, 1993), abalone (Owen *et al.*, 1971; Talmadge, 1977; Fujino *et al.*, 1980; Sasaki *et al.*, 1980; Brown, 1995), snow crabs (Karinen and Hoopes, 1971; Johnson, 1976; Grant *et al.*, 1977) and cetaceans (Cocks, 1887; Spilliaert *et al.*, 1991). Hybrids of lesser economic importance involve the capture and breeding of fishes for collectors, for example, the butterfly fishes (Randall *et al.*, 1977) and the angel-fishes (Pyle and Randall, 1994). It is more than likely that many of these examples of hybridization would not have been described if it was not for the fact that at least one of the taxa involved is commercially important. Consequently, very large sample sizes are available for examination, permitting the identification of hybrids which often occur naturally at very low frequencies.

The data presented in Table 1 (page 70) indicate that hybridization in the sea is not a rare event. There are 108 documented cases of hybridization listed, of which at least 34 involve hybrid zones or swarms, spanning the algae, higher plants, invertebrates and vertebrates. Thus, all major eukaryotic groups are represented (see later section for details). In their extensive review article, Barton and Hewitt (1989) examined over 170 non-marine hybrid zones, indicating that such hybrid zones outnumber marine hybrid zones by about 5 : 1. Approximately 1.4 million species have been identified from the earth, of which approximately 250 000 are marine (Winston, 1992 and references therein), giving a ratio of approximately 4.6 : 1 for non-marine to marine species. This is so close to the ratio of 5 : 1 quoted above, that it seems reasonable to conclude that hybridization (at least when measured in terms of the number of hybrid zones and hybrid swarms) in the sea occurs at the same frequency as hybridization in other environments. It must, however, be recognized that estimating the number of species in any environment is fraught with difficulty and unknown levels of inaccuracy derived from large-scale extrapolations of data collected from small-scale areas (for example see Briggs, 1991 and associated correspondence: Chaloner, 1992; Grassle, 1992; Briggs, 1992a and b).

In conclusion, whilst it is difficult to determine accurately the number of species in any given environment, the best data presently available indicate that rates of hybridization in the sea are surprisingly similar to those in, for example, the terrestrial environment. This fact has not been recognized previously, and it should not be overlooked when patterns and processes of hybridization are being investigated.

6. HYBRIDIZATION AND THE TIMING OF REPRODUCTION

Many authors have studied species-specific reproductive cycles to determine whether or not natural hybridization can possibly occur based upon overlap in the timing of reproductive activity. Such work usually involves an analysis of gonad index or histological examination of gonad maturity. This can be supplemented by plankton analysis which provides an indication of the similarity of spawning time of the two potentially hybridizing taxa based upon the co-occurrence of larval forms of the two taxa. This approach has been applied to populations of fucoid algae

(Burrows and Lodge, 1951), echinoderms (Chia, 1966; Hagström and Lonning, 1967; Schopf and Murphy, 1973; Strathmann, 1981), molluscs (Owen *et al.*, 1971; Seed, 1971; Dalton and Menzel, 1983; Dillon and Manzi, 1989; Gardner and Skibinski, 1990a), crustaceans (Jones and Naylor, 1971; Bert, 1985 cited in Bert and Harrison, 1988), corals (Miller and Babcock), flatfish (Hubbs and Kuronuma, 1941) and seerfish (Devaraj, 1986; Srinivasa Rao and Lakshmi 1993; but see Collette, 1994) to provide evidence supporting the occurrence of natural hybridization.

In other cases, investigation of natural reproductive cycles has indicated that there is in fact little or no overlap in the timing of reproductive activity, which would reduce substantially the opportunity for natural hybridization (e.g. Knight and Parke, 1950 for fucoid algae; Vasseur, 1952; Falk-Petersen and Lonning, 1983 for echinoderms; Dillon, 1992 for hard shell clams).

Clearly, the geographic locality and timing of sampling are of great importance. Periods of overlap in spawning activity between the sea urchins *Strongylocentrotus droebachiensis* and *S. pallidus* along the coast of Norway have been reported from some locations (Hagström and Lonning, 1967), but not at others (Falk-Petersen and Lonning, 1983), just as periods of reproductive overlap for the fucoid algae *Fucus vesiculosus* and *F. serratus* were recorded by Burrows and Lodge (1951) but not by Knight and Parke (1950) from different regions of the Irish Sea coast of England. Such reports reveal the importance of small-scale (in the order of kilometres) as well as larger-scale (hundredss of kilometres) spatial variability. Localized environmental conditions can dictate for two taxa the timing of reproductive activity which might be different at one location but overlap at a different location only a few kilometres away. Thus, hybridization is impossible at the first location but possible at the second location, at least within the time frame of the study (temporal variability in hybridization rates is discussed in a subsequent section). This sort of "mosaic" environmental effect upon reproductive cycles might be expected to have greatest effect in temperate regions where seasonality is most pronounced. In tropical regions, where environmental conditions are more stable, it might be expected that organisms are less likely to respond strongly to environmental cues for spawning activity. For example, Randall (1956) noted that the closely related surgeon fishes *Acanthurus achilles* and *A. glaucopareius* at Hull Island in Oceania do not exhibit definite spawning cycles so that specimens of both species collected at the same time have comparably developed gonads (Randall, 1956). Evidence from corals and other reef invertebrates, however, conflicts with this interpretation. Harrison *et al.* (1984) and Babcock *et al.* (1986, 1992) have shown that mass spawning events for these organisms in the Pacific and

Atlantic Oceans are highly predictable with respect to environmental cues. Whatever the situation, the absence of temporal reproductive isolation between closely related species will increase the chances of hybridization occurring.

Pronounced differences exist in the timing of reproductive activity of a species towards the limits of the distribution when compared with the timing of reproduction at the centre of distribution of that species. As a consequence, the amount of hybridization that occurs throughout the species range can be modified. For example, the amount of natural hybridization between the hard shell clams *Mercenaria mercenaria* and *M. campechiensis* in south-east USA is much less at Hamlin Creek, South Carolina, a site towards the northernmost limit of *M. campechiensis*, than it is further south at Indian River Lagoon, Florida, where the amount of spawning overlap is much greater (Dillon and Manzi, 1989; Dillon, 1992).

Finally, differences in size (and therefore also age) can determine the potential for hybridization between taxa, because individuals of a single species may reproduce at different times of the year as a result of size-dependent variation in the allocation of energy to somatic and gonadal growth. Gardner and Skibinski (1990a) have shown that amongst a hybrid population of the blue mussels *Mytilus edulis* and *M. galloprovincialis* in south-west England, different sized mussels of the same taxon spawn at different times. The greatest amount of overlap in the timing of spawning activity (and hence the greatest potential for hybridization) between the different size (= age) classes of the two taxa occurred between the largest *M. galloprovincialis*-like and the smallest *M. edulis*-like mussels. It was concluded that because these two mussel types exhibited greatest overlap in the number and timing of spawning events throughout the year that hybridization was a common occurrence. This finding is in agreement with an extensive allozyme data set which indicates the same (e.g. Seed, 1971; Skibinski, 1983; Gardner and Skibinski, 1988; Beaumont *et al.*, 1989).

7. EXPERIMENTAL HYBRIDIZATION

Laboratory hybridization experiments have been utilized extensively to confirm the probable hybrid nature of certain individuals by demonstrating that two taxa will interbreed when provided with the opportunity to do so, or that gametes from two taxa can be artificially cross-fertilized. Such information provides a valuable contribution to the study of natural marine hybridization. Many authors have demonstrated that hybridization in the laboratory can take place, thus providing evidence in support of the existence of natural hybrids. Such experiments have been carried out for marine algae (Burrows and Lodge, 1951; Bolwell et al., 1977; Sanbonsuga and Neushul, 1978), molluscs (Loosanoff, 1954; Woodburn, 1961; Owen et al., 1971; Leighton and Lewis, 1982; Dalton and Menzel, 1983; Beaumont et al., 1993), echinoderms (Chia, 1966; Hagström and Lonning, 1961, 1967; Strathmann, 1981), crustacea (Solignac, 1976, 1978; Boitard et al., 1980; Sekiguchi and Sugita, 1980) and corals (Willis et al., 1992; Miller and Babcock, submitted).

Laboratory crosses have also demonstrated that there exists considerable potential for hybridization even among marine taxa which are no longer in geographic contact and therefore could not possibly hybridize in nature. Studies involving red algae from Scandinavia and the Gulf of Mexico (Rueness, 1973), kelp from the Canadian and European coasts of the North Atlantic (Luning et al., 1978), from the west and south coasts of South Africa (Bolton and Anderson, 1987), from the North and South Atlantic Ocean (tom Dieck and de Oliveira, 1993), from the Atlantic and Pacific Oceans (tom Dieck, 1992), sea urchins from either side of the Isthmus of Panama (Lessios and Cunningham, 1990), wrasses from the North Atlantic and Mediterranean (Hagström and Wennerberg, 1964), gobies from either side of the Isthmus of Panama (Rubinoff, 1968) and North American and European lobsters (Mangum, 1993) have all demonstrated that despite geographic separation, hybridization is still possible, and in many cases easily attainable. These results demonstrate that relatively high levels of genetic similarity have been maintained and currently exist between these groups of geographically isolated species despite the long periods of time which have passed since the species were physically separated.

The closely related starfish species *Patiriella calcar*, *P. gunnii* and *P. exigua* occur sympatrically at numerous sites in New South Wales, Australia. Based upon the developmental outcome of laboratory hybridization crosses, Byrne and Anderson (1994) predict that viable natural hybrids between *P. calcar* and *P. gunnii* probably exist in wild populations along these coasts, but that interspecific hybrids involving *P. exigua* are unlikely to occur. Whilst there are no data to indicate the existence of such natural hybrids at present, the use of the laboratory hybridization technique can provide a powerful means by which the presence of natural hybrids may be predicted. This approach is most appropriate for cases in which the hybrid offspring are morphometrically very similar to one parental type, with the result that such hybrids are not easily recognizable and may therefore be overlooked.

8. TEMPORAL VARIABILITY AND STABILITY IN HYBRIDIZATION RATES

It is often assumed that hybrid zones are stable in time and space because theory indicates that this can be the case (Harrison, 1990). The number of studies which have looked at temporal variability in natural marine hybridization is small, but those that have can provide some valuable insights into this area.

Hybridization between the snow crabs *Chionecetes opilio* and *C. bairdi* varies both temporally and according to sex (Karinen and Hoopes, 1971). In 1969, hybrid males constituted 2.8% and hybrid females 1.6% of the populations in Bristol Bay in the south-east Bering Sea, but in 1970 these values had dropped to 1.0% and 0.4% decreases of 64% and 60%, respectively. Even larger changes in the percentage of hybrids have been reported for hybrids of the seerfishes *Scomberomorus commerson* and *S. guttatus* from the south-east coast of India (Devaraj, 1986; but see Collette, 1994). From Palk Bay and the Gulf of Mannar in 1967, hybrids represented 68% and 59% of the total populations, but in 1968 they represented zero and 5.6% of the catch (decreases of 100% and almost 91%, respectively). These two examples illustrate how much variability can exist in the dynamics of marine hybrid zones on short temporal scales. In neither case do the authors suggest why such pronounced temporal differences in the rate of hybridization might exist. It seems likely, however, that these differences reflect natural variation in ecological conditions resulting in, for example, variation in recruitment (especially since the seerfish hybrid zone is associated with up-welling which exhibits temporal variability in its onset), but may to some extent also reflect fishing pressure. They do, however, demonstrate the point that hybridization can be an ephemeral event when measured over short time-scales, for example, from one breeding season to the next.

Not all marine hybrid zones exhibit this degree of short-term temporal instability. Skibinski (1983) examined allozyme variation in two hybrid mussel populations (*Mytilus edulis* and *M. galloprovincialis*) from south-west England in 1980–1 and observed a pronounced length-dependent change in allele frequencies. In this case, alleles at highest frequency in allopatric *M. galloprovincialis* occurred at highest frequency amongst the largest mussels of the hybrid populations. When Gardner and Skibinski (1988) re-examined allozyme variation in these same two populations in 1986–7, they observed almost exactly the same pattern of variation as first described by Skibinski (1983). These findings were interpreted as evidence of short-term temporal stability because during this period six new generations had been added to each population. Maximum individual longevity was subsequently estimated to be about 12 years (Gardner *et al.*, 1993), but such longevity is achieved by very few mussels. In terms of mussel numbers, the turnover during this period would represent nearly all individuals (>99%). Thus, the populations examined by Skibinski (1983) had effectively been replaced by new mussels by the time they were examined by Gardner and Skibinski (1988).

Initial reports of natural hybridization between marine species in the early literature provide a means of estimating short-term periods over which hybridization is known to have occurred. In some cases it is possible to estimate the short-term (in)stability of hybridization events by comparing data between these various accounts. For example, Verrill (1909) reported that the starfish *Asterias epichlora* and *A. hexactis* (now *Leptasterias epichlora* and *L. hexactis*) hybridize freely in the vicinity of Puget Sound, on the Pacific coast of North America. Kwast *et al.* (1990) have shown that this is still the case. Hence, we can infer that this hybrid zone is relatively stable in the short term because, despite the addition of approximately 90 new generations, and the complete replacement of population numbers many times over during this period, the extent of hybridization between these two species is much as described at the start of this century.

Several reports exist of short-term hybridization amongst fishes, with hybrids being observed over periods of many decades, within the same geographical vicinity. Hubbs and Kuronuma (1941) reported collections of an intergeneric hybrid flatfish (*Platichthys stellatus* × *Kareius bicoloratus*) made in 1929 from Oshara on the east coast of Hokkaido, Japan. Fujio (1977) noted the presence of these same hybrids almost five decades later. The "hybrid flounder" *Inopsetta ischyra* (= *Parophrys vetulus* × *Platichthys stellatus*) has been reported from several locations off the Pacific coast of the USA and Canada from 1880 onwards (Norman, 1934; Schultz and Smith, 1936; Herald, 1941; Hart, 1973 and references therein), and the hybrid seerfish *Scomberomorus lineolatus* (= *S. commerson* × *guttatus*) was first described in 1831 from the Bay of Bengal (Srinivasa Rao and Lakshmi, 1993). There is, however, some uncertainty as to the hybrid or specific status of this species (Collette, 1994).

The earliest confirmed report of marine hybridization that I am aware of involves the cetaceans *Balaenoptera musculus* (the blue whale) and *B. physalus* (the fin whale) off north Norway (Cocks, 1887). This report of 11 putative hybrids stems from the days of extensive whaling activity in the northern Atlantic. The application of modern molecular techniques has permitted the identification of a similar hybrid individual from waters off west Iceland (Spilliaert *et al.*, 1991) and suggests that hybridization between these two species might be wide-ranging in time and space.

It is extremely rare that we have a good idea of the time of the initial hybridization event between any two species. However, the example of hybridization between the sea grasses *Spartina maritima* and *S. alterniflora* provides a notable exception. The latter species was accidentally introduced into southern England sometime around 1830 and since about 1880 hybrids have been observed continually, indicating considerable stability in this hybrid zone over almost its entire duration (reviewed by Gray *et al.*, 1991).

The estimates of short-term temporal stability for hybridization gained by comparing reports from the literature encompass a wide variety of different organisms, including algae, invertebrates, fish and cetaceans, which suggests that short-term temporal stability of hybridization rates is the norm for most marine organisms.

Longer-term estimates of hybridization exist for two bivalve mollusc hybrid zones, one in western Europe, the other in North America. The first report of *M. galloprovincialis* in south-west England is by Donovan (1802): since *M. galloprovincialis* and *M. edulis* currently hybridize so readily in this region (reviewed by Gardner, 1994) one might assume that hybrids were likely to be present at the time of Donovan's report. Estimates of a time of genetic divergence between these two taxa are of the order of 1–1.5 million years as determined from biochemical variation (Skibinski *et al.*, 1980), suggesting that a long history of hybridization is likely. Probable hybridization between the hard shell clams *M. mercenaria* and *M. campechiensis* from south-east USA can be dated back even further. The fossil record indicates that both species were established in the Miocene, and have survived, apparently unchanged, for over 5 million years (Humphrey and Crenshaw, 1985; Dillon, 1992). Dillon (1992) suggested that these two clam species should be considered amongst the oldest of any extant species known to hybridize anywhere in the world. Harrison (1990), however, made the very telling point that simply because the environment in which present-day hybridization occurs has remained relatively stable for long periods of time does not necessarily mean that hybridization has been occurring for the same length of time. The stability of the environment does not tell us anything about the past distributions of the parental species, and whether or not they interbred. We must be cautious about inferring long-term stability for hybrid zones based only on information pertaining to environmental conditions. Unfortunately, this "paucity of real evidence will not be easy to remedy" (Harrison, 1990, p. 88). Given the caveat proposed by Harrison (1990) about inferring long-term stability of hybridization from environmental data alone, the examples outlined here provide the best evidence of long-term stability in marine hybrid zone location and composition. However, one method by which estimates of long-term hybridization rates may be obtained is by the application of multivariate statistical techniques to morphometric analysis of (sub)fossil specimens, where such a technique has been shown to be appropriate for living specimens. This would work well for both the *Mytilus* and the *Mercenaria* hybrid zones because the (sub)fossil records of these molluscs are relatively good. This approach would provide a meaningful way to address the question of long-term stability of hybridization, when this has proved to be so difficult to do for other species in other environments.

9. DISTURBANCE AND HYBRIDIZATION

9.1. Introduction

Disturbance, often as a man-made phenomenon, is recognized as an important factor contributing to hybridization (Stebbins, 1950; Mayr, 1979; Harrison, 1990). Of the 108 cases of marine hybridization recorded in Table 1 (page 70), 28 have reported the involvement of some form of disturbance (it is likely that more than this number involve disturbance, but reference to such an event is not made in every paper).

9.2. Human Disturbance

Deliberate or accidental introductions of taxa represent perhaps the ultimate example of human-mediated disturbance. Increasing awareness of the problems (both ecological and economic) associated with the introduction of "exotics" has opened the eyes of the scientific community to many of the risks that are now faced, particularly with the loss of genetic integrity of a native species resulting from hybridization and introgression with the exotic species.

One of the biggest threats to the integrity of native marine fauna and flora is the inadvertent transport of species over great distances, often in ballast water (Carlton and Geller, 1993). Laboratory experiments frequently demonstrate the potential for hybridization between species which are geographically isolated (and as a result could never hybridize in nature). It is likely that we will see rapid growth in the number of instances in which an accidental introduction results in hybridization between an exotic species and a native species.

A striking example of hybridization in the marine realm resulting from the accidental introduction of a species involves marsh grasses of the genus *Spartina* (reviewed by Gray *et al.*, 1991). *S. maritima* ($2n = 60$) is a common salt marsh plant in western Europe and elsewhere, and *S. alterniflora* ($2n = 62$) is a common marsh plant in North America. It is now thought that *S. alterniflora* was accidentally introduced into southern England from the

USA sometime around 1830. Hybridization between the two species gave rise to *S. x. townsendii* (2n with a mode of 62) which is a sterile hybrid. At an unknown date this event was followed by a chromosome doubling to produce a fertile amphidiploid form known as *S. anglica* (2n ranges from 120 to 127, but is usually 120, 122 or 124). These grasses play an important role in stabilizing marsh and estuary sediments. As a result, they have been introduced from southern England to many different localities including France, Spain, China, Germany, Denmark, Australia and New Zealand. Because of the nature of the extensive coexistence of the various *Spartina* species within the salt marsh grass beds of southern England, it is quite possible that most, if not all, forms of these grasses have been transplanted to these other countries. Unfortunately, no information is available concerning the occurrence, structure and dynamics of introduced *Spartina* hybrid zones from any of these locations. This is an area for research which could prove particularly rewarding if pursued.

The opening up of new shipping ways such as the Suez Canal provides for considerable movement of organisms between previously isolated regions. Although in the Suez Canal significant salinity changes provide barriers to dispersal, movements of fauna through the canal are well documented. There has been extensive migration from the Red Sea to the Mediterranean Sea (the Lessepsian migration), but much less movement of organisms in the opposite direction (Briggs, 1974). At the moment no record of hybridization resulting specifically from migration via the Suez Canal or the Panama Canal has been reported, but such a phenomenon is a very real possibility. Hybridization has, however, been recorded in association with the Kiel Canal, or Nordostseekanal, which links the River Elbe in Germany to the Baltic Sea. Solignac (1976) has reported the presence of a hybrid swarm of *Jaera praehirsuta* × *ischiosetosa* (Crustacea: Isopoda) in the canal. It seems likely that the establishment of the canal has resulted in the hybridization of these two isopod species, but pre-existing natural hybridization is difficult to rule out because several other hybrid zones exist in this same geographical location.

Perhaps the most striking example of hybridization having deleterious effects upon species integrity resulting from the deliberate introduction of one or more species, involves the introduction of two Japanese oysters (*Crassostrea sikamea* and *C. gigas*) into California, USA (Hedgecock *et al.*, 1993). This example is dealt with in more detail in Section 19.

9.3. Natural Disturbance

Natural disturbance is recognized as an important factor contributing to hybridization (Hubbs, 1955). Two relatively common themes linking examples of marine hybridization involving natural disturbance are the influences of freshwater input (this can be in a fully saline context with occasional freshwater run-off, can take place in a permanent estuarine setting, or can be associated with the removal of a freshwater barrier) and up-welling.

Examples of marine hybridization which are known to be associated with freshwater disturbance include flatfish (Hubbs and Kuronuma, 1941), and isopods (Jones and Naylor, 1971; Solignac, 1976). Members of the isopod *Jaera albifrons* species complex in western Europe are known to hybridize infrequently at several sites, and frequently at only a small number of sites. It is possible to find up to four species at the same location, each separated from the others by ecological preferences, usually according to tidal height (Solignac, 1976). Jones and Naylor (1971) reported finding a single hybrid individual (*J. albifrons* × *ischiosetosa*) at Milford Haven, South Wales, in the bed of a dried stream leading to the intertidal zone. The temporary natural disturbance caused by the drying of the stream is thought to have permitted the migration of *J. ischiosetosa* into the microhabitat occupied by *J. albifrons*, and resulted in limited hybridization (the single hybrid represented 0.085% of the 1179 *albifrons* and *ischiosetosa* collected). Many other examples of hybridization are associated in some way with the effect of fresh water, often because one of the hybridizing taxa has an ecological preference for lower salinity than the other species involved (refer to Table 1, page 70). This can be viewed as a form of natural disturbance involving freshwater influence because the effect is often highly variable in time and space.

It is interesting to note that many cases of natural hybridization are reported from the Pacific coast of North America (especially southern California and Puget Sound) which is an area of permanent and extensive up-welling. Examples include kelp (Coyer and Zaugg-Haglund, 1982; Coyer *et al.*, 1992), echinoderms (Verrill, 1909; Swan, 1953; Chia, 1966; Strathmann, 1981; Kwast *et al.*, 1990), and molluscs (Owen *et al.*, 1971; McDonald *et al.*, 1991; Sarver and Foltz, 1993). It should be noted, however, that it is difficult to disentangle the effects of up-welling from other effects which might cause hybridization. Srinivasa Rao and Lakshmi (1993) have suggested that hybridization between the seerfish *Scomberomorus commerson* and *S. guttatus* in the Bay of Bengal, is coincident with the timing of the local up-welling event from February to April, which results in intermediate salinity between the low of December and the high of May. It is noted that the environmentally "intermediate conditions obtaining in January/February induce the ready-to-spawn adults of both putative parents, which may not normally hybridize, to breed irrespective of availability of proper mates, resulting in hybridization between the two species" (Srinivasa Rao and Lakshmi, 1993, p. 487; but see Collette,

1994). It is presently impossible to differentiate between two possibilities which might be responsible for the high level of hybridization observed in such regions, namely that the disturbance effect of the up-welling event itself, or that the increased primary productivity, which is thought to permit greater niche partitioning (and therefore ultimately speciation), is the causative factor.

Burrows and Lodge (1951) suggested that the removal of pure adult fucoid algae following the artificial clearance of intertidal regions at Port St Mary, Isle of Man, permited the establishment of competitively inferior hybrid sporelings which could not establish in the presence of the parental species. This hypothesis received support when a natural hybrid swarm of *Fucus spiralis* × *F. vesiculosus* from the River Mersey, north-west England, was recognized. This hybrid swarm exists on very soft sandstone which constantly and naturally erodes away once the weight of the established fucoids is too great to be maintained by the soft rock. This results in continual recolonization of the substratum with an associated short-term decrease in intraspecific competition from pure parental types which permits the establishment of the hybrid plants. Without this natural erosion it is unlikely that the hybrid individuals could become established.

Natural disturbance plays an important role in promoting hybridization of abalone species along the coast of California (Owen et al., 1971). The authors note that "areas where an excessive number of hybrids were found had one thing in common, the presence of large populations of the sea urchin *Strongylocentrotus franciscanus* and a resulting destruction of nearly all algal cover, including the giant kelp, *Macrocyctis pyrifera*" (Owen et al., 1971, p. 32). Over a 4 year period at a site south-west of Point Conception, California, sea urchin grazing completely destroyed a substantial area of kelp forest (kelp fragments provide an important food source, and the kelp forest habitat acts as a refuge and breeding ground for some abalone species). During this 4 year period the number of abalones decreased tenfold whilst the number of hybrids remained the same. The authors concluded that hybridization was most prevalent in areas subjected to habitat disturbances, probably because different species which are not normally in close contact could intermingle and interbreed.

The single phenomenon bring about the greatest incidence of hybridization in the sea is likely to be natural disturbance involving geological upheaval; for example, events such as the formation of the Florida peninsula or Pleistocene cooling associated with the last Ice Age. Such events tend to break up continuous biotic distributions into two or often many more fully or partially isolated subpopulations. Over time, these newly established allopatric populations accumulate genetic differences resulting from stochastic processes (genetic drift) and/or adaptation to the new localized environment (natural selection). When the barrier between the populations is removed and secondary contact is established the newly evolved races, subspecies, semi-species or full species are occasionally able to interbreed. This scenario for the formation of many hybrid zones (especially in temperate and terrestrial regions) has been invoked extensively and much evidence supports it (Harrison, 1990). Inspection of the data of Table 1 (page 70) indicates that such events do indeed have a profound effect upon the establishment of not just isolated hybridization events, but also upon hybrid zones and hybrid swarms. For example, many of the best-known marine hybridization events occur along the coasts of Florida, in north-west Europe and at the boundary between the North and Baltic Seas, areas which are known to have been substantially disturbed by geological upheaval, sea level regressions associated with Ice Age cooling, and salinity changes resulting from the formation of the new environment.

One of the greatest collections of marine hybrid zones is found along the coasts of the Florida peninsula and into the Gulf of Mexico. This region is known to represent a major discontinuity in the genetic composition of many species, some of which show continuous distributions around the peninsula (Reeb and Avise, 1990). The geological formation of the peninsula has been extensively documented (Walters and Robins, 1961; Avise, 1992) and the consequent effects upon biotic distributions have been profound. Clearly, the geological disturbance associated with the formation of the peninsula has been particularly important in the establishment of many marine hybrid zones.

In north-west Europe, the Pleistocene epoch is thought to have had a profound effect upon biotic distributions and the presence of many hybrid zones, both terrestrial and marine. Marine hybrid zones involving *Jaera* (Isopoda, Crustacea) (Jones and Naylor, 1971; Solignac, 1976, 1981), *Xantho* (Decapoda, Crustacea) (Almaça, 1972) and *Mytilus* (*M. edulis* × *galloprovincialis*) (Bivalvia, Mollusca) (Seed, 1971; Skibinski, 1983; Gardner and Skibinski, 1988; Gardner, 1994) are all likely to have originated as a result of secondary contact following the ice sheet regression. Other hybrid zones listed in Table 1 probably arose in this way, too.

Several hybrid zones and abrupt discontinuities in biotic distributions have been recorded from the region between the low-salinity Baltic Sea and the more fully saline North Sea. The present-day Baltic Sea is thought to have been established approximately 6000 years ago (Briggs, 1974) and consequently any unusual biotic effects have arisen since that time. Associated with the Belt and Baltic Seas are hybrid zones involving *Jaera* (Isopoda, Crustacea) (Solignac, 1976, 1981), *Macoma* and *Mytilus* (Bivalvia, Mollusca) (Varvio *et al.*, 1988; Väinölä and Varvio, 1989; McDonald *et al.*, 1991; Väinölä and Hvilsom, 1991) and *Platyichthys* × *Pleuronectes*, *Pleuronectes* × *Glyptocephalus* and *Scophthalmus* (Pleuronectidae, Pisces) (Norman, 1934).

Hewitt (1988) suggests that most hybrid zones are likely to have originated as a result of secondary contact. This is clearly an important mechanism by which hybrid zones in any environment, including the sea, can be generated. Many of the examples of marine hybrid zones given in Table 1 are likely to have arisen in this fashion, but may now be maintained by factors such as ecological differences between the taxa involved.

10. RELATIVE PARENTAL ABUNDANCE AND HYBRIDIZATION

It is often the case that hybridization is reported from areas where one parental type occurs at very low density. This situation would at first sight appear to mitigate against hybridization, but under certain circumstances it can in fact promote it. Individuals which occur at very low density, often beyond the usual distributional range of their species, will have difficulty in finding a conspecific mate. In such cases, Mayr (1979, p. 127) noted that in "the absence of adequate stimuli, that is, stimuli from conspecific individuals, they [lone individuals] are apt to respond to inadequate stimuli, that is to individuals belonging to a different species. Many of the known hybrids of animal species are found at the margin of the normal geographic range of one of the two parental species or even beyond it." This scenario applies equally to species with highly developed mate recognition systems as well as to species which are free-spawning and employ gamete recognition systems (or similar) as part of their reproductive biology.

Owen *et al.* (1971) have reported 12 of the 15 possible hybrid crosses between six abalone species from California. In several of these crosses, one parental type occurs at very low frequency at the limit of the species distribution. Amongst hybridizing stone crabs (*Menippe mercenaria* × *M. adina* from Florida, Bert and Harrison (1988) suggested that a large inequality in the numbers of the parental forms may have resulted in unusual patterns of carapace coloration and allozyme allele frequencies. Dillon (1992) reported hybrids at Hamlin Creek, South Carolina, between the southern hard shell clam *M. campechiensis* which is rare at this site and the northern clam *M. mercenaria* which is abundant. From approximately 10 000 *M. mercenaria* he identified 27 pure *M. campechiensis* and six hybrids and individuals of mixed ancestry, based upon electrophoretic criteria.

Randall *et al.* (1977) observed that the large majority of chaetodontid fishes involved in hybridization can be classified according to their behaviour as either solitary or occurring in pairs. Very little evidence was found of hybridization involving gregarious species which form spawning masses. These authors pointed out that at any given location it is expected that if a conspecific mate is unavailable then a solitary fish will seek to mate with an individual of a closely related species, rather than not reproducing at all: this is exactly what is observed. In this context, then, the occurrence of a single individual or small number of individuals outside the normal range of that species increases the probability of hybridization, provided that isolating barriers are not so well developed as to completely prevent interbreeding.

The social structure of the species in question may also be of importance in determining why individuals of one species interbreed with another species. For example, female angel-fishes of the genus *Centropyge* form harems and, within each harem, females are ranked according to their dominance. The most dominant females monopolize the male during optimal spawning times, often forcing subordinate females to spawn at suboptimal times (Pyle and Randall, 1994). It has been suggested (Pyle and Randall, 1994) that this might make a subordinate female choose to spawn with a heterospecific male at the optimal spawning time, rather than with a conspecific male at a suboptimal time. This does not explain why the male would want to mate with a subordinate (and therefore less fit) heterospecific female, but it does illustrate that under certain conditions, interspecific hybridization might be promoted by the breeding behaviour of at least one of the species involved.

11. THE FITNESS OF HYBRID INDIVIDUALS

There is a large body of evidence from detailed examinations of natural hybrid zones and from laboratory experiments indicating that hybridization often, if not usually, results in the production of infertile offspring, or individuals which possess a greatly reduced reproductive potential compared to the parental types. This can occur for a variety of reasons, but many of the best-documented examples involve hybrid dysgenesis, a phenomenon in which the hybrid individual is, amongst other things, unable to develop fully functional gonads (Kidwell, 1982; Lozovskya *et al.*, 1990; Heath and Simmons, 1991; Khadem and Krimbas, 1993). It is, however, possible that hybrid offspring are produced which are fertile but possess a much reduced reproductive potential compared to both parental types. This can result from the inability of the hybrid to attract a mate of either parental species if pre-mating behaviour is involved, because the hybrid individual produces a much reduced number of gametes, or because the gametes which are produced are not fully functional.

In the strictest sense, individual fitness means reproductive success only, but since many parameters influence reproductive success, fitness can also be used in a wider sense to mean any aspect of the biology of the individual concerned which affects reproduction. Examples of hybrid unfitness are plentiful in the marine biology literature and include slower rates of development (in echinoderms: Hagström and Lonning, 1961, 1967; in molluscs: Fujino *et al.*, 1980), decreased fertilization success (in echinoderms: Hagström and Lonning, 1961; Strathmann, 1981; in crustaceans: Solignac, 1976), lower individual fecundity or complete sterility (in marsh grass: Gray *et al.*, 1991; in crustaceans: Karinen and Hoopes, 1971; Johnson, 1976; in echinoderms: Swan, 1953; in flatfish: Holt, 1893; Norman, 1934), increased mortality rates (in algae: Coyer and Zaugg-Haglund, 1982; in echinoderms: Hagström and Lonning, 1961; in molluscs: Owen *et al.*, 1971), increased susceptibility to gonadal neoplasia (in molluscs: Hesselman *et al.*, 1988; Bert *et al.*, 1993), decreased competitive ability (in algae: Burrows and Lodge, 1951), increased morphological variability (in echinoderms: Kwast *et al.*, 1990), the absence of hybrid genotypes predicted from the presence of parental genotypes (in crustacea: Bert and Harrison, 1988), lower body size or weight (in echinoderms: Kwast *et al.*, 1990), highly skewed sex ratios (in fish: Turner, 1969; Dahlberg, 1970), and higher rates of parasitism (in fish: Turner, 1969).

Given that many hybrids are morphologically intermediate in one form or another, it is not surprising that hybrid fitness is also frequently reported as being intermediate between those of the parental types, or that the fitness parameter under investigation is no different from one or both of the parental types. Examples from the marine realm include developmental stability and developmental rate (in echinoderms: Chia, 1966; Strathmann, 1981; in molluscs: Gardner 1995), growth rate (in molluscs: Gardner *et al.*, 1993), body size (in echinoderms: Kwast *et al.*, 1990; in cetaceans (Cocks, 1887), fertility and fecundity (in marsh grass: Gray *et al.*, 1991; in crustaceans: Solignac, 1976, 1981; in echinoderms: Schopf and Murphy, 1973; in molluscs: Gardner and Skibinski, 1990a), viability (in crustaceans: Solignac, 1976; in molluscs: Menzel, 1985; Gardner and Skibinski, 1991a; Gardner *et al.*, 1993; Willis and Skibinski, 1992), chromosomal structure (in molluscs: Menzel and Menzel, 1965), allozyme thermostability (in molluscs: Gardner and Skibinski, 1990b) and resistance to parasites (in molluscs: Coustau *et al.*, 1991b).

In a smaller number of cases it has been reported that hybrids have a fitness advantage compared with both parental types; this can be viewed as a form of hybrid vigour (heterosis). Examples include increased feeding ability (in echinoderms: Hägstrom and Lonning, 1964), increased growth rate (in molluscs: Menzel and Menzel, 1965; Menzel, 1985; in fish: Srinivasa Rao and Ganapati, 1977; Srinivasa Rao and Lakshmi, 1993; but see Collette, 1994), increased fecundity (in fish: Srinivasa Rao and Lakshmi, 1993; but see Collette, 1994), increased longevity expressed as marketable shelf-life (in molluscs: Menzel, 1985), and increased adaptation to certain environmental conditions (in fucoid algae: Burrows and Lodge, 1951).

On balance, these data indicate that hybridization in the sea results in a decrease in individual fitness, whether this is expressed in a narrow sense (reproductive success) or a wider sense (any factor affecting reproductive success), about as frequently as it results in intermediate or increased hybrid fitness when compared with the parental types. Thus, in the sea, hybrid unfitness does not usually or inevitably result from hybridization, as it has often been considered to do in other environments (Moore, 1977; Hewitt, 1988; Barton and Hewitt, 1989; Harrison, 1990). However, a recent review by Arnold and Hodges (1995) of predominantly non-marine examples has indicated that hybrids possess, on average, either similar fitness to the parental types or higher levels of fitness than at least one parental type. Data from the marine environment indicate that hybrid unfitness does not usually or inevitably result from hybridization in the sea ($n = 21$ cases in which hybrids have decreased fitness versus $n = 19$ cases in which hybrids have equal or greater fitness than the parental types). These data support the thesis that hybrid genotypes are not uniformly unfit but do not indicate that "the general pattern is that hybrids demonstrate either equal fitness to the two parental taxa or higher levels of fitness than at least one of the parents" (Arnold and Hodges, 1995, p. 70).

The fitness of hybrids is one area of great importance which is in need of considerably more attention before a fundamental property of many hybrid zones can be elucidated. It has been demonstrated that many hybrid zones are maintained by a balance between immigration and selection against hybrid individuals, and that such hybrid zones are better defined as tension zones (Barton and Hewitt, 1989). The basis of the selection against the hybrids is thought to be the decreased fitness that such individuals possess as a consequence of the break-up of the co-adapted parental gene complexes resulting from hybridization. Thus, the relative fitness of the hybrids can be important in determining how the hybrid zone is maintained.

In some of the above-cited examples of hybridization in the sea it appears that the tension zone definition is the most appropriate one. In others, however, it appears that this is not the case. For example, the hybrid zone between the blue and Mediterranean mussels (*Mytilus edulis* and *M. galloprovincialis*) in north-west Europe is one of these (reviewed by Gardner, 1994). In brief, there is very good evidence from a number of studies that the numerous hybrids in this extensive region of hybridization on average have a fitness (both in the narrow and wide senses outlined above) that is comparable to one parent (*M. galloprovincialis*) and superior to the other parent (*M. edulis*). All the evidence indicates that this hybrid zone is maintained not by immigration of the parental types and selection against the subsequently produced hybrids, but by massive immigration of *M. edulis* spat and much smaller scale immigration of *M. galloprovincialis* spat, followed by selection against the *M. edulis* and not the hybrid mussels. Because of the paucity of information available for most other marine hybrid zones, it is unclear if this is a unique example (this seems unlikely given the propensity of this genus for hybridization everywhere in the world where two taxa coexist), or whether other marine hybrid zones may be maintained in a similar fashion.

12. REPRODUCTIVE BIOLOGY AND HYBRIDIZATION FREQUENCY

For sexually reproducing animal species it would seem likely that the method of fertilization (internal versus external), as well as the presence or absence of mating behaviour or cues such as colour patterns associated with reproduction, should play an important role in determining the frequency with which hybridization occurs between taxa of these different groups.

Many marine invertebrates have relatively simple reproductive strategies and their gametes are shed directly into the sea. There is thus no visual cue or behavioural component to reproductive activity for many invertebrate taxa. It is therefore not surprising that there is a high frequency of hybridization involving taxa which exhibit such a simple reproductive strategy. In contrast, certain marine invertebrates, notably many crustaceans, employ elaborate mating behaviour which often involves display signals generated by the waving of chelae or the use of antennae for mate recognition. Such behavioural patterns are usually also associated with copulation and mate guarding. In such a case one might predict that this pre-copulatory behaviour and internal fertilization would reduce substantially the frequency with which hybridization occurs. Despite the fact that some crustacean species hybridize (12 examples are given in Table 1), the evidence indicates that mating behaviour and internal fertilization reduce the chances of hybridization; the hybridization rate amongst invertebrates which employ no mating behaviour and external fertilization is approximately twice that of the invertebrate species which exhibit pre-mating courtship behaviour and copulation (data in Table 1).

This difference may be more apparent than real because the majority of marine invertebrates employ the former rather than the latter reproductive strategy. When these figures are standardized according to the number of species in each of the two groups, the frequency of hybridization may be higher in the group employing courtship behaviour and copulation than in the group employing neither of these strategies. Such a situation is counter-intuitive, but might be explained by recent studies of gamete incompatibility and sperm/egg recognition chemistry, which suggest the existence of barriers to interspecific hybridization. For taxa with no mating behaviour and which spawn ripe gametes directly into the sea at the same time as other taxa (such mass spawning activity can involve tens and up to 100 species, e.g. Harrison et al., 1984; Hodgson, 1988; Babcock et al., 1986, 1992; Palumbi and Metz, 1991) the only way to preserve specific identity is for the sperm and eggs to be biochemically incompatible with those of other taxa.

There is thus strong interest in the occurrence of taxon-specific recognition systems that facilitate intrataxon and reduce intertaxon fertilization (recent developments are described and discussed by O'Rand, 1988; Palumbi and Metz, 1991; Palumbi, 1992, 1994; and Vacquier et al., 1990). The underlying mechanism preventing or reducing hybridization is therefore biochemical (or in some cases physical) and in many respects can be considered to be beyond the control of the individuals involved. By comparison, for marine invertebrates which employ species-specific courtship behaviour and/or internal fertilization, less emphasis might be placed upon gamete incompatibility because this represents just one component of preventing hybridization. After all, by the time that interspecific gametes come into contact in such organisms, the courtship ritual should have ensured or maximized the probability that only an individual of the same taxon is contributing the gametes in question. Such a reproductive strategy is thus more in the hands of the participants since active mate choice is usually an integral component of mating behaviour. If for some reason this breaks down (e.g. Wilbur (1989) has shown that male *Menippe mercenaria* stone crabs are 1.6 times more likely to mate with female *M. adina* than by chance alone) then hybridization might result because the gamete recognition system of the two hybridizing taxa is not as strongly developed as that of species which possess no pre-spawning mating behaviour. Whilst the above explanation must remain speculative for the time being, it is interesting to note that studies of fertilization biology tend to be very strongly biased towards those invertebrates with no mating behaviour (see, for example, references in Palumbi, 1994). This is clearly an area of research which requires more attention and will doubtless yield many valuable insights into the cause and prevention of hybridization.

In many fishes, especially the flatfishes (families Bothidae and Pleuronec-tidae), mating behaviour is often reduced or absent, gametes are spawned directly into the sea, and mass matings take place which involve several species in close proximity. Given these conditions it is hardly surprising that hybridization is often reported amongst these fishes. In contrast, the majority of fish taxa employ visual cues derived from colour patterns as well as mating behaviour which often involves species-specific displays before gametes are spawned directly into the sea. This kind of mating behaviour should act to reduce intertaxon fertilization. Inspection of the data in Table 1 (p. 70) indicates that the highest frequency of hybridization occurs amongst those fishes which employ visual cues and/or mating behaviour. Since the majority of fishes are likely to fall into this latter group, the standardized rate of hybridization appears to be highest in the group that does not employ species-specific cues for reproductive behaviour. This situation contrasts sharply with that described for the invertebrates, where hybridization frequency was observed to be highest amongst those in-dividuals which do exhibit species-specific cues for reproductive behaviour. An explanation for this difference is hard to advance. If gamete incom-patibility plays an important role in preventing cross-fertilization in the free-spawning invertebrates (as clearly it must) then one would expect that the same would be true amongst the free-spawning fishes. This is not what is observed from the data presently available.

13. HYBRID ZONE STRUCTURE

13.1. Hybrid Zone Width

One point which has consistently come out of reviews of the literature is that the width of the hybrid zone is narrow when compared to the dispersal ability of the taxa involved (Hewitt, 1988; Barton and Hewitt, 1989; Harrison, 1990; Barton and Gale, 1993). However, a major problem for all biologists is the accurate estimation of the mean dispersal of individuals.

For many animal hybrid zones from the terrestrial environment, mark-recapture methods are frequently used to estimate dispersal poten-tial. There are two main problems associated with estimating dispersal potential for most marine invertebrates (these problems are analogous to the problems faced by plant biologists in estimating plant dispersal ability). First, the adults are sedentary: it is the larval or pre-settlement life-history phase which is the dispersive stage. For many marine invertebrates, the larval stage is very small (often < 1 mm) making mark-recapture methodology impractical. Further, most marine invertebrates (and many marine vertebrates) produce vast quantities of gametes (often millions of gametes per individual) and larval mortality rates are high. This means that even if larvae could be marked, many millions would have to be marked to permit the recovery of only a few newly settled individuals, provided one knew where to look for these randomly dispersed propagules. Clearly, this approach is impractical. The second problem is associated with the unpredictability of the physical environment. Larval dispersal occurs as a function of the local tidal and current regime, and the duration of the dispersive stage in temperate regions is often of the order of 2–4 weeks (e.g. Chipperfield, 1953). This kind of time period for migration and consequent gene flow is used to explain high levels of genetic homogeneity observed in many species over macrogeographic scales (e.g. Skibinski et al., 1983). However, whether a larva is dispersed over 100s of kilometres or over only a few kilometres in the space of 4 weeks will depend almost entirely upon localized hydrographic conditions. For example, certain embayed areas are characterized by entrained circulation, meaning that larval dispersal into and out of such areas is much reduced compared with larval dispersal from other sites. This can be reflected in the genetic composition of the population(s) of the embayment when compared to populations of the same species in other areas, and can explain a high proportion of the genetic heterogeneity of a species with such a dispersive stage (Pande, A., Eyles, R. F., Wear, R. G. and Gardner, J. P. A., data for the greenshell mussel *Perna canaliculus* in New Zealand). Thus, estimating the dispersal potential of the larval stage of many marine invertebrates is particularly difficult, and estimates that are obtained are highly site-specific. What is clear, however, is that the theoretical dispersal potential of a long-lived larval stage is high, and can be as high as thousands of kilometres (a more realistic average figure is probably somewhere around 100–500 km).

In certain circumstances, mark-recapture methods can be applied successfully to motile marine invertebrates such as crustaceans or cephalopods. Such techniques are routinely used by fisheries scientists to estimate population sizes and migration routes. The same is true of mark-recapture techniques as applied to marine vertebrates. However, whilst adult dispersal can be estimated in this way, many of the problems outlined above still exist because many fish species are characterized by external fertilization and a relatively long-lived larval stage which is assumed to represent the time of major dispersal for the species concerned.

The width of most marine hybrid zones is unknown. However, zone width is highly variable when compared between hybrid zones where the width has been estimated. For example, in the Belt Sea, a region which is characterized by a large number of hybrid zones, the width of most zones is very narrow compared to dispersal potential (e.g. for blue mussels: Varvio *et al.*, 1988; McDonald *et al.* 1991; Väinölä and Hvilsom, 1991; for clams: Väinölä and Varvio, 1989; and for a variety of flatfish: Norman, 1934). This probably reflects the pronounced change in environmental conditions from the low salinity Baltic Sea to the more saline North Sea, and is therefore largely determined by physical and not biological parameters. In this case, dispersal potential is determined by environmental factors, because gene flow between the hybridizing mussels *Mytilus edulis* and *M. trossulus*, and the hybridizing races of the soft shell clam *Macoma balthica* in this region is limited by genetic adaptation to salinity differences which exist across the Belt Sea. Given the considerable theoretical dispersal potential of the larval stage (as discussed above) the hybrid zone between these two mussel taxa and the two clam races might well be the narrowest hybrid zones (relative to dispersal potential) known to occur anywhere.

In other regions, marine hybrid zones of considerable width have been described. The zone between the mussels *Mytilus edulis* and *M. galloprovincialis* in north-west Europe apparently extends from south-west England and southern Ireland to the Franco-Spanish border, a distance of approximately 1000 km. Similarly, the hybrid zone between the hard shell clams *Mercenaria mercenaria* and *M. campechiensis* extends from about South Carolina to southern Florida, a distance of approximately 1100 km, and the hybrid zone between members of the *Jaera albifrons* species complex extends from northern Europe to southern France, a distance of approximately 1200 km. These are some of the widest hybrid zones to occur anywhere, largely as a result of the high dispersal ability of the taxa involved, as well as the relative environmental homogeneity of the areas concerned.

Knowlton (1993) has suggested that, although hybrid zones in the sea are wider in absolute terms than those on land, when the extensive dispersal potential of most marine organisms is taken into consideration, then hybrid zones in the sea are relatively narrow too. Within the framework of the caveats outlined above, this is a reasonable suggestion but, in the absence of data pertaining to actual dispersal distances travelled by larvae or juveniles of most of the species listed in Table 1, this will have to await future corroboration.

13.2. Hybrid Zone Shape

Many hybrid zones can be recognized by the shape of the cline, or transition in the characters from one parental type to those of the other parent, which is a sigmoidal curve for each trait being assayed (behavioural, morphological, colour patterns, genetic, etc.; see, for example, Hewitt and Barton, 1980; Hewitt, 1989). Thus, many different traits often show coincidence in their geographic location within the narrow hybrid zone (Barton and Hewitt, 1989). This linkage disequilibrium between traits results from the constant influx of parental gene combinations into the area of the hybrid zone, followed by their breakdown by recombination when hybridization occurs. The position of such hybrid zones is often independent of ecological factors, and probably reflects the differences in fitness between the two species and their hybrids where their distributions overlap. This sort of structure is most often associated with tension zones (Barton and Hewitt, 1989).

In the marine context, it is hard to determine how many of the cases of hybridization might be described by simple but smooth clines. The main problem is that most of the examples have not been described sufficiently well using a variety of approaches to permit a judgement to be made. In many of the examples listed in Table 1, morphological differences are the only criterion to have been used to identify putative hybrids. Whilst identification of hybrids in the sea is important, this tells us nothing about how or why hybridization occurs, and whether or not it is of evolutionary importance. There is clearly a need for a multifaceted approach to be taken to provide as much information as possible about the structure of the marine hybrid zone, thus permitting a fuller evaluation and understanding of the nature of hybridization and speciation in the sea. This is probably the one area in which the greatest discrepancy is observed between our understanding of marine and terrestrial hybrid zones. Whereas many terrestrial hybrid zones are now well characterized, very few marine examples have been investigated to similar levels.

Mosaic hybrid zones were first described by Harrison and Rand (1989) and Rand and Harrison (1989) for crickets (genus *Gryllus*) of the north-eastern USA. These crickets have preferences for certain soil types so that the distributional differences of the two species reflect the underlying environmental mosaic. An equivalent marine mosaic hybrid zone exists between the mussels *Mytilus edulis* and *M. galloprovincialis* in north-west Europe (Gardner, 1994). Within the zone of contact the distributions of the mussels are explained primarily by differences in their preferences and tolerances for salinity variation and wave action. *M. edulis* tends to occur in less saline, more sheltered regions, whereas *M. galloprovincialis* tends to occur in waters with higher salinity and greater wave action. Thus, the distributions of the two types on the scale of only a few kilometres corresponds to the underlying environmental conditions, producing a marine mosaic hybrid zone. Whether or not this is the case for the other *Mytilus* hybrid zones remains to be determined, but preliminary data indicate that the *Mytilus edulis* and *M. trossulus* hybrid zone in eastern Canada is at least partly influenced by physiological differences in salinity tolerance between the two types (Gardner, J. P. A. and Thompson, R. J., unpublished data), consistent with the differences observed for the genus in north-west Europe.

What is apparent from the review of hybridization in the sea is that, whereas the shape of terrestrial hybrid zones is relatively well understood, too few data are presently available to draw conclusions about the shape of marine hybrid zones. Where it is possible to do so, it seems that the standard cline is the most frequent shape encountered, but it is apparent that other types or shapes of hybrid zone are also encountered in the sea. A much more rigorous approach to the description of marine hybrid zones is called for, before meaningful comparisons can be made, either within the marine environment, or between the sea and the land.

14. INTROGRESSION

Introgression, the incorporation of the genes of one taxon into the genome of another taxon as a result of hybridization, has been reported in many studies of non-marine hybrid zones (Harrison, 1990). Introgression is reported in 21 cases presented in Table 1 from the 25 cases from which its presence or absence could be determined. This indicates that rates of introgression in examples of marine hybridization are relatively high. Consistent with findings from other environments, introgression has been reported from a wide variety of marine organisms, including plants (but not algae), crustaceans, echinoderms, molluscs, flatfish and finfish.

Introgression often results in the breakdown of species boundaries or the erosion of differences between species. For example, Kwast *et al.* (1990) have shown that with increasing introgression between the starfish *Leptasterias epichlora* and *L. hexactis* in the Puget Sound region, intrapopulation genetic variation increases, whereas interpopulation genetic variation decreases. Not surprisingly, these two species are more genetically similar to one another at sites with high levels of introgression, than at sites with no such introgression or with much lower levels of interspecific gene flow. The same is true of morphological variation in echinoderms (Schopf and Murphy, 1973) and molluscs (Skibinski, 1983; Gardner, 1995).

Introgression, however, does not always result in the breakdown or erosion of species boundaries. Solignac (1976) reports introgressive hybridization between members of the isopod genus *Jaera* in only a limited number of contact zones. It is pointed out that despite the large extent of the geographical distribution of *Jaera*, these introgressive situations remain localized and do not affect the genetic integrity of the species. Further to this, Solignac (1981) has noted that at localities where introgression occurs between *J. albifrons* and *J. praehirsuta*, this is not associated with hybrid breakdown, but at localities where there is no introgression there is hybrid breakdown.

Levels of mean individual heterozygosity are also usually higher in regions of introgression, reflecting new combinations of alleles from the two previously differentiated parental types (Kwast *et al.*, 1990; Gardner, unpublished data from the *M. edulis*/*M. galloprovincialis* hybrid zone in south-west England). The evolutionary importance of such new combinations of genes has long been of interest because it is largely unknown if such novelties can be translated into hybrids with sufficient fitness to be competitive with the parental types. The presence of hybrid swarms suggests that, at least within the localized region of the swarm, these new recombinant genotypes are sufficiently fit to be maintained within the zone of interaction (marine hybrid swarms have been reported for fucoid algae: Knight and Parke, 1950; Burrows and Lodge, 1951; isopods: Solignac, 1976, 1981; and molluscs: Gardner and Skibinski, 1988; Coustau *et al.*, 1991a). The new genotypes are therefore of evolutionary importance because they can contribute to subsequent generations, as opposed to those recombinant genotypes which experience much reduced fitness compared to the parents, and are highly unlikely to pass on their genes to subsequent generations.

How introgression plays a role in the maintenance of hybrid zones, is exemplified by the hybrid clams (*Mercenaria mercenaria* × *M. campechiensis*) from the Indian River Lagoon, Florida. These hybrid clams have an elevated incidence of gonadal neoplasia compared to the parental types. Bert *et al.* (1993) have proposed that this might result from the individual's inability to receive or process signals from tumour-suppressing genes because of the mixing of genes from the two separate species (introgression) and the correspondingly high mutation rates. Thus, introgression is considered to play an important role in promoting hybrid unfitness in this particular hybrid zone.

It has been suggested that under certain circumstances, introgression can be promoted in cases where hybridization rates are low (Grant, 1971). This results from the relative rarity of the hybrids which increases their chances of mating with a parental genotype at high frequency, rather than with another hybrid individual also at low frequency within the population. In this way, genes can most easily cross species boundaries. The data of Table 1 do not indicate that any association exists between the presence or absence of introgression and either low or high rates of hybridization. Such a test will require a larger data set.

15. HYBRIDIZATION BIAS: DOES IT EXIST?

Systematic bias has been recognized from terrestrial examples of hybridization for a long time (Mayr, 1979). For example, terrestrial hybridization amongst plants is often considered to be more prevalent than amongst animals. Hybridization amongst birds is noted as being common, whilst hybridization amongst mammals is much less frequent. Reptiles exhibit much lower rates of natural hybridization than do, for example, amphibians and fishes. With this sort of hybridization bias having been recorded from the terrestrial environment, it is of interest to determine whether such bias exists in the sea and if the proposed explanations for the bias in the terrestrial environment are applicable to the marine environment.

It has long been recognized that certain marine genera exhibit elevated rates of hybridization. Hybridization bias is most obvious within and between invertebrate phyla and between fish families (insufficient data are available from the algae for any such conclusions to be made). Comparison of algal and plant versus animal hybridization rates in the sea provides sharp contrast with such a comparison from the terrestrial environment.

Amongst the recent invertebrate species which hybridize, 50 of the 51 cases involve only three phyla: the Echinodermata ($n = 11$), Mollusca ($n = 27$) and Arthropoda ($n = 12$). Thus, the large majority of phyla with marine representatives are presently not known to hybridize. Within each phylum where hybridization has been recorded there is evidence of hybridization bias. For example, within the echinoderms the classes Echinoidea (sea urchins) and Stellaroidea (starfish and brittle stars) account for all known examples of hybridization, meaning that the classes Crinoidea (sea lilies) and Holothuroidea (sea cucumbers) are not represented. Even within the class Stellaroidea, it is the subclass Asteroidea (the starfish) but not the subclass Ophiuroidea (brittle stars) that account for all the known cases of hybridization.

Within the Mollusca the classes Bivalvia and Gastropoda are well represented, but there is an absence of recorded hybridization in five other classes (the Aplacophora, Cephalopoda, Monoplacophora, Polyplacophora and Scaphopoda). Certain groups of molluscs, notably the abalones (gastropods) and mussels (bivalves), are particularly well represented on a worldwide scale suggesting that hybridization between (sub)species of these shellfish is a frequent and widespread phenomenon.

Of the phylum Arthropoda, only two classes, the Decapoda and the Isopoda, are presently known to exhibit interspecific hybridization, meaning that there is no record of hybridization for the many other marine classes of Arthropods.

Many other phyla which occur extensively or exclusively in the marine realm are presently not known to hybridize. These include, for example, the Annelida, Porifera, Ctenophora and all classes of the Lophophorates.

The findings of apparent hybridization bias in recent examples is largely consistent with the limited palaeontological evidence (discussed later). Of the four known examples of fossil hybridization, two involve heart urchins (Class Echinoidea, Phylum Echinodermata), the third involves sea lilies (Class Crinoidea, Phylum Echinodermata), and the fourth involves trilobites (Subphylum Trilobitomorpha, Phylum Arthropoda).

The reason for hybridization bias is unclear. The reproductive biology of many marine invertebrates involves the absence of mating behaviour, broadcast spawning of gametes followed by external fertilization and a larval stage. It is expected that such a mechanism for reproduction (similar in many respects to terrestrial plants) would be conducive to hybridization in the absence of temporal barriers to interbreeding. Since most marine invertebrates reproduce in this fashion, there is no reason to expect that certain groups of animals will exhibit hybridization whereas other groups will not. Indeed, amongst those invertebrates which do display some form of pre-mating courtship behaviour and/or mate choice, the opportunity for hybridization might be expected to be much diminished, but the decapod crustaceans (crabs, lobsters, etc.), many of which do possess complex mating behaviour, are well represented in Table 1 (p. 70). In short, the presently available data do not provide any help in explaining which groups exhibit high levels of hybridization and why this should be so.

The hybridization bias that is observed might be best explained in terms of study effort spent on the respective groups of animals (this in turn is at least in part related to the economic value of the species concerned). In other words, groups of animals which have received a lot of scientific attention are much more likely to have hybridization events recorded, thus creating an apparent bias in hybridization. This situation can of course only be rectified by more intensive study of the poorly represented groups. However, as a counter to this argument it should be noted that there are currently no reports of hybridization within or between species of certain groups of commercially important invertebrates, for example, the cephalopods.

Fishes have long been recognized as a group of animals which are predisposed towards hybridization. However, within this group, certain biases are quite clear. As far as I am aware, there are no records of hybridization amongst the agnathan (jawless) fishes (hagfishes and lampreys) and, more strikingly, hybridization amongst the Chondrichthyes (cartilaginous fishes such as sharks, rays and skates) is also unreported. This latter is a major group of fishes from which hybridization is noticeably absent.

Hubbs and Kuronuma (1941) have suggested that flatfishes of the families Bothidae and Pleuronectidae resemble the freshwater fishes in the frequency of hybridization between species. This is in line with ecological considerations, because in many areas several species of flounders often coexist in abundance over rather uniform habitats of sand and mud bottom, in shallow to moderate depths. The breeding seasons tend to be rather long, and are synchronous or overlapping for the different species. Whilst the data of Table 1 do indicate that the flatfishes are well represented, there is a higher representation of finfishes (there are 9 cases of hybridization involving flatfish, and 49 involving finfish). Of these 49 cases, 13 families are involved, and the representation ranges from single, isolated instances to 10 cases each being recorded for the families Chaetodontidae and Pomacanthidae.

Hybridization amongst marine mammals is apparently restricted to one species pair, *Balaenoptera musculus* (the blue whale) × *B. physalus* (the fin whale) (Cocks, 1887; Spilliaert *et al.*, 1991). Because of the economic and conservation importance of marine mammals, one might expect that if more cases of hybridization existed, then they would have been recognized by now. As this is not the case, it is possible that the social lifestyle and breeding biology of marine mammals is a natural mechanism by which hybridization can be reduced or prevented.

Perhaps the greatest bias in the frequency of hybridization is between algal/plant species and animal species. Of the 108 cases listed in Table 1 only 6 involve algae and plants. This bias towards animal hybridization is surprising given that in the terrestrial environment the bias is in the other direction (Mayr, 1979), although this imbalance might have been some-what redressed more recently (Hewitt, 1988). Algal species all over the world are of significant economic importance both directly as food and indirectly as sources of chemical compounds, fouling organisms, stabilizers of shifting substrata, etc., so it seems unlikely that the apparent hybridization bias exists simply because the animal species have received more attention than the algal and plant species. There are also more similarities in reproductive biology between marine algae/plants and many marine invertebrates than there are differences (e.g., the absence of pre-zygotic reproductive barriers such as mating behavior, the direct spawning of gametes into the sea followed by external fertilization, etc.) which suggests that the hybridization bias does not result from any profound differences in reproductive biology between these two groups, but more likely reflects a greater propensity of marine animal species towards hybridization than marine plant and algal species. Any explanation for this difference will be speculative until further data are collected which confirm its existence.

It is interesting to note that Mayr (1979) explains the greater hybridization rate of terrestrial plants compared with animals in terms of "the ecological needs and their reproductive mechanisms favor[ing] hybridization" (Mayr, 1979, p. 129) as a result of "the logical consequences of the known differences in ecology and reproduction prevalent in the two kingdoms" (Mayr, 1979, pp. 129–130). The underlying biological explanation for the difference in the terrestrial environment (non-mobility of plants resulting in increased selection for non-genetic plasticity as well as genetic variability which is enhanced by hybridization and is therefore favoured (Mayr, 1979)) applies to many marine invertebrate species, just as it applies to marine algal and plant species. This explanation does not appear to explain the observed bias in hybridization rates between marine plants/algae and marine invertebrates.

The high motility of marine vertebrate species (primarily the fishes) should, according to the arguments proposed by Mayr (1979), mitigate against a high rate of natural hybridization, but the fishes in particular exhibit frequent natural hybridization in the sea, a finding which is consistent with the hybridization frequency of freshwater fishes (Mayr, 1979), suggesting that the motility explanation does not apply to this group of animals either.

In conclusion, the apparently lower rate of hybridization amongst marine plants and algae compared to marine animals must await further examination before an explanation can be advanced, but provides a most interesting comparison with the situation in the terrestrial environment. The explanation, based on differences in motility, advanced by Mayr (1979) to account for differences in the rates of hybridization between terrestrial plants and animals, does not explain the biases observed in the marine realm.

16. INTERGENERIC versus INTERSPECIFIC HYBRIDIZATION

The degree of genetic similarity that exists between two potentially interbreeding individuals will, to a large extent, determine the ability of these individuals to hybridize, all other factors such as behavioural differences and the timing of reproductive activity, being similar. Thus, races, sub-, semi- or incipient species are more likely to successfully interbreed than are members of different genera. This is indeed what is observed (Table 1, p. 70). The number of intergeneric hybridization events listed in Table 1 is only 13 of 108, but interestingly, 9 of these 13 events involve fish (2 cases involve algal species and account for one-third of all algal/plant examples, and two involve echinoderms). Whilst the distinction between interspecific and intergeneric hybridization to some extent reflects the current state of taxonomy and systematics, there is apparently a much greater potential amongst the fishes for intergeneric hybridization to occur than there is amongst any other group of marine organisms. It is hard to advance a satisfactory explanation for this phenomenon. The fish taxa involved in these hybridization events live in close proximity with each other and reproduce by spawning gametes directly into the water column, thereby enhancing the opportunity for cross-reproduction. However, the same can be said of many of the other marine species for which hybridization has been recorded. This suggests, therefore, that the fishes are genuinely predisposed at some genetic level to intergeneric hybridization, a phenomenon not observed in other marine groups.

17. THE BIOGEOGRAPHY OF MARINE HYBRIDIZATION

17.1. Geographical Distribution of Marine Hybridization

Figure 1 shows the geographical distribution of the cases of hybridization reviewed in this paper, within the context of the biogeographic zones recognized by Briggs (1974). It is apparent from this map that hybridization in the sea tends to occur more frequently in some areas than in others. In regions such as around the Florida peninsula, along the Pacific coast of North America (especially in the vicinity of Puget Sound), in western Europe, and in the Belt Sea (at the boundary between the Baltic Sea and the North Sea), there is considerable coincidence of hybridization involving many different species and phyla (these four regions account for 55 of the 108 hybridization events recorded in Table 1). An apparent coincidence of hybridization has been noted for certain regions in the terrestrial environment, but it remains unclear as to whether or not hybrid zones occur at higher than expected frequency in certain "hot spots" (Harrison, 1990).

It might be argued that the locations of many marine hybrid zones reflect the locations of marine research laboratories. There is probably some truth to this, but there are sufficient exceptions to this generalization to consider that genuine "hot spots" for hybridization do exist. For example, no cases of hybridization have been reported from the Mediterranean Sea or from around the coasts of South Africa, yet both of these two regions are well represented in the marine literature. Further, the three main "hot spots"

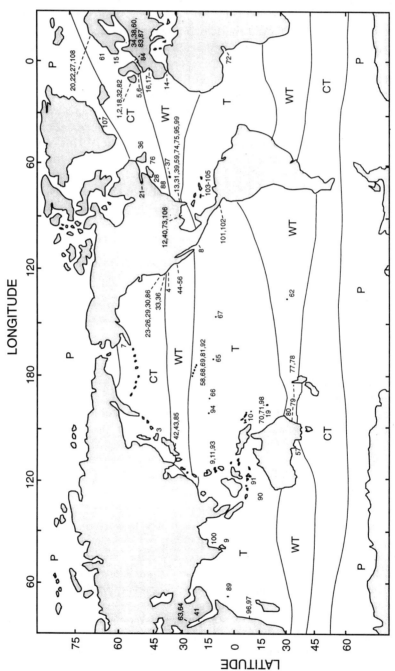

Figure 1. Worldwide distribution of marine hybridization (numbers correspond to the listing in the left-hand column of Table 1). P, polar; CT, cold temperate; WT, warm temperate; T, tropical. (Biogeographic zones are from Briggs, 1974.)

(Florida, Puget Sound and the Belt Sea) are all relatively small areas in terms of the number of kilometres of coastline, compared with, for example, the thousands of kilometres of Atlantic coastline of North America from northern Labrador to South Carolina (with its many marine laboratories) where only five cases of marine hybridization have been reported.

At present, there are no recorded examples of marine hybridization in polar regions, 1 case at the boundary between polar and cold temperate regions, 32 cases in the cold temperate zone, 26 cases at the boundary of cold temperate and warm temperate regions, 1 case in warm temperate regions, 14 cases at the warm temperate to tropical boundary and 33 cases in tropical regions. There is no discernible pattern to this distribution; hence hybridization in the sea, when viewed across all species, genera and phyla, is independent of sea water temperature.

Certain groups of organisms display patterns or trends in the location of hybridization with respect to sea water temperature. For the algal/plant examples, all cases occur in or around the cold temperate region, but with only six cases to refer to it is difficult to draw any definitive conclusions about the distribution of these hybrid zones with respect to water temperature. All examples of echinoderm hybridization are from the cold temperate region, yet echinoderms are a well-represented group in all regions of the world. Amongst the flatfish (families Bothidae and Pleuronectidae) all recorded hybridization events have been from cold temperate regions, yet economically important flatfish fisheries exist outside this biogeographic zone. Amongst the other families of fishes, the majority of cases (35 of 40) occur in warm temperate or tropical areas. Explanations for these examples are hard to find, other than to say that for the fishes, the number of cases of hybridization per zone probably reflects an underlying difference in the distributions of the two types of fishes (more flatfish in cooler waters, more finfish in warmer waters).

17.2. Ecological Differentiation

In this review it has previously been pointed out that the sea has a relatively continuous distribution, and sharp ecotonal differences between areas of the type observed on land are usually absent from the sea. The most pronounced changes in the physical nature of the sea often occur at the boundaries between biogeographic zones (ecotones). Associated with this is a high proportion (41 of 108 cases) of hybridization events. Given that such boundaries occupy only a small percentage of the sea, this high representation of marine hybridization at ecotones is of interest.

Hewitt (1988) estimated that at least 25% of all hybrid zones occurred at or near an environmental transition, and Harrison (1990) observed that many interacting taxa of the hybrid zones that he reviewed show habitat associations which arise as a consequence of preferences expressed by the taxa for different habitats, or from differential survivorship within the different habitats. Examples of the former are uncommon, those of the latter more frequent. Consistent with Harrison's (1990) findings, many of the hybridization events described in Table 1 are associated with ecological differences between the hybridizing taxa, and not with habitat preference choice. Not surprisingly, salinity, depth and temperature are the most frequently cited factors which affect distributions.

A high proportion of marine hybridization events occur at the boundaries between biogeographic zones (38%), suggesting that marine hybrid zones are non-randomly distributed in space. Further, analysis of the long-term stability of the majority of marine hybridization events indicates that they are reasonably constant in time and space (as far as we can measure—see Section 8). This suggests that many marine hybrid zones are associated with the environment, and lie in their present position because of ecological differences between the taxa involved. By contrast, tension zones are not static and their position is unrelated to ecological patterns of distribution, reflecting primarily low population density (density troughs), and dispersal and fitness differences between the genotypes. Thus, tension zones are expected to be randomly distributed in space, so the occurrence of more than one-third of the examples of marine hybridization at ecotones is unexpected, and best explained by the ecological preferences and fitness differences of genotypes associated with the habitat differences on either side of the ecotone.

18. THE FOSSIL RECORD

Whilst hybridization amongst recent biota is now accepted to be a fairly common phenomenon, hybridization as recorded in the fossil record is much less frequent. Geary (1992) suggests that this results for two main reasons: first, the fossil record lacks the resolution required to interpret speciation on a meaningful time-scale (10^2–10^4 years); and secondly, it is particularly difficult to disentangle genetic from environmental influences. A third explanation involves the incomplete representation of the fossil record itself. We know from present-day hybridization events that hybridization can be short-lived on geological time-scales, and is often restricted in space to small geographic areas of sympatry or parapatry. Based upon knowledge of recent hybridization events, the number of hybrid offspring produced can be extremely small when compared to the number of parental individuals present in the population. All of these factors will mitigate against hybrid individuals being preserved in the fossil record. Despite these shortcomings, reports of interspecific hybridization involving marine biota do exist and provide valuable information for comparative purposes with examples of recent hybridization.

I am aware of four cases of possible marine hybridization in the fossil record. Three involve echinoderms, adding new records of marine hybridization for two groups of animals apparently unrepresented amongst examples of recent hybridization: the heart urchins and the crinoids. The fourth involves putative hybridization amongst trilobites. In all four cases the species involved in the hybridization are extinct, but in three of the cases the group of animals is still extant (the heart urchins and the crinoids). Whilst the case for hybridization in the fossil record can never be definitive because morphological species do not necessarily represent biological species, it can be strong when based upon morphological intermediacy between what are recognized as "good" or well-defined species. The following examples all appear to meet this requirement.

Nichols (1959) described heart urchins which were morphometrically intermediate between *Micraster coranguinum* and *M. senonensis* (Phylum Echinodermata) from Kent, southern England, as well as individuals which were intermediate between *M. glyphus* and *M. stolleyi* from Norfolk, eastern England. Both sets of putative hybrid heart urchins were collected from chalk beds dating from the Upper Cretaceous (approximately 64–100 million years BP). Kermack (1954) had previously suggested, based upon differences in test shape and structure, that *M. coranguinum* and *M. senonensis* occupied different niches in the same area of distribution (the former is postulated to be a burrower whilst the latter is postulated to be a surface dweller). The numerous morphologically intermediate forms found between *M. coranguinum* and *M. senonensis* (putative hybrids represent approximately 21% of the total number of specimens examined) indicate extensive interbreeding between the two taxa, and may even constitute a hybrid swarm. If this is the case then this is likely to be one of the oldest, if not the oldest, hybrid swarm known to have existed.

Three specimens which were morphological intermediates between the well-defined crinoid species *Eretmocrinus magnificus* and *E. praegravis* (phylum Echinodermata) have been described from the Lower Mississippian (325–345 million years BP) Fort Payne Formation of south-central Kentucky, USA, by Ausich and Meyer (1994). These three individuals were recognized using multivariate statistical techniques which clearly separated the two parental types whilst at the same time revealing the intermediate nature of the putative hybrids. These hybrids represented

0.8% of the total sample size, a figure which is consistent with present-day records of the occurrence of hybrids amongst echinoderm taxa (Table 1). The authors suggested that the Fort Payne Formation occurs at the extremes of the environmental ranges of the respective species (but where both species were reasonably successful). This indicates the possibility of both environmental stress and a reduction in the number of prospective conspecific partners for reproduction as explanations for this example of interspecific hybridization (see previous sections for more detailed discussion of these phenomena).

The final example of marine hybridization in the fossil record involves trilobite species of the Viola group (*Cryptolithoides* and *Cryptolithus* of the phylum Arthropoda). Shaw (1991) described morphologically intermediate trilobites collected from the Ordovician period (435–500 million years BP) of Oklahoma, USA. The convergence of genera is interpreted as "hybridization of mere subspecies after several million years of incomplete geographic separation" (Shaw, 1991, p. 919). *Cryptolithoides* (the western form) and *Cryptolithus* (the eastern form) are assumed to be closely related descendants of an ancestor that migrated to the Ordovician North American continent (Laurentia) from the rim of Gondwana. The two subspecies were geographically isolated until the marine transgression of the late Middle Ordovician, at which time present-day Oklahoma was the area of hybridization between the two taxa, which rarely penetrated further into each other's territory. As with the fossil crinoid example, this trilobite example suggests the possibility of both environmental stress and relative absence of conspecific partners for reproduction as explanations for the occurrence of hybridization between these two forms. Dating from approximately 450–470 million years BP, this example is probably the oldest case of (postulated) hybridization so far recorded.

Care should be taken when attempting to extrapolate from a limited number of studies of fossil organisms which exhibit possible hybridization to the more recent situation. With this caveat in mind we can, however, ask the question: "What does hybridization between marine species as recorded in the fossil record tell us about hybridization events in general and their relevance to speciation?" First, although being relatively few in number, these examples confirm what we might suspect—that hybridization is not a new phenomenon and has been going on for millions of years. Indeed, the example of the putative hybridization between trilobite taxa is probably the oldest account of hybridization presently known. Because it is generally believed that metazoan animals did not invade the land until about the late Silurian (395–435 million years BP) (Stormer, 1977; Seldon, 1990), the trilobite hybridization event must pre-date any such terrestrial event. Secondly, the basic processes resulting in hybridization appear to be similar for both fossil and recent taxa. These include fundamental processes such as geographic isolation with the development of genetic/ morphological differences followed by an event (such as a marine transgression) which establishes secondary contact. At least two of the records of fossil marine hybridization specifically mention that the hybridization event occurred at the boundaries of the distributions of the taxa involved (?ecotones), consistent with what we know from recent examples of hybridization. In accordance with findings from recent hybridization events, the roles of environmental stress and the relative absence of a reproductive partner of the same taxon are both emphasized by these examples as possible mechanisms resulting in hybridization.

19. HYBRIDIZATION AS A FISHERIES AND CONSERVATION PROBLEM

For the individual organism, hybridization results in a mixed and reorganized genome, whilst for the population, hybridization constitutes the mixing of two previously isolated gene pools. As such, hybridization adversely affects the management of organisms as discrete stocks, populations or species. Furthermore, hybridization has the potential to eradicate rare and endangered species (Campton, 1987). For these reasons, hybridization is viewed as being of importance to fisheries science and conservation.

The point has already been made that many examples of natural marine hybridization are economically important (e.g. the use of *Spartina* for stabilizing sediments) or involve commercial fisheries (see Table 1), both vertebrate and invertebrate. Indeed, 51 of the 108 cases listed in Table 1 have some economic relevance (this excludes those fish hybrids which are collected by hobbyists and aquaria). If it was not for this commercial importance many of these examples would probably have remained unrecorded. Significantly, hybridization between marine species can act to both the benefit and the detriment of fisheries biology, as described subsequently.

Much effort has been directed by fisheries biologists towards obtaining experimental hybrid crosses between various species with the intention of maximizing productivity, usually as a result of hybrid vigour (heterosis). Thus, genetic traits associated with natural hybridization are exploited in an experimental setting to maximize the potential return of the fishery. There has been, for example, a relatively long history of studies of aspects of experimental and natural hybridization of the hard clam (*Mercenaria mercenaria* and *M. campechiensis*) fishery off the south-east coast of the United States (e.g. Loosanoff, 1954; Chestnut *et al.*, 1956; Haven and

Andrews, 1956; Porter and Chestnut, 1960; Menzel, 1962, 1977, 1985; Menzel and Menzel, 1965; Humphrey and Crenshaw, 1985; Hesselman *et al.*, 1988; Dillon and Manzi, 1989; Dillon, 1992; Bert *et al.*, 1993). Studies of these two naturally hybridizing bivalves have shown that in Virginia (Haven and Andrews, 1956), North Carolina (Chestnut *et al.*, 1956) and Florida (Menzel, 1962, 1977) hybrids have better growth rates than *M. mercenaria* (the northern, colder water form) and as good as those of *M. campechiensis* (the southern, warmer water form). Menzel (1977) experimentally hybridized *M. mercenaria* with *M. campechiensis* to obtain F1, F2, F3 and backcross clams (with no apparent loss of fertility) which exhibited percentages of successful rearing through metamorphosis comparable to *M. mercenaria* (85%) and greater than *M. campechiensis* (50%). These hybrid progeny had growth rates equivalent to *M. campechiensis* and better than *M. mercenaria* in warmer waters, as well as having as good a shelf-life as *M. mercenaria* and better than *M. campechiensis*. Menzel (1985) concluded by recommending the F1 hybrid as the preferred type for use by mariculturists in southern waters.

Products of natural hybridization events can be of considerable economic benefit within certain fisheries. Menhaden fishes (*Brevoortia smithi* the yellowfin menhaden and *B. tyrannus* the Atlantic menhaden: family Clupeidae) from the east coast of Florida constitute an important fish oil and crab bait industry. In the vicinity of Indian River Lagoon, on the Atlantic coast of Florida, these two species interbreed extensively and form abundant hybrids. The natural seasonal fat cycles and rates of spoilage as bait of these two species were investigated by Dahlberg (1969a), who showed that *B. tyrannus* lays down greater fat stores than *B. smithi* at most times of the year and that *B. tyrannus* attracts more blue crabs when used as bait. However, *B. smithi* has a much slower spoilage rate than *B. tyrannus* when used as crab bait. The natural hybrids exhibit intermediate fat storage levels and spoilage rates when compared with the parental types and thus, from the point of view of the blue crab fishery, represent a natural compromise of considerable value.

In an experimental setting, F1 hybrids are very useful for gene mapping because they are heterozygous at many loci. This can be exploited by crossing such hybrids with individuals which are multiply homozygous to permit an analysis of gene linkage amongst the offspring (Campton, 1990). This in turn provides valuable information about the quantitative genetics of the species in question, and such knowledge can be usefully applied to the selective breeding of strains for restocking of depleted (overexploited) populations or seeding of presently unoccupied areas. Other aspects of experimental hybridization which are of considerable utility to fisheries biologists include ploidy manipulation, gene regulation and genetic tagging (Campton, 1990).

Natural hybridization can, however, be a problem for managers of commercial fisheries. For example, hybridization between the tanner crabs *Chionoecetes opilio* and *C. bairdi* in the Bering Sea presents difficulties "in resource assessment studies and is particularly problematical in commercial catch regulation because it is often difficult to distinguish the hybrid from its parent species" (Grant *et al.*, 1977, p. 127). Furthermore, female hybrid snow crabs have reduced fecundity (many are not gravid at maturity or have abnormally small egg clutches with large numbers of dead eggs) when compared to the parental types (male hybrid fecundity was not investigated) (Johnson, 1976). Natural hybrids of the hard shell clams *M. mercenaria* and *M. campechiensis* from the Indian River Lagoon, Florida, experience elevated levels of gonadal neoplasia, a tumorous condition which interferes with and possibly suppresses gametogenesis, and is probably fatal (Hesselman *et al.*, 1988; Bert *et al.*, 1993). The ratio of *M. mercenaria* : *M. campechiensis* : hybrid amongst non-diseased clams was 1.00 : 0.11 : 0.42, but amongst diseased individuals this ratio changed to 1.00 : 0.22 : 1.63. In total, 21.6% of hybrids, 6.5% of *M. mercenaria* and 11.8% of *M. campechiensis* exhibited gonadal neoplasia (species designations were based on electrophoretic criteria) (Bert *et al.*, 1993). This example of pronounced hybrid unfitness probably arises as a result of the break-up during hybridization of the co-adapted gene complexes (including proto-oncogenes and tumour suppressing genes) of the parental types since the neoplasia is not viral in origin and is apparently unrelated to localized environmental conditions. Hybridization apparently disrupts the genetic mechanisms of cell transformation and differentiation, but not the gene associations controlling cell proliferation (Bert *et al.*, 1993).

One of the fundamental aspects of fisheries biology is the conservation of stocks from over-exploitation and from the introduction of species which might adversely affect native species by, for example, hybridization (Campton, 1990). This requires that populations of native species which are threatened by hybridization and introgression be identified and isolated or eradicated to preserve the integrity of the gene pool of the native species. However, in many instances, a non-native species has been introduced into a region with the explicit purpose of establishing or improving upon a fishery. It is often the case that little attention is paid to the possibility of cross-breeding between the introduced and native stock, or when this is the specific intention of the introduction, too little attention has been paid to preventing the spread of introduced genes throughout a much larger range than was originally envisaged.

Hedgecock *et al.* (1993) document the example of two Japanese oysters introduced to the west coast of America for the aquaculture industry. The Pacific oyster (*Crassostrea gigas*) and the Kumamoto oyster (*Crassostrea sikamea*) were found to hybridize in Washington and California, and a genetic survey of the Kumamoto oyster at these sites revealed that these stocks had been contaminated by stocks of the Pacific oyster (Pacific oyster sperm can fertilize Kumamoto oyster eggs). It was thus concluded that hybridization and introgression pose very serious problems to the maintenance of the genetic integrity of *C. sikamea* in America, and that steps should be taken to eradicate *C. gigas* and interspecific hybrids in American Kumamoto oyster brood stocks to prevent the loss of genetic diversity within these brood stocks. Ironically perhaps, the only Kumamoto oysters now known to exist are the brood stocks introduced into America: attempts to identify *C. sikamea* from the Kumamoto area of Japan have revealed only *C. gigas* individuals (Hedgecock *et al.*, 1993 and references therein). This of course reinforces the importance of the preservation of the genetic integrity of the Kumamoto stocks in America. Hybridization and potential introgression with *C. gigas* represent a serious threat to the conservation of this species on a worldwide scale.

20. HYBRIDIZATION AND SPECIATION IN THE SEA

It is often suggested that hybridization is an important source of new genetic variation in certain environments (Lewontin and Birch, 1966). Indeed, Woodruff (1989) has coined the term "hybrizyme" for alleles which are unique to regions of interspecific hybridization involving the pulmonate land snail *Cerion*. Examination of the literature shows that the occurrence of "hybrizymes" is relatively widespread in many hybrid zones, although all examples, as far as I am aware, are restricted to non-marine environments. This just as likely reflects the under-representation of the study of hybrid zones in the sea, as the absence of hybrizymes from marine hybrid zones.

If hybrid individuals display consistently lower fitness than parental types, and are selected against, then hybridization may be viewed as an interesting anomaly, but of very little evolutionary importance. However, if hybrid individuals possess intermediate fitness or fitness that is greater than both parental types, then the evolutionary impact of the mixing of the two parental genomes is of much greater interest and importance, because such hybrid individuals will not be swamped by pure genotypes. There is sufficient evidence from studies of hybridization among a wide variety of different taxa in the sea to indicate that hybrids do not always experience reductions in parameters related to fitness (see Section 11). This suggests that hybridization will play an important role in the speciation process in the sea. As pointed out by Arnold and Hodges (1995), the first generations of hybrids in any population may be at a fitness disadvantage compared to the parental types and thus represent a bottleneck to the production of subsequent hybrids, but at the same time they may be an important step in the production of future hybrids (and individuals of mixed ancestry) which have relatively fit genotypes (e.g. in hybrid swarms). This in turn could result in the generation of individuals which are adapted preferentially (compared to the parental types) to the environment in which the hybrid zone is located (the bounded hybrid superiority hypothesis of Moore, 1977), and ultimately perhaps to speciation. As noted earlier, the existence of several marine hybrid swarms indicates that individuals of mixed ancestry can coexist with parental types in the region of the hybrid zone, i.e. they are not preferentially selected against. Hence the novelties created by the recombination of previously separated genomes can be of significance to the speciation process.

Reproductive isolating mechanisms between taxa can, in theory, be developed and perfected by hybridization if hybridization is rare enough and the hybrids are inviable (Mayr, 1979). This results from the selective advantage enjoyed by individuals who mate preferentially with their own taxon, and in so doing do not "waste" their gametes by mating with another taxon which would produce inviable offspring. Thus, speciation by reinforcement should result in the preferential mating of individuals of one taxon with other such individuals, and not with other taxa. The evidence available from hybridization events in the sea does not permit the recognition of the establishment of speciation by reinforcement. Many of the examples listed in Table 1 involve the production of either a relatively high frequency of hybrid individuals (possibly as a hybrid swarm) or isolated instances of hybridization between well-established species. Neither of these two cases will prove to be particularly helpful in the recognition of speciation by reinforcement. Of the cases of hybridization between closely related species, subspecies or semi-species which involve the production of a small number of hybrids, insufficient data exist to be able to judge whether or not such hybrid offspring are in fact inviable. This is not to say that speciation by reinforcement is absent from the sea, but reflects the greater ease of its recognition on land (particularly among insects) than in the sea. The decapod crustaceans, many of which employ elaborate pre-copulatory mating behaviour as a species recognition device, might well prove particularly valuable model organisms for the demonstration of speciation by reinforcement amongst marine biota.

Whilst hybridization between coral species might be an important source of evolutionary novelties (for example, polyploid individuals—see later), it was pointed out that hybridization could just as easily "be retarding divergence among evolutionary lineages and contributing to the high morphological variability characteristic of many species" (Wallace and

Willis, 1994, p. 255). The constant erosion of genetic and non-genetic differences between taxa is a logical result of hybridization, and may play an important role in reducing the tempo of speciation between races and species complexes.

Richmond (1990) has discussed the potential importance of hybridization as a mechanism resulting in speciation in the marine environment for corals. Several observations support the concept of hybridization as a potential mechanism in coral speciation. For example, scleractinian corals and many marine invertebrates occur in very high densities and diversities, and in many species reproduction occurs via mass spawnings (Harrison *et al.*, 1984; Babcock *et al.*, 1986, 1992). It is easy to envisage the formation of a hybrid larva as a result of a synchronous multispecies spawning event. The larva is viable, settles, metamorphoses and grows. Because many coral species have long generation times (permits opportunities for backcrossing) and can reproduce asexually as well as sexually, this hybrid individual can contribute new recruits to the population. Since many coral species boundaries are based on morphological variation (morphospecies) the hybrid could easily be recognized as a new species by morpho-taxonomists. On balance though, the majority of mass spawning corals are not presently known to be able to hybridize, and there are many species for which barriers to hybridization are known to exist (e.g. Hodgson, 1988; Wallace and Willis, 1994).

A recent review of the systematics of the coral genus *Acropora* has also highlighted the role that hybridization can play in promoting speciation. Laboratory crosses have demonstrated that hybrids between certain coral species are viable (Wallace and Willis, 1994; Miller and Babcock, submitted), and it has been suggested that unusual chromosome counts in some species of *Acropora* may be evidence of polyploid individuals generated by natural hybridization (cited in Wallace and Willis, 1994). Thus, Wallace and Willis (1994, p. 247) concluded that "it seems likely that hybrid corals could be a significant component of adult coral assemblages . . .". One is left to wonder at just how many new, unreported coral species, owing their existence to polyploidy, are yet to be identified in one of the world's most speciose environments.

An excellent example of speciation resulting from allopolyploidy following on from natural hybridization involves the marsh grass *Spartina*. This example has been well documented (Gray *et al.*, 1991), and the general details have been noted previously in this review (Section 9.2). This is one of the best-documented examples of speciation resulting from hybridization in any environment, and supports theoretical considerations which suggested that such an event was possible if preceded by chromosomal doubling.

It is clear that hybridization in the sea can, and does, result in speciation.

It is also clear that we are largely ignorant of how often this mechanism plays a role in the generation of a new species.

21. CONCLUSIONS AND PROSPECTS FOR THE FUTURE

More than anything else, by writing this review I hope to have raised the profile of natural hybridization in the sea, a topic which has been largely overlooked, probably because it was considered to occur so infrequently. It should be apparent that this is not the case: hybridization in the sea occurs at a rate almost identical to that in terrestrial systems. Examination of the body of data collected from studies of marine hybridization indicates that most of the properties and processes of hybridization in the sea are very similar to those described for terrestrial hybrid zones (why would they not be?).

It is apparent that marine ecologists and marine evolutionary biologists have much progress to make in the description, analysis and modelling of individual marine hybrid zones. Only a multidisciplinary approach involving biochemical, mitochondrial and nuclear genetic markers, supported by data from studies of behaviour, morphology, reproductive ecology and fertilization potential (where appropriate), will provide the kind of detail that is required to answer questions about the origin, maintenance, evolutionary and, in many cases, economic importance of hybridization in the sea. This kind of information is presently available for less than 10 marine hybrid zones.

One area of interest in particular that studies of marine hybridization might be able to address, is the stability of hybridization in time and space. Many marine invertebrates possess hard body parts which are particularly well preserved, hence the fossil record of many marine species is relatively good. Marine hybrid zones are the widest known from any ecosystem, so it should be possible to use multivariate morphometric studies of (sub)fossil material to look at the duration and geographic distribution of hybridization. This will go some way to addressing the question of long-term temporal and geographic stability of hybrid zones, at least in the sea.

As far as prospects for future study are concerned, the good news is that hybridization and hybrid zones are common in the sea. Certain "hot spots" can be recognized where previously unreported hybrids might be found. Many of these hybrid zones occur in highly accessible regions such as the intertidal and shallow subtidal zones, and exist in groups of organisms which are widely distributed and usually occur at relatively high densities. All of these factors aid future study and provide excellent

Table 1 Summary table of recent hybridization in the sea.

Taxa involved	Location	Biogeographic zone	Characters examined	Hybrids fertile	Introgression	Hybrid unfitness	Disturbance	% hybrids	Ecological differences	Economic importance	Reference
Kingdom Plantae											
Division Phaeophyceae											
Family Fucales											
1. *Fucus serratus* × *F. vesiculosus*	NW Scotland to SW England NE Atlantic	CT	M	—	—	Y	Y	Hybrid swarm	Height on shore	N	10, 56
2. *Fucus spiralis* × *F. vesiculosus*	Isle of Man Irish Sea	CT	M	—	—	Y	Y	65%	Height on shore	N	10
Family Laminariales											
3. *Alaria crassifolia* × *Laminaria angustata*	N Japan NW Pacific	CT	M	—	—	—	—	$N=1$	—	Y	96
4. *Macrocystis pyrifera* × *Pelagophycus porra*	S California E Pacific	CT/WT	M	N	?N	?Y	U	3% ($N=6$) to "relatively high frequency"	—	Y	15, 16, 69
Division Tracheophyta											
Family Gramineae											
5. *Spartina maritima* × *S. alterniflora* = *S. anglica*	S England NE Atlantic	CT	A, M, K	Y	Y	N	H	Can be common	—	Y	43
6. *Spartina maritima* × *S. alterniflora* = *S. x. townsendii*	SE England NE Atlantic	CT	A, M, K	N	Y	Y	H	Can be common	—	Y	43
Kingdom Animalia											
Phylum Arthropoda											
Subphylum Crustacea											
Order Decapoda											
Family Majidae											
7. *Chionocetes opilio* × *C. bairdi*	Bristol Bay SE Bering Sea	CT	A, M	Y	—	Y	—	0.4–20%	—	Y	42, 53, 55
Family Ocypodidae											
8. *Uca princeps princeps* × *U. p. monilifera*	Mexico E Pacific	WT/T	M	—	—	—	—	$N=2$	—	N	17
9. *Uca lactea perplexa* × *U. l. annulipes*	SE India to Philippines	T	M	—	—	—	—	Rare to common	—	N	17
10. *Uca vocans vomeris* × *U. v. pacificensis*	New Guinea W Pacific	T	M	—	—	—	—	—	—	N	17
11. *Uca vocans vocans* × *U. v. pacificensis*	Philippines W Pacific	T	M	—	—	—	—	—	—	N	17

Taxa involved	Location	Biogeographic zone	Characters examined	Hybrids fertile	Introgression	Hybrid unfitness	Disturbance	% hybrids	Ecological differences	Economic importance	Reference
Family Xanthidae											
12. *Menippe adina* × *M. mercenaria*	Florida Gulf of Mexico	WT/T	A, C	Y	Y	?Y^A/?N^B	—	34%	Depth and habitat type	Y	5, 6, 105
13. *Menippe adina* × *M. mercenaria*	S Carolina to Florida W Atlantic	WT/T	A, C	Y	Y	Y	—	—	Depth and habitat type	Y	5, 6
14. *Xantho incisus incisus* × *X. i. granulicarpus*	Portugal E Atlantic	WT	M	—	—	—	—	—	Depth	N	1
Order Isopoda											
Family Asellota											
15. *Jaera praehirsuta* × *J. ischiosetosa*	Nordost-Seekanal Baltic Sea	CT	M	—	—	—	—	Hybrid swarm	Height on shore	N	87, 88
16. *Jaera albifrons* × *J. praehirsuta*	NW France NE Atlantic	CT/WT	M	—	Y	—	—	20% and hybrid swarm	Height on shore	N	87, 88
17. *Jaera forsmani* × *J. praehirsuta*	W France NE Atlantic	CT/WT	M	—	—	—	—	Hybrid swarm	Height on shore	N	87, 88
18. *Jaera albifrons* × *J. ischiosetosa*	S Wales NE Atlantic	CT	M	—	—	—	S	0.085% ($N = 1$)	Height on shore	Y	54
Phylum Cnidaria											
Class Anthozoa											
Family Faviidae											
19. *Platygyra* spp. (7 morphological species)	Australia SW Pacific	T	M, A	?Y	—	N	—	Hybrids assumed to be common	Probably not	N	66–68
Phylum Echinodermata											
Class Echinoidea											
Family Echinidae											
20. *Echinus acutus* × *E. esculentus*	Norway NE Atlantic	CT	M, S	N	N	Y	—	10–20%	?	N	45
Family Strongylocentrotidae											
21. *Strongylocentrotus droebachiensis* × *S. pallidus*	Gulf of St Lawrence NW Atlantic	CT	C, M	—	—	—	—	—	Depth	Y	94
22. *Strongylocentrotus droebachiensis* × *S. pallidus*	Norway NE Atlantic	CT	C, M	—	—	—	—	0.53%	Depth	Y	102
23. *Strongylocentrotus droebachiensis* × *S. pallidus*	Puget Sound NE Pacific	CT	C, M	—	—	Y	U	—	Depth	Y	92
24. *Strongylocentrotus droebachiensis* × *S. purpuratus*	Puget Sound NE Pacific	CT	C, M	?Y	—	?Y	U	—	Very small, if any	?Y	93

Taxa involved	Location	Biogeographic zone	Characters examined	Hybrids fertile	Introgression	Hybrid unfitness	Disturbance	% hybrids	Ecological differences	Economic importance	Reference
25. *Strongylocentrotus droebachiensis* × *S. franciscanus*	Puget Sound NE Pacific	CT	M	—	—	—	U	—	Very small, if any	?Y	93
26. *Strongylocentrotus purpuratus* × *S. franciscanus*	Puget Sound NE Pacific	CT	M	—	—	—	U	—	Very small, if any	?Y	93
27. *Strongylocentrotus droebachiensis* × *Psammechinus miliaris*	Norway NE Atlantic	CT	M	—	—	—	—	—	?N	N	45
Class Stelleroidea											
Family Asteriidae											
28. *Asterias forbesi* × *A. vulgaris*	Cape Cod NW Atlantic	CT	A, M	Y	?N	—	—	1.4% (N=295)	Depth	Y*	63, 83
29. *Leptasterias epichlora* × *Pisaster ochraceus*	Puget Sound NE Pacific	CT	M	—	—	—	U	—	Temperature	N	103
30. *Leptasterias epichlora* × *L. hexactis*	Puget Sound NE Pacific	CT	A, C, M	?Y	?Y	?Y[C]/N[D]	U	38%	—	N	11, 58, 103
Phylum Mollusca											
Class Bivalvia											
Family Mytilidae											
31. *Geukensia demissa* × *G. granosissima*	Florida W Atlantic	WT/T	A, M	—	—	—	—	19%	N	N	79
32. *Mytilus edulis* × *M. galloprovincialis*	W Europe NE Atlantic	CT/WT	A, C, K, M, Mt	Y	Y	N	—	Hybrid swarm	Salinity Temperature Wave action	Y	4, 13, 14, 27, 32–36, 41, 85, 86, 106
33. *Mytilus trossulus* × *M. galloprovincialis*	California and Oregon E and NE Pacific	CT/WT	A, M	—	—	—	U	12.6%	Salinity Temperature	?Y	62, 79
34. *Mytilus edulis* × *M. trossulus*	Belt Sea Baltic Sea	CT	A, M	Y	Y	—	—	Hybrids common	Salinity Temperature	N	62, 99, 101
35. *Mytilus edulis* × *M. trossulus*	Canada NW Atlantic	CT	A, C, M	Y	Y	—	—	Hybrids common	Salinity Temperature	Y	3, 37–39, 57, 62
Family Ostreoidea											
36. *Crassostrea gigas* × *C. sikamea*	Pacific coast, USA	CT/WT	A, M, Mt	Y	—	—	H	5%	Y	Y	47
Family Tellinidae											
37. *Tellina magna* × *T. laevigata*	Bermuda W Atlantic	WT/T	C, M	—	—	—	—	10% (N=1)	—	N	8

Taxa involved	Location	Biogeographic zone	Characters examined	Hybrids fertile	Introgression	Hybrid unfitness	Disturbance	% hybrids	Ecological differences	Economic importance	Reference
38. *Macoma balthica* (North Sea race × Baltic Sea race)	Belt Sea, Baltic Sea	CT	A, C	Y	Y	?Y	—	—	Salinity	N	100
Family Veneridae											
39. *Mercenaria mercenaria* × *M. campechiensis*	S Carolina to Florida, W Atlantic	WT/T	A, K, M	Y	Y	$Y^E N^F$	—	0.06–88%	Depth, Salinity, Temperature	Y	7, 23, 24, 49, 64, 65, 72
Class Gastropoda											
Family Buccinidae											
40. *Stramonita haemastoma floridana* × *S. h. canaliculata*	Florida, Gulf of Mexico	WT/T	A, M	—	—	—	—	1.6–9.5%	N	N	59
Family Cypraeidae											
41. *Cypraea tigris* × *C. pantherina*	Gulf of Aden, Red Sea	T	C, M	?Y	—	—	—	30%	—	N	82
Family Haliotidae											
42. *Haliotis discus discus* × *H. d. hannai*	N, E and S Japan, W Pacific	CT/WT	A, M	—	—	?Y	—	22–46%	Y	Y	30, 81
43. *Haliotis discus discus* × *H. gigantea*	S and W Japan, W Pacific	CT/WT	A, M	—	—	—	—	33–57%	—	Y	81
44. *Haliotis rufescens* × *H. k. kamtschatkana*	S California, E Pacific	CT/WT	C, M	—	—	—	U	N=1	—	Y	95
45. *Haliotis rufescens* × *H. sorenseni*	S California, E Pacific	CT/WT	C, M	Y	—	—	U, Y	N>1000	Y	Y	71
46. *Haliotis rufescens* × *H. corrugata*	S California, E Pacific	CT/WT	C, I, M	—	—	—	U, Y	N=70	Y	Y	71
47. *Haliotis rufescens* × *H. k. assimilis*	S California, E Pacific	CT/WT	C, M	—	—	—	U, Y	N=40	Y	Y	71
48. *Haliotis rufescens* × *H. walallensis*	S California, E Pacific	CT/WT	C, M	—	—	—	U, Y	N=2	Y	Y	71
49. *Haliotis rufescens* × *H. fulgens*	S California, E Pacific	CT/WT	C, M	—	—	—	U, Y	N=2	Y	Y	71
50. *Haliotis corrugata* × *H. fulgens*	S California, E Pacific	CT/WT	C, M	—	—	—	U, Y	N=25	Y	Y	71
51. *Haliotis corrugata* × *H. walallensis*	S California, E Pacific	CT/WT	C, M	—	—	—	U, Y	N=24	Y	Y	71
52. *Haliotis corrugata* × *H. sorenseni*	S California, E Pacific	CT/WT	C, M	—	—	—	U, Y	N=7	Y	Y	71
53. *Haliotis corrugata* × *H. k. assimilis*	S California, E Pacific	CT/WT	C, M	—	—	—	U, Y	N=1	Y	Y	71
54. *Haliotis k. assimilis* × *H. sorenseni*	S California, E Pacific	CT/WT	C, M	—	—	—	U, Y	N=4	Y	Y	71

Taxa involved	Location	Biogeographic zone	Characters examined	Hybrids fertile	Introgression	Hybrid unfitness	Disturbance	% hybrids	Ecological differences	Economic importance	Reference
55. *Haliotis k. assimilis* × *H. sorenseni*	S California E Pacific	CT/WT	C, M	—	—	—	U, Y	N=1	Y	Y	71
56. *Haliotis sorenseni* × *H. walallensis*	S California E Pacific	CT/WT	C, M	—	—	—	U, Y	N=1	Y	Y	71
57. *Haliotis rubra* × *H. laevigata*	S Australia Tasman Sea	CT/WT	A, M	Y	?Y	—	—	N=17 Hybrids rare	Y	Y	9
Phylum Chordata Class Osteichthyes Family Acanthuridae											
58. *Acanthurus achilles*** × *A. glaucopareius*	Hawaii E Pacific	T	—	—	—	—	—	—	—	N	75
Family Atherinidae											
59. *Menidia menidia* × *M. beryllina*	Florida W Atlantic	WT/T	M	Y	Y	—	—	6.3–15.0%	Y	—	40
Family Bothidae											
60. *Scophthalmus maximus* × *S. rhombus*	Kattegat Baltic Sea	CT	M	?Y	—	?Y	—	N=4 examined, but hybrids common	—	Y	70
61. *Scophthalmus maximus* × *S. rhombus*	NE England North Sea	CT	C, M	—	—	?Y	—	N=3 examined, but hybrids common	—	Y	51
Family Chaetodontidae											
62. *Chaetodon auriga* × *C. ephippium*	Tuamotu Pacific Ocean	T	C, M	—	—	—	—	N=1	—	N	75
63. *Chaetodon auriga* × *C. fasciatus*	Northern Red Sea	T	C	—	—	—	—	N=1	—	N	74
64. *Chaetodon auriga* × *C. lunula*	Northern Red Sea	T	C, M	—	—	—	—	—	—	N	75
65. *Chaetodon ephippium* × *C. semeion*	Marshall Is. Pacific Ocean	T	C, M	—	—	—	—	N=1	—	N	75
66. *Chaetodon kleini* × *C. unimaculatus*	Enewetak Pacific Ocean	T	C, M	—	—	—	—	N=1	—	N	75
67. *Chaetodon meyeri*** × *C. ornatissinus*	Palau Pacific Ocean	T	—	—	—	—	—	—	—	N	75
68. *Chaetodon miliaris* × *C. tinkeri*	Hawaii E Pacific	T	C, M	—	—	—	—	N=3	—	N	75
69. *Chaetodon miliaris*** × *C. multicinctus*	Hawaii E Pacific	T	—	—	—	—	—	N=2	—	N	75
70. *Chaetodon rainfordi* × *C. aureofasciatus*	NE Australia SW Pacific	T	C, M	—	—	—	—	N=1	—	N	75

Taxa involved	Location	Biogeographic zone	Characters examined	Hybrids fertile	Introgression	Hybrid unfitness	Disturbance	% hybrids	Ecological differences	Economic importance	Reference
71. *Chaetodon petewensis*** × *C. punctatofasciatus*	NE Australia SW Pacific	T	—	—	—	—	—	—	—	N	75
Family Citharinidae											
72. *Citharidium ansorgii* × *Citharinus distichodoides*	Cameroon E Atlantic	T	C, M	—	—	—	—	$N=3$	—	—	18
Family Clupeidae											
73. *Brevoortia patronus* × *B. smithi*	Florida Gulf of Mexico	WT/T	M	Y	Y	Y	—	37–74%	Temperature	Y	19, 21, 50, 97, 98
74. *Brevoortia smithi* × *B. tyrannus*	Florida W Atlantic	WT/T	B, M	?Y	?Y	$N^{G,H}Y^J$	—	1.2–51.4%	Inshore vs. offshore	Y	19–21
Family Cyprinodontidae											
75. *Fundulus majalis* × *F. similis*	NE Florida W Atlantic	WT/T	A	—	—	—	—	—	Salt marsh vs. mangrove marsh	N	25
Family Gadidae											
76. *Gadus morhua* × *Melanogrammus aeglefinus*	Nova Scotia NW Atlantic	CT	M	—	—	—	—	$N=2$	—	Y	26
Family Labridae											
77. *Notolabrus celidotus* × *N. fucicola*	NE New Zealand Tasman Sea	CT/WT	B, M	—	—	—	—	0.2% ($N=4^K$) $N=2^L$	Y	N	2, 77
78. *Notolabrus celidotus* × *N. inscriptus*	NE New Zealand Tasman Sea	CT/WT	C, M	—	—	—	—	$N=1$	—	N	2
79. *Notolabrus fucicola* × *N. inscriptus*	SE Australia and/or NE New Zealand Tasman Sea	CT/WT	C, M	—	—	—	—	$N=1$	?	N	77
80. *Notolabrus fucicola* × *N. tetricus*	SE Australia Tasman Sea	CT/WT	C, M	—	—	—	—	—	?Y	N	77
81. *Thalassoma lutescens* × *T. duperrey*	Hawaii E Pacific	T	M	—	—	—	—	$N=1^M$ Hybrids common^N	—	N	2, 78
Family Pleuronectidae											
82. *Limanda limanda* × *Platichthys flessus*	SW England NE Atlantic	CT	M	—	—	—	—	$N=1$	—	Y	70
83. *Platichthys flessus* × *Pleuronectes platessa*	Kattegat Baltic Sea	CT	M	—	—	—	—	$N=2$ examined. but hybrids common	—	Y	70
84. *Platichthys flesus* × *Pleuronectes platessa*	SE England NW Atlantic	CT	M	—	—	—	—	$N=2$ examined, but hybrids common	—	Y	70

Taxa involved	Location	Biogeographic zone	Characters examined	Hybrids fertile	Introgression	Hybrid unfitness	Disturbance	% hybrids	Ecological differences	Economic importance	Reference
85. *Plaichthys stellatus* × *Kareius bicoloratus*	Japan NW Pacific	CT	A, M	Y	Y	—	—	Hybrids common	Depth	Y	31, 52
86. *Plaichthys stellatus* × *Parophrys vetulus*	N California to Bering Sea E and NE Pacific	CT	M	?Y	—	—	—	$N=12^O$; $N=1^P$	Salinity	Y	46, 48, 84
87. *Pleuronectes platessa* × *Glyptocephalus cynoglossus*	Kattegat Baltic Sea	CT	M	—	—	—	—	$N=1$	—	Y	70
88. *Pseudopleuronectes americanus* × *Limanda ferruginea*	New York NW Atlantic	CT	M	—	—	—	—	$N=1$	—	Y	70
Family Pomacanthidae											
89. *Apolemichthys xanthurus* × *A. trimaculatus*	Seychelles and Maldives Indian Ocean	T	C, M	—	—	—	—	$N=1$	—	N	73
90. *Centropyge eibli* × *C. flavissimus*	Christmas Is. and Cocos-Keeling Is. Indian Ocean	T	C	—	—	—	—	$N=4$	—	N	73
91. *Centropyge eibli* × *C. vrolikii*	N Indonesia W Pacific	T	C, M	—	—	—	—	Hybrids common	—	N	73
92. *Centropyge loriculus* × *C. potteri*	Hawaii E Pacific	T	C, M	—	—	—	—	$N=3$	—	N	73
93. *Centropyge venustus* × *C. multifasciatus*	Philippines W Pacific	T	C, M	—	—	—	—	$N=5$	—	N	73
94. *Centropyge vrolikii* × *C. flavissimus*	Marshall Is., Guam, Kosrae and Ryukyus Is. Pacific Ocean	T	C, M	—	—	—	—	Hybrids common (10%)	—	N	73
95. *Holacanthus isabelita*† × *H. ciliaris*	Florida W Atlantic	WT/T	B, M	?Y	N	?N	—	22%	Y	N	28
96. *Pomacanthus chrysurus* × *P. maculosus*	Kenya W Indian Ocean	T	C, M	—	—	—	—	$N=2$	—	N	73
97. *Pomacanthus maculosus* × *P. semicirculatus*	Kenya and S. Oman Indian Ocean	T	C	—	—	—	—	$N=2$	—	N	73
98. *Pomacanthus sextriatus* × *P. xanthometapon*	N Australia W Pacific	T	C, M	—	—	—	—	$N=1$	—	N	73

Taxa involved	Location	Biogeographic zone	Characters examined	Hybrids fertile	Introgression	Hybrid unfitness	Disturbance	% hybrids	Ecological differences	Economic importance	Reference
Family Pomacentridae											
99. *pomacentrus planifrons* × *P. leucosticus*	Florida Keys and W Bahamas W Atlantic	WT/T	C, M	—	—	—	—	3.7% (N=6)	Depth Microhabitat	N	76
Family Scombridae											
100. *Scomberomorus commerson* × *S. guttatus*	SE India Indian Ocean	T	C, M	Y	—	N	U	59-68% in 1967, 0-5.6% in 1968^Q	Oceanic vs. coastal	Y	22, 90, 91
Family Serranidae											
101. *Hypoplectrus aberrans* × *H. nigricans*	Panama W Atlantic	T	C, M	?Y	?Y	?Y	—	N=4	?Y	N	29
102. *Hypoplectrus unicolor* × *H. puella*	Panama W Atlantic	T	C, M	?Y	?Y	?Y	—	N=1	?Y	N	29
103. *Hypoplectrus unicolor* × *H. puella*	Jamaica Caribbean	T	C, M	?Y	?Y	?Y	—	N=1	?Y	N	29
104. *Hypoplectrus aberrans* × *H. puella*	Jamaica Caribbean	T	C, M	?Y	?Y	?Y	—	N=2	?Y	N	29
105. *Hypoplectrus puella* × *H. indigo*	Jamaica Caribbean	T	C, M	?Y	?Y	?Y	—	N=2	?Y	N	29
Family Triglidae											
106. *Prionotus alatus* × *P. paralatus*	Gulf of Mexico	WT/T	M	—	—	—	—	25.6%	—	N	60
Class Mammalia											
Order Cetacea											
Family Balaenopteridae											
107. *Balaenoptera musculus* × *B. physalus*	W Iceland N Atlantic	P/CT	C, M, MH MT, N	Y	—	—	—	N=1	—	Y	89
108. *Balaenoptera musculus* × *B. physalus*	N Norway NE Atlantic	CT	M	Y	—	N^R	—	N=11	—	Y	12

Biogeographic zones: CT, cold temperate; P, polar; T, tropical; WT, warm temperate.
Characters examined: A, allozymes; B, behaviour; C, colour; K, karyotype; M, morphology; MH, molecular hybridization; Mt, mitochondrial DNA; N, nuclear DNA.
Disturbance: H, human; S, salinity; U, upwelling; Y, yes (some other cause).

References: 1 Almaça (1972); 2 Ayling (1980); 3 Bates (1992); 4 Beaumont *et al.* (1989); 5 Bert (1986); 6 Bert and Harrison (1988); 7 Bert *et al.* (1993); 8 Boss (1964); 9 Brown (1995); 10 Burrows and Lodge (1951); 11 Chia (1966); 12 Cocks (1887); 13 Coustau *et al.* (1991a); 14 Coustau *et al.* (1991b); 15 Coyer and Zaugg-Haglund (1982); 16 Coyer *et al.* (1992); 17 Crane (1975); 18 Daget (1963); 19 Dahlberg (1969a); 20 Dahlberg (1969b); 21 Dahlberg (1970); 22 Devaraj (1986); 23 Dillon (1992); 24 Dillon and Manzi (1989); 25 Duggins *et al.* (1995); 26 Dymond (1939); 27 Edwards and Skibinski (1987); 28 Feddern (1968); 29 Fischer (1980); 30 Fujino *et al.* (1980); 31 Fujio (1977); 32 Gardner and Skibinski (1988); 33 Gardner and Skibinski (1990a); 34 Gardner and Skibinski (1990b); 35 Gardner and Skibinski (1991a); 36 Gardner and Skibinski (1991b); 37 Gardner and Thompson (unpubl.); 38 Gartner-Kepkay *et al.* (1980); 39 Gartner-Kepkay *et al.* (1983); 40 Gosline (1948); 41 Gosling and Wilkins (1981); 42 Grant *et al.* (1977); 43 Gray *et al.* (1991); 44 Hagström and Lonning (1961); 45 Hagström and Lonning (1964); 46 Hart (1973); 47 Hedgecock *et al.* (1993); 48 Herald (1941); 49 Hesselman *et al.* (1988); 50 Hettler (1968); 51 Holt (1893); 52 Hubbs and Kuronuma (1941); 53 Johnson (1976); 54 Jones and Naylor (1971); 55 Karinen and Hoopes (1971); 56 Knight and Parke (1950); 57 Koehn *et al.* (1984); 58 Kwast *et al.* (1990); 59 Liu *et al.* (1991); 60 McClure and McEachran (1992); 61 McDonald and Koehn (1988); 62 McDonald *et al.* (1991); 63 Menge (1986); 64 Menzel (1985); 65 Menzel and Menzel (1965); 66 Miller (1994); 67 Miller and Babcock (unpubl.); 68 Miller and Benzie (unpubl.); 69 Neushul (1962); 70 Norman (1934); 71 Owen *et al.* (1971); 72 Porter and Chestnut (1960); 73 Pyle and Randall (1994); 74 Randall and Fridman (1981); 75 Randall *et al.* (1977); 76 Rivas (1960); 77 Russell (1988); 78 Sale (1991); 79 Sarver and Foltz (1993); 80 Sarver *et al.* (1992); 81 Sasaki *et al.* (1980); 82 Schilder (1962); 83 Schopf and Murphy (1973); 84 Schultz and Smith (1936); 85 Skibinski (1983); 86 Skibinski (1985); 87 Solignac (1976); 88 Solignac (1981); 89 Spilliaert *et al.* (1991); 90 Srinivasa Rao and Ganapati (1977); 91 Srinivasa Rao and Lakshmi (1993); 92 Strathmann (1981); 93 Swan (1953); 94 Swan (1962); 95 Talmadge (1977); 96 Tokida *et al.* (1958); 97 Turner (1967); 98 Turner (1969); 99 Väinölä and Hvilsom (1991); 100 Väinölä and Varvio (1989); 101 Varvio *et al.* (1988); 102 Vasseur (1952); 103 Verrill (1909); 104 Victor (1986); 105 Wilbur (1989); 106 Willis and Skibinski (1992).

Superscripts: A, the authors suggest hybrid unfitness (ref. 6); B, the author suggests no hybrid unfitness (ref. 103); C, hybrids have lower wet weight (but intermediate size) and greater morphological variability than parents (ref. 58); D, laboratory studies indicate no abnormalities among hybrids (ref. 11); E, high incidence of neoplasia in hybrids (ref. 49); F, hybrids have increased growth rate and better shelf life than parents (ref. 64); G, hybrids exhibit intermediate rates of parasitism (ref. 19); H, hybrids exhibit intermediate fat reserves (ref. 20); I, hybrids observed to spawn (ref. 21); J, sex ratio of hybrids dominated by males (ref. 21); K, ref. 2; L, ref. 75; M, ref. 2; N, ref. 76; O, ref. 82; P, ref. 48; Q, ref. 22; R, hybrids, on average, greater length than at least one parent (ref. 12); *, these species economically important as shellfish predators; **, these hybrids mentioned but not described; †, now known as *H. bermudensis* (ref. 71).

opportunities for new research. Given that the sea is the single largest ecosystem on Earth, the continued and expanded study of marine hybridization will provide confirmation of patterns and processes already described from work in other environments, as well as new insights into what might be collectively termed "hybrid zone theory".

ACKNOWLEDGEMENTS

I thank a variety of colleagues and friends, especially from the University of Wales (Swansea) UK, University of Leicester, UK, Memorial University of Newfoundland, Canada, and Victoria University of Wellington, New Zealand, for helpful discussions about hybridization and hybrid zones. Thanks to Dr Peter Castle for assistance with fish taxonomy and systematics, Dr Karen Miller for permission to cite unpublished data, and Drs L. Brown, R. Pyle and J. Randall for making their work available to me. In particular I would like to thank my colleagues Drs Geoff Chambers and Karen Miller for their critical reading of the manuscript and their many suggestions for improvements.

REFERENCES

Abbott, R. J. (1992). Plant invasions, interspecific hybridization and the evolution of new plant taxa. *Trends in Ecology and Evolution* 7, 401–405.

Almaça, C. (1972). Le littoral portugais, zone d'intergradation entre *Xantho incisus incisus* (Leach) et *X. incisus granulicarpus* (Forest). *Thalassia Jugoslavia* 8, 59–61.

Arnold, M. (1992). Natural hybridization as an evolutionary process. *Annual Review of Ecology and Systematics* 23, 237–261.

Arnold, J. (1993). Cytonuclear disequilibria in hybrid zones. *Annual Review of Ecology and Systematics* 24, 521–554.

Arnold, M. L. and Hodges, S. A. (1995). Are natural hybrids fit or unfit relative to their parents? *Trends in Ecology and Evolution* 10, 67–71.

Ausich, W. I. and Meyer, D. L. (1994). Hybrid crinoids in the fossil record (Early Mississippian, Phylum Echinodermata). *Paleobiology* 20, 362–367.

Avise, J. C. (1992). Molecular population structure and the biogeographic history of a regional fauna: a case history with lessons for conservation biology. *Oikos* 63, 62–76.

Avise, J. C. and Ball, R. M. (1990). Principles of genealogical concordance species concepts and biological taxonomy. *Oxford Surveys in Evolutionary Biology* 7, 45–67.

Ayling, A. M. (1980). Hybridization in the genus *Pseudolabrus* (Labridae). *Copeia* 1980, 176–180.

Babcock, R. C., Bull, G. D., Harrison, P. L., Heyward, A. J., Oliver, J. K., Wallace, C. C. and Willis, B. L. (1986). Synchronous spawnings of 105 scleractinian coral species on the Great Barrier Reef. *Marine Biology* **93**, 379–394.

Babcock, R. C., Mundy, C., Keesing, J. and Oliver, J. (1992). Predictable and unpredictable spawning events: *in situ* behavioural data from free-spawning coral reef invertebrates. *Invertebrate Reproduction and Development* **22**, 213–228.

Barton, N. H. and Gale, K. S. (1993). Genetic analysis of hybrid zones. In "Hybrid Zones and the Evolutionary Process" (R. G. Harrison, ed.), pp. 13–45. Oxford University Press, Oxford, UK.

Barton, N. H. and Hewitt, G. M. (1985). Analysis of hybrid zones. *Annual Review of Ecology and Systematics* **16**, 113–148.

Barton, N. H. and Hewitt, G. M. (1989). Adaptation, speciation and hybrid zones. *Nature, London* **341**, 497–503.

Bates, J. A. (1992). Genetic and morphological variation in the mussel *Mytilus* in Newfoundland. MSc. Thesis, Memorial University of Newfoundland, Canada, pp. 123.

Battaglia, B. (1957). Ecological differentiation and incipient intraspecific isolation in marine copepods. *Année Biologique* **33**, 259–268.

Beaumont, A. R., Seed, R. and Garcia-Martinez, P. (1989). Electrophoretic and morphometric criteria for the identification of the mussels *Mytilus edulis* and *M. galloprovincialis*. In "Proceedings of the 23rd European Marine Biology Symposium, University of Wales, Swansea" (J. S. Ryland and P. A. Tyler, eds), pp. 251–258. Olsen and Olsen, Fredensborg, Denmark.

Beaumont, A. R., Abdul-Matin, A. K. M. and Seed, R. (1993). Early development, survival and growth in pure and hybrid larvae of *Mytilus edulis* and *M. galloprovincialis*. *Journal of Molluscan Studies* **59**, 120–123.

Bert, T. M. (1986). Speciation in western Atlantic stone crabs (genus *Menippe*): the role of geological patterns and climatic events in the formation and distribution of species. *Marine Biology* **93**, 157–170.

Bert, T. M. and Harrison, R. G. (1988). Hybridization in western Atlantic stone crabs (genus *Menippe*): evolutionary history and ecological context influence species interactions. *Evolution* **42**, 528–544.

Bert, T. M., Hesselman, D. M., Arnold, W. S., Moore, W. S., Cruz-Lopez, H. and Marelli, D. C. (1993). High frequency gonadal neoplasia in a hard clam (*Mercenaria* spp.) hybrid zone. *Marine Biology* **117**, 97–104.

Boitard, M., Guillaumin, M., Lefebvre, J. and Solignac, M. (1980). Application du D^2 de Mahalanobis a l'étude du dimorphisme sexuel, de la variabilité entre populations et des hybrides interspecifique dans le complexe *Jaera albifrons* (Crustacés, Isopodes). *Archives de Zoologie Expèrimentale et Génèrale* **121**, 115–136.

Boitard, M., Lefebvre, J. and Solignac, M. (1982). Analyse en composantes principales de la variabilité de taille, de croissance et de conformation des espèces du complexe *Jaera albifrons* (Crustaces, Isopodes). *Cahiers de Biologie Marine* **23**, 115–142.

Bolton, J. J. and Anderson, R. J. (1987). Temperature tolerances of two southern African *Ecklonia* species (Alariaceae: Laminariales) and of hybrids between them. *Marine Biology* **96**, 293–297.

Bolwell, G. P., Callow, J. A., Callow, M. E. and Evans, L. V. (1977). Cross-fertilisation in fucoid seaweeds. *Nature, London* **268**, 626–627.

Boss, K. J. (1964). Notes on a hybrid *Tellina* (Tellinidae). *Nautilus* **78** 18–21.

Briggs, J. C. (1974). "Marine Zoogeography". McGraw-Hill, New York, USA.

Briggs, J. C. (1991). "Global species diversity". *Journal of Natural History* **25**, 1403–1406.

Briggs, J. C. (1992a). "Global species diversity": a reply to W. G. Chaloner. *Journal of Natural History* **26**, 455.

Briggs, J. C. (1992b). "Global species diversity": a reply to J. F. Grassle. *Journal of Natural History* **26**, 455–456.

Brock, V. (1978). Morphological and biochemical criteria for the separation of *Cardium glaucum* (Bruguière) from *Cardium edule* (L.). *Ophelia* **17**, 207–214.

Brock, V. (1979). Habitat selection of two congeneric bivalves, *Cardium edule* and *C. glaucum* in sympatric and allopatric populations. *Marine Biology* **54**, 149–156.

Brown, L. D. (1995). Genetic evidence for hybridization between *Haliotis rubra* and *H. laevigata*. *Marine Biology* **123**, 89–93.

Burrows, E. M. and Lodge, S. (1951). Autecology and the species problem in *Fucus*. *Journal of the Marine Biological Association of the UK* **30**, 161–176.

Byrne, M. and Anderson, M. J. (1994). Hybridization of sympatric *Patiriella* species (Echinodermata: Asteroidea) in New South Wales. *Evolution* **48**, 564–576.

Campton, D. E. (1987). Natural hybridization and introgression in fishes. Methods of detection and genetic interpretations. *In* "Population Genetics and Fishery Management" (N. Ryman and F. Utter, eds), pp. 161–192. University of Washington Press, Seattle, Washington, USA.

Campton, D. E. (1990). Application of biochemical and molecular genetic markers to analysis of hybridization. *In* "Electrophoretic and Isoelectric Focusing Techniques in Fisheries Management" (D. H. Whitmore, ed.), pp. 241–263. CRC Press, Boca Raton, Florida, USA.

Campton, D. E. and Utter, F. M. (1985). Natural hybridization between steelhead trout (*Salmo gairdneri*) and coastal cutthroat trout (*Salmo clarki clarki*) in two Puget Sound streams. *Canadian Journal of Fisheries and Aquatic Sciences* **42**, 110–119.

Carlton, J. T. and Geller, J. B. (1993). Ecological roulette: the global transport of nonindigenous marine organisms. *Science N.Y.* **261**, 78–81.

Carson, H. L. (1975). The genetics of speciation at the diploid level. *American Naturalist* **109**, 83–92.

Chaloner, W. G. (1992). "Global species diversity": a reply to J. C. Briggs. *Journal of Natural History* **26**, 265.

Chestnut, A. F., Fahy, W. E. and Porter, H. J. (1956). Growth of young *Venus mercenaria*, *Venus campechiensis*, and their hybrids. *Proceedings of the National Shellfisheries Association* **47**, 50–56.

Chia, F.-S. (1966). Systematics of the six-rayed sea star, *Leptasterias*, in the vicinity of San Juan Island, Washington. *Systematic Zoology* **15**, 300–306.

Chipperfield, P. N. J. (1953). Observations on the breeding and settlement of *Mytilus edulis* (L.) in British waters. *Journal of the Marine Biological Association of the UK* **32**, 449–476.

Cocks, A. H. (1887). The finwhale fishery of 1886 on the Lapland coast. *Zoologist* **11**, 207–222.

Collette, B. B. (1994). *Scomberomorus lineolatus* is a valid species of Spanish mackerel, not an interspecific hybrid: a reply. *Journal of Natural History* **28**, 1205–1208.

Coustau, C., Renaud, F. and Delay, B. (1991a). Genetic characterization of the hybridization between *Mytilus edulis* and *M. galloprovincialis* on the Atlantic coast of France. *Marine Biology* **111**, 87–93.

Coustau, C., Renaud, F., Maillard, C., Pasteur, N. and Delay, B. (1991b). Differential susceptibility to a trematode parasite among genotypes of the *Mytilus edulis/galloprovincialis* complex. *Genetical Research* **57**, 207–212.

Coyer, J. A. and Zaugg-Haglund, A. C. (1982). A demographic study of the elk kelp, *Pelagophycus porra* (Laminariales, Lessoniaceae), with notes on *Pelagophycus* × *Macrocystis* hybrids. *Phycologia* **21**, 399–407.

Coyer, J. A., Engle, J. M. and Zimmerman, R. C. (1992). Discovery of a fertile *Pelagophycus* × *Macrocystis* (Phaeophyta) putative hybrid and subsequent production of F2 sporophytes in the laboratory. *Journal of Phycology* **28**, 127–130.

Cracraft, J. (1983). Species concepts and speciation analysis. *Current Ornithology* **1**, 159–187.

Cracraft, J. (1989). Speciation and its ontology: the empirical consequences of alternative species concepts for understanding patterns and processes of differentiation. *In* "Speciation and its Consequences" (D. Otte and J. A. Endler, eds), pp. 28–59. Sinauer Associates, Sunderland, Massachusetts, USA.

Crane, J. (1975). "Fiddler Crabs of the World". Princeton University Press, Princeton, New Jersey, USA.

Daget, J. (1963). Sur plusieurs cas probables d'hybridization naturelle entre *Citharidium ansorgii* et *Citharinus distichodoides*. *Memoire d'Institut Français d'Afrique Noire* **68**, 81–83.

Dahlberg, M. (1969a). Fat cycles and condition factors of two species of menhaden, *Brevoortia* (Clupeidae), and natural hybrids from the Indian River Lagoon. *American Midland Naturalist* **82**, 117–126.

Dahlberg. M. (1969b). Incidence of the isopod, *Olencira praegustator*, and copepod, *Lernaeenicus radiatus*, in three species and hybrid menhaden (*Brevoortia)* from the Florida coasts, with five new host records. *Transactions of the American Fisheries Society* **98**, 111–115.

Dahlberg, M. D. (1970). Atlantic and Gulf of Mexico Menhadens, genus *Brevoortia* (Pisces: Clupeidae). *Bulletin of the Florida State Museum* **15**, 91–162.

Dalton, R. and Menzel, W. (1983). Seasonal gonadal development of young laboratory-spawned southern (*Mercenaria campechiensis*) and northern (*Mercenaria mercenaria*) quahogs and their reciprocal hybrids in northwest Florida. *Journal of Shellfish Research* **3**, 11–17.

Devaraj, M. (1986). Maturity, spawning and fecundity of the streaked seer, *Scomberomorus lineolatus* (Cuvier and Valenciennes), in the Gulf of Mannar and Palk Bay. *Indian Journal of Fisheries* **33**, 293–319.

Dillon, R. T. Jr (1992). Minimal hybridization between populations of the hard clams, *Mercenaria mercenaria* and *Mercenaria campechiensis*, co-occurring in South Carolina. *Bulletin of Marine Science* **50**, 411–416.

Dillon, R. T. Jr and Manzi, J. J. (1989). Genetics and shell morphology of hard clams (Genus *Mercenaria*) from Laguna Madre, Texas. *Nautilus* **103**, 73–77.

Donovan, E. (1802). "The Natural History of British Shells". F. and C. Rivington, London, UK.

Duggins, C. F. Jr, Karlin, A. A., Mousseau, T. A. and Relyea, K. G. (1995). Analysis of a hybrid zone in *Fundulus majalis* in a northeastern Florida ecotone. *Heredity* **74**, 117–128.

Dymond, J. R. (1939). Cod × haddock hybrids? *Canadian Field Naturalist* **53**, 91.

Edwards, A. J. and Skibinski, D. O. F. (1987). Genetic variation of mitochondrial DNA in mussel (*Mytilus edulis* and *M. galloprovincialis*) populations from south west England and south Wales. *Marine Biology* **94**, 547–556.

Endler, J. A. (1977). "Natural Selection in the Wild". Princeton University Press, Princeton, New Jersey, USA.

Falk-Peterson, I.-B, and Lonning, S. (1983). Reproductive cycles of two closely related sea urchin species, *Strongylocentrotus droebachiensis* (O. F. Müller) and *Strongylocentrotus pallidus* (G. O. Sars). *Sarsia* **68**, 157–164.

Feddern, H. A. (1968). Hybridization between the western Atlantic angelfishes, *Holacanthus isabelita* and *H. ciliaris*. *Bulletin of Marine Science* **18**, 351–382.

Fischer, E. A. (1980). Speciation in the Hamlets (*Hypoplectrus*: Serranidae)—a continuing enigma. *Copeia* **1980**, 649–659.

Freeman, K. R., Perry, K. L. and DiBacco, T. G. (1992). Morphology, condition and reproduction of two co-occurring species of *Mytilus* at a Nova Scotia mussel farm. *Bulletin of the Aquaculture Association of Canada* **3**, 8–10.

Fujino, K., Sasaki, K. and Wilkins, N. P. (1980). Genetic studies on the Pacific abalone. III. Differences in electrophoretic patterns between *Haliotis discus* Reeve and *H. discus hannai* Ino. *Bulletin of the Japanese Society of Scientific Fisheries* **46**, 543–548.

Fujio, Y. (1977). Natural hybridization between *Platichthys stellatus* and *Kareius bicoloratus*. *Japanese Journal of Genetics* **52**, 117–124.

Gage, J. D. and Tyler, P. A. (1991). "Deep-Sea Biology: A Natural History of Organisms at the Deep-sea Floor". Cambridge University Press, Cambridge, UK.

Gardner, J. P. A. (1992). *Mytilus galloprovincialis* (Lmk) (Bivalvia, Mollusca): The taxonomic status of the Mediterranean mussel. *Ophelia* **35**, 219–243.

Gardner, J. P. A. (1994).The structure and dynamics of naturally occurring hybrid *Mytilus edulis* Linnaeus, 1758 and *Mytilus galloprovincialis* Lamarck, 1819 (Bivalvia, Mollusca) populations: review and interpretation. *Archiv für Hydrobiologie, Monographische Beiträge* **99**, 37–71.

Gardner, J. P. A. (1995). Developmental stability is not disrupted by extensive hybridization and introgression among populations of the marine bivalve molluscs *Mytilus edulis* (L.) and *M. galloprovincialis* (Lmk.) from southwest England. *Biological Journal of the Linnean Society* **54**, 71–86.

Gardner, J. P. A. (1996). The *Mytilus edulis* species complex in SW England: the extent of hybridization and introgression and their effects upon interlocus associations and morphometric variation. *Marine Biology* **125**, 385–399.

Gardner, J. P. A. and Skibinski, D. O. F. (1988). Historical and size dependent genetic variation in hybrid mussel populations. *Heredity* **61**, 93–105.

Gardner, J. P. A. and Skibinski, D. O. F. (1990a). Genotype-dependent fecundity and temporal variation of spawning in hybrid mussel populations. *Marine Biology* **105**, 153–162.

Gardner, J. P. A. and Skibinski, D. O. F. (1990b). Thermostability differences of allozyme loci in *Mytilus edulis*, *M. galloprovincialis* and hybrid mussels. *Marine Ecology Progress Series* **64**, 99–105.

Gardner, J. P. A. and Skibinski, D. O. F. (1991a). Biological and physical factors influencing genotype-dependent mortality in hybrid mussel populations. *Marine Ecology Progress Series* **71**, 235–243.

Gardner, J. P. A. and Skibinski, D. O. F. (1991b). Mitochondrial DNA and

allozyme covariation in a hybrid mussel population. *Journal of Experimental Marine Biology and Ecology* **149**, 45–54.

Gardner, J. P. A., Skibinski, D. O. F. and Bajdik, C. D. (1993). Shell growth and viability differences between the marine mussels *Mytilus edulis* (L.), *M. galloprovincialis* (Lmk.) and their hybrids from two sympatric populations in SW England. *Biological Bulletin* **185**, 405–416.

Gartner-Kepkay, K. E., Dickie, L. M., Freeman, K. R. and Zouros, E. (1980). Genetic differences and environments of mussel populations in the Maritime Provinces. *Canadian Journal of Fisheries and Aquatic Sciences* **37**, 775–782.

Gartner-Kepkay, K. E., Zouros, E., Dickie, L. M. and Freeman, K. R. (1983). Genetic differentiation in the face of gene flow: a study of mussel populations from a single Nova Scotian embayment. *Canadian Journal of Fisheries and Aquatic Sciences* **40**, 443–451.

Geary, D. H. (1992). An unusual pattern of divergence between two fossil gastropods: ecophenotypy, dimorphism, or hybridization? *Paleobiology* **18**, 93–109.

Gosline, W. A. (1948). Speciation in the fishes of the genus *Menidia*. *Evolution* **2**, 306–311.

Gosling, E. M. (1980). Gene frequency changes and adaptation in marine cockles. *Nature, London* **286**, 601–602.

Gosling, E. M. and Wilkins, N. P. (1981). Ecological genetics of the mussels *Mytilus edulis* and *Mytilus galloprovincialis* on Irish coasts. *Marine Ecology Progress Series* **4**, 221–227.

Grant, V. (1971). "Plant Speciation". Columbia University Press, New York, USA.

Grant, W. S., Bartlett, L. and Utter F. M. (1977). Biochemical genetic identification of species and hybrids of the Bering Sea tanner crab, *Chionoecetes bairdi* and *C. opilio*. *Proceedings of the National Shellfisheries Association* **67**, 127.

Grassle, J. F. (1992). "Global species diversity": a reply to J. C. Briggs. *Journal of Natural History* **26**, 265–266.

Gray, A. J., Marshall, D. F. and Raybould, A. F. (1991). A century of evolution in *Spartina anglica*. *Advances in Ecological Research* **21**, 1–62.

Hagström, B. E. and Lonning, S. (1961). Morphological and experimental studies on the genus *Echinus*. *Sarsia* **4**. 21–31.

Hagström, B. E. and Lonning, S. (1964). Morphological variation in *Echinus esculentus* from the Norwegian west coast. *Sarsia* **17**, 39–46.

Hagström, B. E. and Lonning, S. (1967). Experimental studies of *Strongylocentrotus droebachiensis* and *S. pallidus*. *Sarsia* **29**, 165–176.

Hagström, B. E. and Wennerberg, C. (1964). Hybridization experiments with wrasses (Labridae). *Sarsia* **17**, 47–54.

Harrison, R. G. (1990). Hybrid zones: windows on evolutionary process. *Oxford Surveys in Evolutionary Biology* **7**, 69–128.

Harrison, R. G. (1993). "Hybrid Zones and the Evolutionary Process". Oxford University Press, Oxford, UK.

Harrison, R. G. and Rand, D. M. (1989). Mosaic hybrid zones and the nature of species boundaries. *In* "Speciation and its Consquences" (D. Otte and J. A. Endler, eds), pp. 111–133. Sinauer Associates, Sunderland, Massachusetts, USA.

Harrison, P. L., Babcock, R. C., Bull, G. D., Oliver, J. K., Wallace, C. C. and Willis, B. L. (1984). Mass spawning in tropical reef corals. *Science N. Y.* **223**, 1186–1189.

Hart, J. L. (1973). Pacific Fishes of Canada. *Bulletin of the Fisheries Research Board of Canada* **180**, 740 pp.

Haven, D. and Andrews, J. D. (1956). Survival and growth of *Venus mercenaria*, *Venus campechiensis*, and their hybrids in suspended trays and on natural bottoms. *Proceedings of the National Shellfisheries Association* **47**, 43–49.

Heath, E. M. and Simmons, M. J. (1991). Genetic and molecular analysis of repression in the P–M system of hybrid dysgenesis in *Drosophila melanogaster*. *Genetical Research* **57**, 213–226.

Hedgecock, D., Banks, M. A. and McGoldrick, D. J. (1993). The status of the Kumamoto oyster *Crassostrea sikamea* (Amimia 1928) in US commercial broodstocks. *Journal of Shellfish Research* **12**, 215–221.

Herald, E. S. (1941). First record of the hybrid flounder, *Inopsetta ischyra*, from California. *California Fish and Game* **27**, 44–46.

Hesselman, D. M., Blake, N. J. and Peters, E. C. (1988). Gonadal neoplasms in hard shell clams *Mercenaria* spp., from the Indian River, Florida: occurrence, prevalence, and histopathology. *Journal of Invertebrate Pathology* **52**, 436–446.

Hettler, W. F., Jr (1968). Artificial fertilization between yellowfin and Gulf Menhaden (*Brevoortia*) and their hybrid. *Transactions of the American Fisheries Society* **97**, 119–123.

Hewitt, G. M. (1988). Hybrid zones—natural laboratories for evolutionary studies. *Trends in Ecology and Evolution* **3**, 158–167.

Hewitt, G. M. (1989). The subdivision of species by hybrid zones. *In* "Speciation and its Consequences" (D. Otte and J. A. Endler, eds), pp. 85–110. Sinauer Associates, Sunderland, Massachusetts, USA.

Hewitt, G. M. and Barton, N. H. (1980). The structure and maintenance of hybrid zones as exemplified by *Podisma pedestris*. *In* "Insect Cytogenetics" (R. L. Blackman, G. M. Hewitt and N. M. Ashburner, eds), pp. 149–169. Blackwell, Oxford, UK.

Hodgson, G. (1988). Potential gamete wastage in synchronously spawning corals due to hybrid inviability. *Proceedings of the 6th International Coral Reef Symposium, Australia*, **2**, 707–714.

Holt, E. W. L. (1893). Note on some supposed hybrids between the turbot and the brill. *Journal of the Marine Biological Association of the UK* **3**, 292–299.

Hubbs, C. L. (1955). Hybridization between fish species in nature. *Systematic Zoology* **4**, 1–20.

Hubbs, C. L. and Kuronuma, K. (1941). Hybridization in nature between two genera of flounders in Japan. *Papers of the Michigan Academy of Science, Arts and Letters* **27**, 267–306.

Humphrey, C. M. and Crenshaw, J. W. Jr (1985). Clam genetics. *In* "Clam Mariculture in North America. Developments in Aquaculture and Fisheries Science" (J. J. Manzi and M. Castagna, eds), pp. 323–356. Elsevier, Amsterdam, Holland.

Johnson, A. G. (1976). Electrophoretic evidence of hybrid snow crab, *Chionoecetes bairdi* × *opilio*. *Fishery Bulletin US* **74**, 693–694.

Jones, M. B. and Naylor, E. (1971). Breeding and bionomics of the British members of the *Jaera albifrons* group of species (Isopoda: Asellota). *Journal of Zoology* **165**, 183–199.

Karinen, J. F. and Hoopes, D. T. (1971). Occurrence of tanner crabs (*Chionoecetes* sp.) in the eastern Bering Sea with characteristics intermediate between *C. bairdi* and *C. opilio*. *Proceedings of the National Shellfisheries Association* **61**, 8–9.

Kermack, K. A. (1954). A biometrical study of *Micraster coranguinum* and *M. (Isomicraster) senonensis. Philosophical Transactions of the Royal Society of London B* **237**, 375–428.

Khadem, M. and Krimbas, C. B. (1993). Studies of the species barrier between *Drosophila subobscura* and *D. madeirensis*. III. How universal are the rules of speciation? *Heredity* **70**, 353–361.

Kidwell, M. (1982). Hybrid dysgenesis in *Drosophila melanogaster*: a syndrome of aberrant traits inducing mutation, sterility and male recombination. *Genetics* **86**, 813–833.

Kingston, P. (1973). Interspecific hybridization in *Cardium. Nature, London* **243**, 360.

Knight, M. and Parke, M. (1950). A biological study of *Fucus vesiculosus* L. and *F. serratus* L. *Journal of the Marine Biological Association of the UK* **29**, 439–499.

Knowlton, N. (1993). Sibling species in the sea. *Annual Review of Ecology and Systematics* **24**, 189–216.

Knowlton, N. and Jackson, J. B. C. (1994). New taxonomy and niche partitioning on coral reefs: jack of all trades or master of some? *Trends in Ecology and Evolution* **9**, 7–9.

Knox, G. A. (1963). Problems of speciation in intertidal animals with special reference to New Zealand shores. *Systematics Association Publication* **5**, 7–29.

Koehn, R. K., Hall, J. G., Innes, D. J. and Zora, A. J. (1984). Genetic differentiation of *Mytilus edulis* in eastern America. *Marine Biology* **79**, 117–126.

Kolding, S. (1985). Genetic adaptation to local habitats and speciation process within the genus *Gammarus* (Amphipoda: Crustacea). *Marine Biology* **89** 249–255.

Kwast, K. E., Foltz, D. W. and Stickle, W. B. (1990). Population genetics and systematics of the *Leptasterias hexactis* (Echinodermata: Asteroidea) species complex. *Marine Biology* **105**, 477–489.

Leighton, D. L. and Lewis, C. A. (1982). Experimental hybridization in abalones. *International Journal of Invertebrate Reproduction* **5**, 273–282.

Lessios, H. A. and Cunningham, C. W. (1990). Gametic incompatibility between species of the sea urchin *Echinometra* on the two sides of the Isthmus of Panama. *Evolution* **44**, 933–941.

Lewontin, R. C. and Birch, L. C. (1966). Hybridization as a source of variation for adaptation to new environments. *Evolution* **20**, 315–336.

Liu, L. L., Foltz, D. W. and Stickle, W. B. (1991). Genetic population structure of the southern oyster drill *Stramonita* (= *Thais*) *haemostoma. Marine Biology* **111**, 71–79.

Loosanoff, V. L. (1954). New advances in the study of bivalve larvae. *American Scientist* **42**, 607–624.

Lozovskaya, E. R., Scheinker, V. S. and Evgen'ev, M. B. (1990). A hybrid dysgenesis syndrome in *Drosophila virilis. Genetics* **126**, 619–623.

Luning, K., Chapman, A. R. O. and Mann, K. H. (1978). Crossing experiments in the non-digitate complex of *Laminaria* from both sides of the Atlantic. *Phycologia* **17**, 293–298.

McClure, M. R. and McEachran, J. D. (1992). Hybridization between *Prionatus alatus* and *P. paralatus* in the northern Gulf of Mexico (Pisces: Triglidae). *Copeia* **1992**, 1039–1046.

McDonald, J. H. and Koehn, R. K. (1988). The mussels *Mytilus galloprovincialis* and *M. trossulus* on the Pacific coast of North America. *Marine Biology* **99**, 111–118.

McDonald, J. H., Seed, R. and Koehn, R. K. (1991). Allozyme and morphometric characters of three species of *Mytilus* in the Northern and Southern Hemispheres. *Marine Biology* **111**, 323–333.

McKitrick, M. C. and Zink, R. M. (1988). Species concepts in ornithology. *Condor* **90**, 1–14.

Mallet, J. (1995). A species definition for the Modern Synthesis. *Trends in Ecology and Evolution* **10**, 294–299.

Mangum, C. P. (1993). Hemocyanin sub-unit composition and oxygen binding in two species of the lobster genus *Homarus* and their hybrids. *Biological Bulletin* **184**, 105–113.

May, R. M. (1988). How many species are there on Earth? *Science N.Y.* **241**, 1441–1449.

Mayr, E. (1942) "Systematics and the Origin of Species". Columbia University Press, New York, USA.

Mayr, E. (1970). "Population Species and Evolution". Belknap Press, Cambridge, Massachusetts, USA.

Mayr, E. (1979). "Animal Species and Evolution", 6th edn. Belknap Press, Cambridge, Massachusetts, USA.

Menge, B. (1986). A preliminary study of the reproductive ecology of the seastars *Asterias vulgaris* and *A. forbesi* in New England. *Bulletin of Marine Science* **39**, 467–476.

Menzel, R. W. (1962). Seasonal growth of northern and southern quahogs, *Mercenaria mercenaria* and *M. campechiensis*, and their hybrids in Florida. *Proceedings of the National Shellfisheries Association* **53**, 111–119.

Menzel, R. W. (1977). Selection and hybridization in quahog clams. *Proceedings of the World Mariculture Society* **8**, 507–521.

Menzel, R. W. (1985). The biology, fishery and culture of quahog clams, *Mercenaria. In* "Clam Mariculture in North America. Developments in Aquaculture and Fisheries Science" (J. J. Manzi and M. Castagna, eds), pp. 201–242. Elsevier, Amsterdam, Holland.

Menzel, R. W. and Menzel, M. Y. (1965). Chromosomes of two species of quahog clams and their hybrids. *Biological Bulletin* **129**, 181–188.

Miller, K. J. (1994). The *Platygyra* species complex; implications for coral taxonomy and evolution. PhD Thesis, James Cook University, Australia, 164 pp.

Miller, K. J. and Benzie, J. A. H. (1997). No clear genetic distinction between morphological species with the coral genus *Platygyra*. *Bulletin of Marine Science* **60** (in press).

Moore, W. S. (1977). An evaluation of narrow hybrid zones in vertebrates. *Quarterly Review of Biology* **52**, 263–277.

Neushul, M. (1962). Possible intergenic hybrids in giant kelps (Lessoniaceae). *American Journal of Botany* **49**, 672.

Nichols, D. (1959). Changes in the chalk heart-urchins *Micraster* interpreted in relation to living forms. *Philosophical Transactions of the Royal Society of London B*, **242**, 347–437.

Norman, J. R. (1934). "A Systematic Monograph of the Flatfishes (Heterosomata)", Vol. 1: "Psettodidae, Bothidae, Pleuronectidae". Natural History Section of the British Museum, London, UK.

O'Rand, M. G. (1988). Sperm–egg recognition and barriers to interspecies fertilization. *Gamete Research* **19**, 315–328.

Owen, B., McLean, J. H. and Meyer, R. J. (1971). Hybridization in the eastern Pacific abalones (*Haliotis*). *Bulletin of the Los Angeles County Museum of Natural History* **9**, 1–37.

Palumbi, S. R. (1992). Marine speciation on a small planet. *Trends in Ecology and Evolution* **7**, 114–118.

Palumbi, S. R. (1994). Genetic divergence, reproductive isolation, and marine speciation. *Annual Review of Ecology and Systematics* **25**, 547–572.

Palumbi, S. R. and Metz, E. (1991). Strong reproductive isolation between closely related tropical sea urchins (genus *Echinometra*). *Molecular Biology and Evolution* **8**, 227–239.

Paterson, H. E. H. (1985). The recognition concept of species. *Transvaal Museum Monograph* **4**, 21–29.

Porter, H. J. and Chestnut, A. F. (1960). The offshore clam fishery in North Carolina. *Proceedings of the National Shellfisheries Association* **53**, 103–109.

Price, P. W. (1996). "Biological Evolution". Saunders College Publishing, Fort Worth, Texas, USA.

Pyle, R. L. and Randall, J. E. (1994). A review of hybridization in marine angelfishes (Perciformes: Pomacanthidae). *Environmental Biology of Fishes* **41**, 127–145.

Rand, D. M. and Harrison, R. G. (1989). Ecological genetics of a mosaic hybrid zone: mitochondrial, nuclear, and reproductive differentiation of crickets by soil type. *Evolution* **43**, 432–449.

Randall, J. E. (1956). *Acanthurus rackliffei*, a possible hybrid surgeonfish (*A. achilles* × *A. glaucopareius*) from the Phoenix Islands. *Copeia* **1956**, 21–25.

Randall, J. E. and Fridman, D. (1981). *Chaetodon auriga* × *Chaetodon fasciatus*, a hybrid butterflyfish from the Red Sea. *Revue français Aquariologie* **7**, 113–114.

Randall, J. E., Allen, G. R. and Steene, R. C. (1977). Five probable hybrid butterfly fishes of the genus *Chaetodon* from the central and western Pacific. *Records of the Western Australian Museum* **6**, 3–26.

Reeb, C. A. and Avise, J. C. (1990). A genetic discontinuity in a continuously distributed species: mitochondrial DNA in the American oyster, *Crassostrea virginica*. *Genetics* **124**, 397–406.

Richmond, R. H. (1990). Relationships among reproductive mode, biogeographic distribution patterns and evolution in scleractinian corals. *Advances in Invertebrate Reproduction* **5**, 317–322.

Rivas, L. R. (1960). The fishes of the genus *Pomacentrus* in Florida and the western Bahamas. *Quarterly Journal of the Florida Academy of Science* **23**, 130–162.

Rubinoff, I. (1968). Central American sea-level canal: possible biological effects. *Science N.Y.* **161**, 857–861.

Rueness, J. (1973). Speciation in *Polysiphonia* (Rhodophyceae, Ceramiales) in view of hybridization experiments: *P. hemisphaerica* and *P. boldii*. *Phycologia* **12**, 107–109.

Russell, B. C. (1988). Revision of labrid genus *Pseudolabrus* and allied genera. *Records of the Australian Museum* (Suppl.) **9**, 1–72.

Sale, P. F. (1981). "The Ecology of Fishes on Coral Reefs". Academic Press, San Diego, California, USA.

Sanbonsuga, Y. and Neushul, M. (1978). Hybridization of *Macrocystis*

(Phaeophyta) with other float-bearing kelps. *Journal of Phycology* **14**, 214–224.

Sara, M. (1985). Ecological factors and their biogeographic consequences in the Mediterranean ecosystems. *In* "Mediterranean Marine Ecosystems" (M. Moratou-Apostolopoulou and V. Kiortsis, eds), pp. 1–17, Plenum Press, New York, USA.

Sarver, S. K. and Foltz, D. W. (1993). Genetic population structure of a species complex of the blue mussels (*Mytilus* spp.). *Marine Biology* **117**, 105–112.

Sarver, S. K., Landrum, M. C. and Foltz, D. W. (1992). Genetics and taxonomy of ribbed mussels (*Geukensia* spp.). *Marine Biology* **113**, 385–390.

Sasaki, K., Kanazawa, K. and Fujino, K. (1980). Zymogram differences among five species of the abalones from the coasts of Japan. *Bulletin of the Japanese Society of Scientific Fisheries* **46**, 1169–1175.

Schilder, F. A. (1962). Hybrids between *Cypraea tigris* Linnaeus, 1758 and *Cypraea pantherina* Solander, 1786 (Mollusca: Gastropoda). *Veliger* **5**, 83–87.

Schopf, T. J. M. and Murphy, L. S. (1973). Protein polymorphism of the hybridizing seastars *Asterias forbesi* and *Asterias vulgaris* and implications for their evolution. *Biological Bulletin* **145**, 589–597.

Schultz, L. P. and Smith, R. T. (1936). Is *Inopsetta ischyra* (Jordan and Gilbert), from Puget Sound, Washington, a hybrid flatfish? *Copeia* **1936**, 199–203.

Schwartz, F. J. (1972). "World Literature to Fish Hybrids with an Analysis by Family, Species and Hybrid". Gulf Coast Research Laboratory Publication, Wilkes Printing Company, Biloxi, Mississippi, USA.

Seed, R. (1971). A physiological and biochemical approach to the taxonomy of *Mytilus edulis* L. and *Mytilus galloprovincialis* (Lmk.). *Cahiers de Biologie Marine* **12**, 291–322.

Sekiguchi, K. and Sugita, H. (1980). Systematics and hybridization in the four living species of horseshoe crab. *Evolution* **34**, 712–718.

Seldon, P. A. (1990). Terrestrialization: invertebrates. *In* "Paleobiology: A Synthesis" (D. E. G. Briggs and P. R. Crowther, eds), pp. 64–68. Blackwell Scientific, Oxford, UK.

Shaw, F. C. (1991). Viola group (Ordovician, Oklahoma) cryptolothinid trilobites: biogeography and taxonomy. *Journal of Paleontology* **65**, 919–935.

Sick, K., Frydenberg, O. and Nielsen, J. T. (1963). Haemoglobin patterns of plaice, flounder and their natural and artificial hybrids. *Nature, London* **198**, 411–412.

Skibinski, D. O. F. (1983). Natural selection in hybrid mussel populations. *In* "Protein Polymorphism: Adaptive and Taxonomic Significance" (G. S. Oxford and D. Rollinson eds), Systematics Association Special Publication, Vol. 24, pp. 283–298. Academic Press, London, UK.

Skibinski, D. O. F. (1985). Mitochondrial DNA variation in *Mytilus edulis* L. and the Padstow mussel. *Journal of Experimental Marine Biology and Ecology* **92**, 251–258.

Skibinski, D. O. F., Cross, T. F. and Ahmad, M. (1980). Electrophoretic investigation of systematic relationships in the marine mussels *Modiolus modiolus* L., *Mytilus edulis* L., and *Mytilus galloprovincialis* Lmk. (Mytilidae; Mollusca). *Biological Journal of the Linnean Society* **13**, 65–73.

Skibinski, D. O. F., Beardmore, J. A. and Cross, T. F. (1983). Aspects of the population genetics of *Mytilus* (Mytilidae: Mollusca) in the British Isles. *Biological Journal of the Linnean Society* **19**, 137–183.

Sokal, R. R. and Crovello, T. J. (1970). The biological species concept: a critical evaluation. *American Naturalist* **104**, 127–153.

Solignac, M. (1976). Demographic aspects of interspecific hybridization. *Oecologia* **26**, 33–52.

Solignac, M. (1978). Genetics of ethological isolating mechanisms in the species complex *Jaera albifrons* (Crustacea, Isopoda). *In* "Marine Organisms: Genetics, Ecology and Evolution" (B. Battaglia and J. A. Beardmore, eds), pp. 637–664. Plenum Publishing, New York, USA.

Solignac, M. (1981). Isolating mechanisms and modalities of speciation in the *Jaera albifrons* species complex (Crustacea, Isopoda). *Systematic Zoology* **30**, 387–405.

Spilliaert, R., Vikingsson, G., Arnason, U., Palsdottir, A., Sigurjonsson, J. and Arnason, A. (1991). Species hybridization between a female blue whale (*Balaenoptera musculus*) and a male fine whale (*B. physalus*): molecular and morphological documentation. *Journal of Heredity* **82**, 269–274.

Srinivasa Rao, K. and Ganapati, P. N. (1977). Description of the post-larvae and juveniles of *Scomberomorus lineolatus* (Cuvier and Valenciennes, 1831) from Indian waters. *Journal of Natural History* **11**, 101–111.

Srinivasa Rao, K. and Lakshmi, K. (1993). *Scomberomorus lineolatus* (Cuvier), an interspecific natural hybrid (*S. commerson* (Lacépéde) × *S. guttatus* (Bloch and Schneider) off Visakhapatnam, India. *Journal of Natural History* **27**, 471–491.

Stebbins, G. L. (1950). "Variation and Evolution in Plants". Columbia University Press, New York, USA.

Stormer, L. (1977). Arthropod invasion of land during late Silurian and Devonian times. *Science N.Y.* **197**, 1362–1364.

Strathmann, R. R. (1981). On the barriers to hybridization between *Strongylocentrotus droebachiensis* (O. F. Muller) and *S. pallidus* (G. O. Sars). *Journal of Experimental Marine Biology and Ecology* **55**, 39–47.

Swan, E. F. (1953). The Strongylocentrotidae (Echinoidea) of the northeast Pacific. *Evolution* **7**, 269–273.

Swan, E. F. (1962). Evidence suggesting the existence of two species of *Strongylocentrotus* (Echinoidea) in the northwest Atlantic. *Canadian Journal of Zoology* **40**, 1211–1222.

Talmadge, R. R. (1977). Notes on a California hybrid *Haliotis*. *Veliger* **20**, 37–38.

Templeton, A. R. (1989). The meaning of species and speciation: a genetic perspective. *In* "Speciation and its Consequences" (D. Otte and J. A. Endler, eds), pp. 3–27. Sinauer Associates, Sunderland, Massachusetts, USA.

Tokida, J., Ohmi, H. and Imashima, M. (1958). A chimaera of *Alaria* and *Laminaria* found in nature. *Nature, London* **181**, 923–924.

tom Dieck (Bartsch), I. (1992). North Pacific and North Atlantic digitate *Laminaria* species (Phaeophyta): hybridization experiments and temperature responses. *Phycologia* **31**, 147–163.

tom Dieck (Bartsch), I. and de Oliveira, E. C. (1993). The Section Digitatae of the genus *Laminaria* (Phaeophyta) in the northern and southern Atlantic; crossing experiments and temperature responses. *Marine Biology* **115**, 151–160.

Turner, W. R. (1967). Biology of Gulf of Mexico menhaden. *US Fisheries and Wildlife Services Circular* **264**, 20–22.

Turner, W. R. (1969). Life history of menhadens in the eastern Gulf of Mexico. *Transactions of the American Fisheries Society* **98**, 216–224.

Vacquier, V. D., Carnier, K. R. and Stout, C. D. (1990). Species specific sequences of abalone lysin, the sperm protein that creates a hole in the egg envelope. *Proceedings of the National Academy of Science of the USA* **87**, 5792–5796.

Väinölä, R. and Hvilsom, M. M. (1991). Genetic divergence and a hybrid zone between Baltic and North Sea *Mytilus* populations (Mytilidea: Mollusca). *Biological Journal of Linnean Society* **43**, 127–148.

Väinölä, R. and Varvio, S.-L. (1989). Biosystematics of *Macoma balthica* in northwestern Europe. *In* "Reproduction, Genetics and Distributions of Marine Organisms" (J. S. Ryland and P. A. Tyler, eds), pp. 309–316. Fredensborg, Denmark.

Varvio, S.-L., Koehn, R. K. and Väinölä, R. (1988). Evolutionary genetics of the *Mytilus edulis* complex in the North Atlantic region. *Marine Biology* **98**, 51–60.

Vasseur, E. (1952). Geographic variation in the Norwegian sea-urchins, *Strongylocentrotus droebachiensis* and *S. pallidus*. *Evolution* **6**, 87–100.

Verrill, A. E. (1909). Remarkable development of sea-stars on the northwest American coast; multiplicity of rays; teratology; problems in evolution; geographical distribution. *American Naturalist* **43**, 542–555.

Victor, B. C. (1986). Duration of the planktonic larval stage of one hundred species of Pacific and Atlantic wrasses (Family Labridae). *Marine Biology* **90**, 317–326.

Wallace, C. C. and Willis, B. L. (1994). Systematics of the coral genus *Acropora*: implications of new biological findings for species concepts. *Annual Review of Ecology and Systematics* **25**, 237–262.

Walters, V. and Robins, C. R. (1961). A new toadfish (Batrachoididae) considered to be a glacial relict in the West Indies. *American Museum Novitates* **2047**, 1–24.

Warwick, T., Knight, A. J. and Ward, R. D. (1990). Hybridization in the *Littorina saxatilis* species complex (Prosobranchia: Mollusca). *Hydrobiologia* **193**, 109–116.

Wilbur, D. H. (1989). Reproductive biology and distribution of stone crabs (Xanthidae, *Menippe*) in the hybrid zone on the northeastern Gulf of Mexico. *Marine Ecology Progress Series* **52**, 235–244.

Wiley, E. O. (1981). "Phylogenetics: The Theory and Practice of Phylogenetic Systematics". John Wiley, New York, USA.

Willis, B. L., Babcock, R. C., Harrison, P. L. and Wallace, C. C. (1992). Experimental evidence of hybridization in reef corals involved in mass spawning events. *In* "Proceedings of the 7th International Coral Reef Symposium" (Abstracts), Guam, June 1992. Published by the University of Guam Marine Laboratory, p. 109.

Willis, G. L. and Skibinski, D. O. F. (1992). Variation in strength of attachment to the substrate explains differential mortality in hybrid mussel (*Mytilus galloprovincialis* and *Mytilus edulis*) populations. *Marine Biology* **112**, 403–408.

Winston, J. E. (1992). Systematics and marine conservation. *In* "Systematics, Ecology and the Biodiversity Crisis" (N. Eldredge, ed.), pp. 144–168. Columbia University Press, USA.

Woodburn, K. D. (1961). Survival and growth of laboratory-reared northern clams (*Mercenaria mercenaria*) and hybrids (*M. mercenaria* × *M. campechiensis*) in Florida waters. *Proceedings of the National Shellfisheries Association* **52**, 31–36.

Woodruff, D. S. (1989). Genetic anomalies associated with *Cerion* hybrid zones: the origin and maintenance of new electromorphic variants called hybrizymes. *Biological Journal of the Linnean Society* **36**, 281–294.

Reproductive Biology of Marine Cladocerans

David A. Egloff[1], Paul W. Fofonoff[2] and Takashi Onbé[3,4]

[1] *Department of Biology, Oberlin College, Oberlin, OH 44074-1082, USA*
[2] *Smithsonian Environmental Research Center, PO Box 28, Edgewater, MD 21037-0028, USA*
[3] *Faculty of Applied Biological Science, Hiroshima University, Higashi-Hiroshima 739, Japan*
[4] *Present address: 2-60-406 Saijo Nishi-Honmachi, Higashi-Hiroshima 739, Japan*

ADVANCES IN MARINE BIOLOGY VOL. 31
ISBN 0-12-026131-6

1. INTRODUCTION

1.1. Taxonomy and Distribution

Marine cladocerans are seasonally abundant and widely distributed in continental shelf waters (Sherman, 1966), estuaries (Barlow, 1955; Bosch and Taylor, 1968, 1973a; Bryan and Grant, 1979), the open ocean (Gieskes, 1971a), the Baltic (Ackefors, 1965), the Mediterranean (Bernard, 1955; Le Tourneau, 1961) and inland seas (Mordukhai-Boltovskoi and Rivier, 1987; Aladin, 1995). They comprise a significant fraction, and at times the dominant component, of mesoplanktonic communities (Bosch and Taylor, 1973a; Platt, 1977; Sherman *et al.* 1987) with densities recorded in excess of 50 000 (Bosch and Taylor, 1973a; Onbé, 1974; Grahame, 1976) and 100 000 individuals m^{N3} (Platt, 1977; Bryan and Grant, 1979). The abundance and cosmopolitan distribution of cladocerans is not matched by their species diversity. Only eight (Table 1) of the approximately 600 described species of cladocerans (Schram, 1986) are distributed widely in neritic and oceanic waters. Each of the eight species in the oceans has a wide distribution and all exhibit a high degree of morphological constancy over their geographic ranges.

Other marine cladocerans appear regularly in estuaries or saline, inland waters but seldom in neritic or oceanic environments (Table 2). Of particular interest are the 24 endemic species of Onychopoda found in the Ponto-Caspian basin. Included also in Table 2 are four freshwater species, in particular, *Polyphemus pediculus*, that will be referenced in this review. Comparative studies are helpful in clarifying the biology of marine species. Table 2 does not include all species that have been reported from estuaries or inland waters. Many freshwater species are introduced regularly into estuaries from freshwater systems but are generally believed not to thrive in saline waters. Many other species omitted from Table 2 occur in inland, athalassic environments (Frey, 1993) at a wide range of salinities.

A general overview of the biology and systematics of marine cladocerans may be found in Rammner (1930). Mordukhai-Boltovskoi (1968b) and Mordukhai-Boltovskoi and Rivier (1987) present detailed reviews of the

Table 1 Cladocerans prevalent in neritic and oceanic systems.

Branchiopoda	Taxa widely distributed in neritic and oceanic systems*
Order Onychopoda Sars, 1865 (replaces superfamily Polyphemoidea Brooks, 1959) Family Podonidae Mordukhai-Boltovskoi, 1968	 *Evadne nordmanni* Lovén, 1836 *Evadne spinifera* P. E. Müller, 1867 *Podon intermedius* Lilljeborg, 1853 *Podon leuckarti* (G. O. Sars, 1862) *Pseudevadne tergestina* Claus, 1877** *Pleopis polyphemoides* (Leuckart, 1859)† *Pleopis schmackeri* Poppe, 1889†
Family Polyphemidae Baird, 1845 Family Cercopagidae Mordukhai-Boltovskoi, 1968	
Order Haplopoda Sars, 1865	—
Order Ctenopoda Sars, 1865 (replaces superfamily Sidoidea Brooks, 1959) Family Sididae Baird, 1850	*Penilia avirostris* Dana, 1852
Order Anomopoda Sars, 1865 (replaces superfamily Chydoroidea, Brooks, 1959) Family Bosminidae Baird, 1845 Family Chydoridae Stebbing, 1902 Family Moinidae Goulden, 1968 Family Daphniidae (Straus, 1820)	—

* In addition to wide distribution in the oceans, all of the above species occur in the Black Sea, except for *Pleopis schmackeri*, and two species have moved, probably via the Volga–Don canal, into the Ponto-Caspian basin. *Pleopis polyphemoides* appeared in the Caspian Sea in 1957 (Mordukhai-Boltovskoi, 1965) and *Podon intermedius* in 1985 (Kurashova *et al.*, 1992).
** *Pseudevadne tergestina* was formerly classified as *Evadne tergestina* (see Section 1.1).
† These *Pleopis* species were formerly classified as *Podon* (see Section 1.1).

biology of the cladocerans of the Caspian Sea and its environs. Other major treatises on marine cladocerans have been published by Lilljeborg (1901), Baker (1938) and Dolgopolskaya (1958). Useful keys or descriptions for the identification of six or seven of the eight species in the oceans may be found in Della Croce (1974), Dolgopolskaya (1958), Flössner

Table 2 Cladocerans found in estuaries, mesohaline or hypersaline systems but not in the neritic or offshore oceanic environments (middle column) and pertinent, other taxa found in freshwater systems (right column). In the middle column, all of the Onychopoda, except the *Podon* sp., are endemic in the Ponto-Caspian basin. Some of the Anomopoda (*B. longispina*, *C. sphaericus*, *M. macrocopa*, *D. atkinsoni* and *D. similis*) are found primarily in freshwater, but have also been reported from saline systems.

Branchiopoda	Taxa found in estuarine and inland mesohaline or hyperhaline systems but not in the open oceans	Pertinent taxa found in freshwater systems
Order Onychopoda Sars, 1865 (replaces superfamily Polyphemoidea Brooks, 1959)		
Family Podonidae Mordukhai-Boltovskoi, 1968	*Caspievadne maximowitschi* (G. O. Sars, 1902)[1,2] *Cornigerius* Mordukhai-Boltovskoi (4 spp.)[1,2,6] *Evadne anonyx* G. O. Sars, 1897[1,2] *Evadne prolongata* Behning, 1938[1,2] *Podonevadne* Gibitz, 1922 (3 spp.)[1,2,4] *Podon* sp. ? (Tibet)[5] *Polyphemus exiguus* G. O. Sars, 1897[1,2]	
Family Polyphemidae Baird, 1845		*Polyphemus pediculus* (Linné) 1761 *Bythotrephes longimanus* Leydig, 1860 *Bythotrephes cederstroemi* Schoedler, 1877*
Family Cercopagidae Mordukhai-Boltovskoi, 1968	*Cercopagis* G. O. Sars, 1897 (13 spp.)[1-4]	
Order Haplopoda Sars, 1865		*Leptodora kindtii* (Focke, 1844)
Order Ctenopoda Sars, 1865 (replaces superfamily Sidoidea Brooks, 1959) Family Sididae Baird, 1850	*Penilia* sp.[7] *Diaphanosoma celebensis* Stingelin, 1900[8] *Diaphanosoma* sp.[9]	

Order Anomopoda Sars, 1865
(replaces superfamily Chydoroidea, Brooks, 1959)

Family Bosminidae Baird, 1845

Bosmina (Eubosmina) longispina Leydig, 1860[10]
= *B. maritima*
= *E. longispina maritima*

Family Chydoridae Stebbing, 1902

Alona taraporevalae Shirgur & Naik, 1977[11]
Chydorus sphaericus (O. F. Müller, 1785)[12]

Family Moinidae Goulden, 1968

Moina eugeniae Olivier, 1954[13]
Moina hutchinsoni Brehm, 1937[13,14]
Moina macrocopa (Straus, 1820)[15]
M. mongolica Daday, 1901[16]
= *M. microphthalma* Sars, 1903[17]
= ?*M. salina* Daday, 1888[18]

Family Daphniidae (Straus, 1820)

Daphnia akinsoni Baird[19]
Daphnia mediterranea Alonso, 1985[20]
Daphnia similis Claus, 1876[21]
Daphniopsis tibetana Sars, 1903[22]
Daphniopsis pusilla Serventy, 1929[23]
Daphniopsis australis Sergeev and Williams, 1985[24]

* *B. cederstroemi* may be a synonym of *B. longimanus* (Zozulya and Mordukhai-Boltovskoi, 1977; Yurista, 1992; Berg and Garton, 1994).

1 Mordukhai-Boltovskoi (1965, 1968a).
2 Mordukhai-Boltovskoi and Rivier (1987).
3 Mordukhai-Boltovskoi and Rivier (1971a).
4 Mordukhai-Boltovskoi and Rivier (1971b).
5 Hutchinson (1967, p. 121).
6 Mordukhai-Boltovskoi (1967a).
7 Egborge (1987).
8 Segawa and Yang (1987, 1990); Korovchinsky (1989).
9 Inoue and Aoki (1971).
10 Vuorinen and Ranta (1987).
11 Shirgur and Naik (1977).
12 Spittler and Schiller (1984).
13 Hutchinson (1937a, 1967).
14 Frey (1993).
15 Young (1924).
16 Aladin (1983).
17 Sukhanova (1971).
18 Goulden (1968).
19 Beadle (1943).
20 Alonso (1985).
21 Moore (1952).
22 Hutchinson (1937b).
23 Sergeev and Williams (1983).
24 Sergeev and Williams (1985).

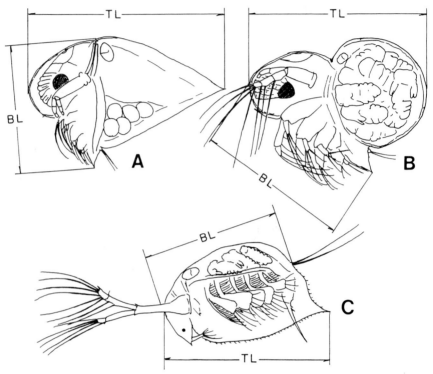

Figure 1 Conventional measurements of body length (BL) and total length (TL) for three principal morphotypes of cladocera found in the oceans: (A) *Evadne* or *Pseudevadne*; (B) *Podon* or *Pleopis*; and (C) *Penilia*. (After Onbé, 1974.)

(1972) and Smirnov and Timms (1983). *Pleopis schmackeri*, the one species not included in any of these keys, is described by Mordukhai-Boltovskoi (1978), Mordukhai-Boltovskoi and Rivier (1987) and Onbé (1983). A system of standard measurements of body size for the three principal morphotypes (Figure 1) was developed by Onbé (1974).

We have adopted the taxonomic system espoused by Fryer (1987a, b), who rejected the Cladocera as a taxonomic unit and assigned ordinal status to each of these four groups, as listed in Table 1. Many other classification systems have been applied to cladocerans in the past, but the one commonly cited in the literature on marine cladocerans was promulgated by Brooks (1959). Brooks divided the order Cladocera into two suborders: the Haplopoda and the Eucladocera; the latter accommodated the three superfamilies Polyphemoidea, Sidoidea and Chydoroidea, corresponding to the orders Onychopoda, Ctenopoda and Anomopoda as delineated by Fryer (1987b).

Relationships among the four orders proposed by Fryer (1987a) are

Table 3 Number of exopodite setae on the thoracopods of Podonidae found in oceans.

	Thoracopod			
	I	II	III	IV
Evadne spinifera[1,2]	2	2	2	1
Evadne nordmanni[1,2]	2	2	1	1
Podon leuckarti[1,2]	1	1	1	2
Podon intermedius[1,2]	2	1	1	2
Pseudevadne tergestina[2,3]	2	3	3	1
Pleopis polyphemoides[1,2]	3	3	3	2
Pleopis schmackeri[4]	4	4	4	2

[1] Gieskes (1971a).   [3] Claus (1877); Baker (1938).
[2] Dolgopolskaya (1958). [4] Mordukhai-Boltovskoi (1978).

uncertain but traditionally the Onychopoda and Haplopoda have been grouped together as "Gymnomera" because of presumed shared ancestral and derived characters that include locomotory naupliar antennae and open-water raptorial habits. Despite significant dissimilarities, the Ctenopoda and Anomopoda have been united as "Calyptomera" because of the bivalve carapace and other shared characteristics. Fryer (1987a, b) argues cogently that each of these four orders is sufficiently distinct to warrant the presumption that each has had a long and separate evolutionary history. Unraveling these relationships is not yet possible but further study of the marine cladocerans may contribute to the clarification and understanding of the relationships within and between cladoceran orders.

Of the eight, truly neritic or oceanic species (Table 1), one—*Penilia avirostris*—is the sole representative of the order Ctenopoda (Lochhead, 1954) and is readily distinguished from all others by its six pairs of phyllopodial limbs, enclosed by a carapace (Dolgopolskaya, 1958); the phyllopodial limbs are adapted ostensibly for filter feeding (Gore, 1980; Meurice, 1982). The other seven species are in the order Onychopoda, family Podonidae. In contrast to the Ctenopoda, the carapace in the Podonidae is reduced to a dorsal pouch and the four pairs of elongate and prehensile trunk appendages are capable of raptorial feeding.

The neritic and oceanic podonids include four genera: *Evadne* (two species), *Podon* (two species), *Pseudevadne* (one species) and *Pleopis* (two species). Definitive identifications of these seven species may be made by noting the number of setae on the exopods of the thoracic appendages (Table 3). *Pleopis polyphemoides* and *P. schmackeri* can be distinguished from *Podon leuckarti* and *P. intermedius* by having more setae on exopods of the thoracic appendages (Table 3), as well as by differences in the morphology of caudal claws and other features. Because of these differences Mordukhai-Boltovskoi (1968a, 1978) and Gieskes (1971c) concluded that the genus *Pleopis* Sars, 1861 should be retained for *P. polyphemoides* and *P. schmackeri*. In contrast, Meurice and Dauby (1983), using numerical taxonomy based on 25 characters, concluded that *P. polyphemoides* did not differ sufficiently from *P. intermedius* and *P. leuckarti* to justify its removal from the genus *Podon*. Similarly, Meurice and Dauby's analysis did not support Di Caporiacco's (1938) and Mordukhai-Boltovskoi's (1978) removal of *Evadne tergestina* to the genus *Pseudevadne*. Mordukhai-Boltovskoi and Rivier (1987) concurred with the earlier appraisal that *Pleopis* should be retained for *P. polyphemoides* and *P. schmackeri*, and *Pseudevadne* for *P. tergestina*. Until new morphological or molecular data are available to support another interpretation, we accept Mordukhai-Boltovskoi and Rivier's (1987) classification.

In contrast to their low species diversity in the oceans, the Onychopoda have become quite diverse in mesosaline, inland waters. All but one of the 24 Onychopoda species listed in Table 2 (middle column) are endemic to the inland seas and estuaries of the Ponto-Caspian region, which includes the Sea of Azov, the Caspian Sea and the Aral Sea; the one exceptional species is an undescribed *Podon* reported from Tibet (Hutchinson, 1967). The Onychopoda in the Ponto-Caspian region include six genera, four of which are endemic. The 24 endemic podonid species are characterized by an extraordinary transformation of limbs and body forms relative to the eight species in the oceans (Mordukhai-Boltovskoi 1965, 1966, 1967a b, 1968a b). The family Podonidae is represented in the Ponto-Caspian region by three endemic genera: *Cornigerius* (four species), *Caspievadne* (one specie) and *Podonevadne* (three or more species), as well as several endemic species in the genus *Evadne* (two species). The family Polyphemidae is represented in the Ponto-Caspian region by *Polyphemus exiguus*, a dwarf species that closely resembles the only other species in this genus, the freshwater *P. pediculus*. Lastly, the family Cercopagidae is represented by an endemic genus, *Cercopagis* (13 species), which includes four species assigned by Sars (1897) to the genus *Apagis* (Mordukhai-Boltovskoi, 1968a); note that the Cercopagidae is represented outside of the Ponto-Caspian area by the freshwater species *Bythotrephes longimanus*. The latter species, often reported by its synonym, *B. cederstroemi* (Zozulya and Mordukhai-Boltovskoi, 1977; Yurista, 1992; Berg and Garton, 1994), is a recent invader of the Great Lakes of North America from Europe (Bur et al., 1986; Lehman, 1991).

Ctenopoda found in hypersaline inland waters and estuaries (Table 2) include a *Penilia* sp. (Egborge, 1987), *Diaphanosoma* sp. (Inoue and Aoki, 1971) and *Diaphanosoma celebensis* (Segawa and Yang, 1990).

Among the Anomopoda found in estuaries are *Bosmina longispina* (Vuorinen and Ranta, 1987), *Alona taraporevalae* (Shirgur and Naik, 1977) and *Chydorus sphaericus* (Spittler and Schiller, 1984). Anomopoda found in hypersaline inland waters are several species of *Moina* (Goulden, 1968), *Daphnia* and *Daphniopsis* (Sergeev and Williams, 1983, 1985). *Moina salina* is planktonic in the Aral Sea and reproduces at salinities up to $97^{\circ}/_{oo}$ in the Kiziltash estuary in the Black Sea (Sukhanova, 1971).

1.2. Horizontal and Vertical Distributions

Cladocerans in the oceans have long been considered as neritic epizooplankton, occurring mainly in surface waters of coastal and embayment regions, but they also occur offshore. Wiborg (1955) found populations of *Evadne nordmanni* in oceanic waters of the North Atlantic. His findings were confirmed for various species in the North Atlantic Ocean and North Sea (Gieskes, 1971a), the eastern (Longhurst and Seibert, 1972) and northwestern (Kim, 1989; Kim and Onbé, 1989b, 1995) Pacific Ocean, the Indian Ocean (Della Croce and Venugopal, 1972) and the Gulf of Mexico and the Caribbean Sea (Della Croce and Angelino, 1987). Cladocerans appear to be transported to oceanic waters by surface currents because the successful hatching of diapaused embryos and migration of neonates to the surface from the muds of deep and cold ocean basins is highly unlikely. Kim (1989) found *Pseudevadne tergestina* widely distributed in offshore waters in the northwestern Pacific, where they had been carried from neritic regions by the Warm Kuroshio Current, with further distribution in warm core rings. Once established, populations are dependent on favorable environmental factors, especially the higher surface temperatures associated with vertical stratification. Gieskes (1970, 1971a) noted that cladocerans are primarily daylight-feeding surface water animals, well adapted to life in shallow neritic waters. They resemble phytoplankton in being adversely affected by vertical turbulence that would take them out of the euphotic zone. Hence in the deeper coastal waters and in the ocean they become abundant only when the water is stratified, i.e. in spring and summer in temperate latitudes.

Vertical profiles confirm the restriction of marine cladocerans to surface waters (Bainbridge, 1958; Trégouboff, 1963; Thiriot, 1968; Longhurst and Seibert, 1972; Onbé, 1974). Although reported migration patterns vary, recent findings in the southern Sea of Japan (Onbé and Ikeda, 1995) confirmed a pattern of *reverse* diel vertical migration within the surface layers. In contrast to their common occurrence in surface waters, *Evadne spinifera* and *Pseudevadne tergestina* were found by Trégouboff (1963) at abyssal depths of 1000–2000 m off Villefranche-sur-Mer in the western

Podonid cladoceran life cycle

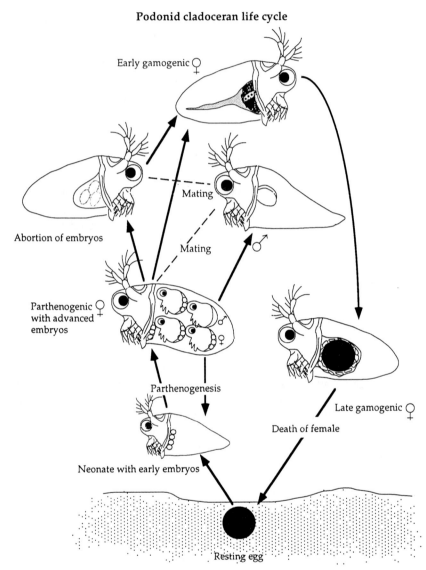

Early gamogenic ♀

Mating

Abortion of embryos

Mating

♂

Parthenogenic ♀
with advanced
embryos

♂
♀

Parthenogenesis

Late gamogenic ♀

Death of female

Neonate with early embryos

Resting egg

Figure 2   Life cycle of a podonid cladoceran illustrated for the marine species, *Evadne nordmanni*. The cycle is initiated in the spring with hatching of parthenogenic females from resting embryos which have been dormant in muds. These females produce subitaneous eggs which develop pædogenically. Note that the advanced-stage embryos in the parental brood chamber, as illustrated for three individuals, have already deposited eggs of the next generation in their own brood chambers. Normally, several generations of parthenogenic females are produced before males and gamogenic (sexual) females appear. Gamogenic females are assumed to be transformed parthenogenic females. Gamogenesis in females is

Fig. 2 (continued)
believed to begin during parthenogenic development of embryos and also following
an abortion of parthenogenic embryos. The timing of mating is not known, but
we assume here that it occurs before or during the deposition of one or two eggs
in the brood chamber. After mating, these gamogenic eggs, surrounded by nurse
cells, develop in the brood pouch. Details of gamogenic embryo development are
derived from Cheng (1947), Rivier (1968) and original observations (Fofonoff,
1994).

Mediterranean. Their appearance at these depths may have been the result
of the submergence of Mediterranean water beneath Atlantic surface water
flowing in through the Strait of Gibraltar, or by vertical mixing as a result
of convective water movements (Onbé and Ikeda, 1995).

1.3. Life Cycles

Marine cladocerans, like most of their freshwater counterparts, exhibit direct development, alternation of gamogenesis (amphimixis) and
diploid parthenogenesis (apomixis), and pædogenesis or the initiation of
reproduction by young individuals during their own early development
(see Section 5.2.1). Figure 2 summarizes the life cycle of a podonid
cladoceran, using *Evadne nordmanni* as an example. Normally, many
generations of all-female broods occur before males are produced.

Parthenogenic reproduction during favorable conditions for growth
typically consists of the production of successive broods of subitaneous
(parthenogenic) eggs, which develop without delay into parthenogenic
females. At other times females produce eggs that develop without
fertilization into diploid males, or one or more haploid eggs that are
fertilized by a male during copulation. These fertilized eggs undergo early
cleavage before development is arrested, at which time the mother may
cease feeding and die, or possibly revert to parthenogenic reproduction.
Production of more than one brood of gamogenic eggs by a single female
has not been reported in marine species. Diapausing embryos ("resting
eggs") are released during ecdysis. In freshwater species these resting
embryos may be attached to vegetation, as in some chydorids and
macrothricids, float and accumulate along the shorelines, or sink out of
the water column. In marine systems where vegetation and shorelines are
often remote, the resting embryos usually sink to the sediments where they
remain until development resumes, often at times correlated with rising
water temperatures.

Anatomically distinct parthenogenic[1] (amictic) and gamogenic[1] (mictic) females characterize the cladocerans. Individual females in many freshwater cladocerans and in the marine *Penilia* (Della Croce and Bettanin, 1969) may produce parthenogenic eggs at one time and gamogenic eggs at another. To accomplish this alternation of reproductive modes, a female must undergo anatomical reorganization, especially with respect to the brood chamber. Podonids may also alternate gamogenic and parthenogenic reproductive modes. Rivier (1968) observed the co-occurrence of late-stage, parthenogenic and gamogenic embryos in individual *Evadne anonyx* females. Fofonoff (1994) made similar observations in laboratory-reared *Pleopis polyphemoides* and observed a few *Evadne nordmanni* with late-stage, parthenogenic embryos in the brood pouch and a gamogenic egg developing in the ovary. Further studies, based on rearing individual females in the laboratory or on cytological studies of field populations, will be necessary to demonstrate whether these observations are exceptional or whether single females of marine cladocerans routinely alternate mictic and amictic reproduction.

Cyclic parthenogenesis (heterogony) is not widespread among the Metazoa, but it occurs also in monogonont rotifers, digenean trematodes and various insects, for example, cynipid wasps (Hymenoptera), aphids (Homoptera), cecidomyids (Diptera) and a beetle, *Micromomalthus debilis* (Coleoptera) (White, 1973; Bell, 1982).

1.4. Physiology and Nutrition

1.4.1. *Physiology*

To persist in saline environments cladocerans require either a tolerance for high salt concentrations or active osmoregulation. The latter is the case for the most successful marine and inland, saline species, namely, *Penilia* and all Podonidae. Adult *Penilia avirostris* osmoregulate (Aladin, 1979) through specialized cells on the epipodites (branchiae) of thoracic legs. Podonids and *Penilia* possess a cervical organ (Meurice and Goffinet, 1990) that has been variously termed a cervical gland (Nackendrüse), neck organ (Nackenorgan, organe nucal), dorsal gland, or an endusium (Sudler, 1899). Claus (1877) erroneously interpreted this structure as an adhesive organ (Haftorgane).

The cervical organ consists of no more than 12 mitochondria-rich cells, arranged radially around a central, permeable area of the exoskeleton (Potts and Durning, 1980; Meurice and Goffinet, 1983). These cells are

---

[1] For this review we elected to use the adjectives parthenogenic and gamogenic, and not their synonyms parthenogenetic and gamogenetic.

polarized with enlarged distal ends suspended in the hemocoel and small apical surfaces through which ions are secreted (Meurice and Goffinet, 1983). Using silver stains Dejdar (1931) demonstrated that the exoskeleton overlying the cervical organ is permeable to salts, and the underlying cells converted reduced (colorless) methylene blue to its oxidized (blue) form. The hypo-osmotic body fluids in the body and brood chambers (Khlebovich and Aladin, 1976; Aladin, 1982) of podonids presumably are maintained by active transport of salts out of the body through the cervical organ.

The cervical organ is fully formed (Meurice and Goffinet, 1983) but not metabolically active in embryonic podonids (Dejdar, 1931). This is not a handicap to embryos because the osmolarity of the fluid in the closed brood chamber (Section 2.1) is equal to that of the adult body fluid (Khlebovich and Aladin, 1976). The cervical gland remains inactive until the embryos are released from the brood chamber (Khlebovich and Aladin, 1976). In contrast the salt-regulating cervical gland in most embryos of freshwater cladocerans is active because the brood pouch is open. This requires that the embryos be osmotically self-sufficient (Rammner, 1931).

Hypo-osmotic regulation, although exacting a metabolic cost, may benefit marine organisms in general, and marine podonids in particular, because of the buoyancy conferred by hypo-osmotic body fluids (Potts, 1959; Potts and Durning, 1980). Experimental work remains to be done to determine if the energy required to maintain hypo-osmotic body fluids, thereby increasing buoyancy, is offset by the decrease in energy required for locomotion. Energy expenditure for swimming in podonids may be high, especially during the day when they typically aggregate near the sea surface.

1.4.2. Nutrition

Anatomically, podonids appear to be raptorial feeders capable of capturing prey in a size range of 20–170 μm (Nival and Ravera, 1979). Most investigators have assumed that podonids are primarily raptorial feeders on animal prey, or on dinoflagellates because of the observed abundance of podonids during dinoflagellate blooms (Bainbridge, 1958; Morey-Gaines, 1979). *Evadne nordmanni* frequently are seen in preserved samples clasping thecate dinoflagellates (*Ceratium*, *Protoperidinium*) and tintinnids (Bainbridge, 1958; Gieskes 1970; Fofonoff, 1994). Similarly, preserved *E. anonyx* (Rivier, 1968) and *Pleopis polyphemoides* (Bosch, 1970) appear to have captured copepod nauplii or copepodites. In most cases the items held by the cladocerans were too small to be retained by the nets used. In concentrated plankton sampler, Sedgwick-Rafter chambers, *P. polyphemoides* has been observed capturing small copepodites, *Eurytemora affinis*, and sucking fluid from them (Fofonoff, 1994). This mechanism of feeding on large prey resembles that described for *Polyphemus pediculus* (Butorina, 1965). Attempts to confirm this type of feeding at more natural prey densities have so far been unsuccessful (Fofonoff, personal observation).

Morey-Gaines (1979), using light microscopy, found thecal plates of *Ceratium* sp. in the gut contents of *Pleopis polyphemoides*. Using scanning electron microscopy (SEM), Jagger *et al.* (1988) found few dinoflagellates and no animal remains in the gut contents of several species of marine cladocerans. Kim *et al.* (1989) confirmed these results and found *Skeletonema costatum* to be a predominant prey item among the recognizable gut contents of *Penilia* and four podonid species, irrespective of the abundance of *Skeletonema costatum* in the phytoplankton. In addition, they found fragments of the diatoms *Chaetoceros*, *Thalassiosira*, *Cyclotella* and *Coscinodiscus*. Mean size of diatom fragments was 16.9 μm with a range of 4–115 μm. Kim *et al.* (1993) suggest that dinoflagellates were undercounted in earlier studies. An obvious shortcoming of studies based on light and scanning electron microscopy is that only organisms with indigestible parts are detected. To detect visually unrecognizable prey in the gut contents, investigators must use biochemical and molecular methods (Calver, 1984; Schulz and Yurista, 1995).

Several investigators have recently re-examined feeding by podonids *in vitro*. Jagger *et al.* (1988) reported clearance rates by *Podon intermedius* up to 0.5 ml animal^{-1} h^{-1} when feeding on the diatom *Rhizosolenia* sp. ($150 \times 10 \mu$m), but found no significant feeding on the flagellate *Tetraselmis chui*, the thecate dinoflagellate *Prorocentrum gracile* ($40 \times 15 \mu$m), or the diatom *Hemiaulus* sp. In Turner and Graneli's (1992) enclosure studies, *Pleopis polyphemoides* exposed to natural assemblages did not feed on ciliates but did feed on microflagellates at rates of up to 6 ml animal^{-1} h^{-1}, *Skeletonema costatum* at rates of up to 2.5 ml animal^{-1} h^{-1}, and *Rhizosolenia* sp. at rates of up to 2.5 ml animal^{-1} h^{-1}. Nielsen (1991) reported that *Evadne nordmanni* and *Podon* spp. from the Kattegat near Helsingør, Denmark, cleared three species of the large dinoflagellate *Ceratium* sp. ($100–200 \mu$m) at rates of 0.1–0.4 ml animal^{-1} h^{-1}. Fofonoff (1994) found that *P. polyphemoides*, grazing on a natural assemblage including *Gymnodinium sanguineum* ($40–60 \mu$m, 27 cells ml^{-1}) and *Skeletonema costatum* ($31\,000$ cells ml^{-1}), fed on the *Gymnodinium* at 0.6 ml animal^{-1} h^{-1} but did not feed detectably on *Skeletonema*. Clearance rates of *P. polyphemoides* on unialgal cultures of *G. sanguineum* decreased (from 5.0 to 0.30 ml animal^{-1} h^{-1}) as food concentration increased from 1 to 100 cells ml^{-1} (7–655 μg C ml^{-1}). Maximum ingestion rates were equivalent to 27% of body carbon h^{-1}. Turner and Graneli's (1992) experiments were run at 9°C, Nielsen's (1991) at "*in situ* conditions" (*c*.15–20°C) while Fofonoff's (1994) were conducted at 18°C.

Penilia's feeding apparatus appears adapted for suspension feeding on particles as small as 2 μm (Turner *et al.*, 1988). In laboratory studies, *Penilia* ingested heterotrophic flagellates 2–7 μm in size (Pavlova, 1959b; Turner *et al.*, 1988), the autotrophic flagellate *Isochrysis galbana* (Paffen-höfer and Orcutt, 1986), and the diatoms *Thalassiosira pseudonana* (4–6 μm) and *T. weissflogii* (10–12 μm) (Turner *et al.*, 1988). Grazing on free-living bacteria was reported by Pavlova (1959b), but not detected by Turner *et al.* (1988). Kim *et al.* (1989), using SEM, found primarily diatom fragments in the guts of *Penilia* taken from the Inland Sea of Japan, yet in experimental studies *Penilia* failed to ingest net-collected diatoms, including *Skeletonema costatum* (Turner *et al.*, 1988).

2. PARTHENOGENESIS

2.1. Reproductive Anatomy of Parthenogenic Females

Ovaries in cladocerans are elongated, spindle-shaped organs located on each side of the intestine. In *Evadne nordmanni* the ovary extends nearly the length of the thorax at maturity (Jorgensen, 1933; Gieskes, 1970). Oogenesis occurs in this species at the narrow anterior end of the ovary (Jorgensen, 1933), while mature eggs accumulate at the broader and more rounded posterior end. A short oviduct connects the posterior end of the ovary to the brood chamber lying dorsal to the ovary (Claus, 1877; Jorgensen, 1933). Neither the oviduct nor the ovary is easily detected histologically in the absence of developing eggs (Jorgensen, 1933). In contrast, oogenesis is initiated in *Penilia avirostris* at the posterior end of the ovary, and mature eggs accumulate at the anterior end (Sudler, 1899). This arrangement requires that mature ova pass through the germarium during ovulation. The position of the germarium—anterior in *Evadne* (Onychopoda) and posterior in *Penilia* (Ctenopoda)—is reversed in the freshwater representatives of these families; for example, in *Bythotrephes* (Onychopoda) it is at the posterior end (Rossi, 1980), while in *Sida* (Ctenopoda) the germarium is at the anterior end of the ovary. The significance of these different orientations is not known.

After release from the ovary, ova in marine podonids are enclosed totally in a brood chamber. A closed brood chamber (Figure 3A) permits nutritive secretions to be retained, whereas the open chambers of most

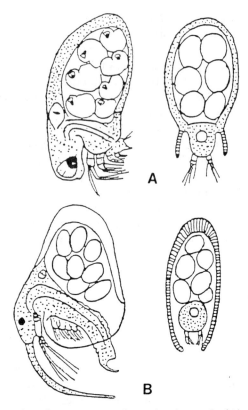

Figure 3 Lateral and transverse schematic views of: (A) the closed brood chamber of most marine cladocerans; and (B) the open brood chamber of most freshwater cladocerans. (After Rammner, 1931.)

freshwater cladocerans (Figure 3B) are flushed continuously with water by the movements of the abdomen (Rammner, 1931).

An anatomical description of the brood chambers in podonids is available only for the Caspian *Evadne anonyx* (Rivier, 1968), but some details are available for the related freshwater onychopods, *Polyphemus pediculus* (Patt, 1947; Butorina, 1968) and *Bythotrephes longimanus* (Zaffagnini, 1964; Rossi, 1980). In these species the brood chamber totally encloses the mature eggs and developing embryos. The brood chamber is bounded by an inner membrane that expands outwards towards the carapace as the eggs develop (Figure 4). Rossi (1980) reported that in *B. longimanus* the double membrane is formed by the cells of the expanding brood chamber sac and the innermost cells of the hypodermis underlying the carapace. During ecdysis and the release of the mature embryos, the

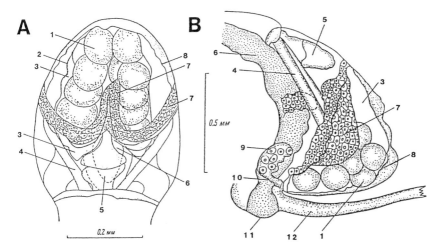

Figure 4 Anatomy of the brood chamber of parthenogenic *Polyphemus pediculus* females: (A) dorsal view of a young specimen; and (B) lateral view of a mature specimen. 1, Parthenogenic egg; 2, expanding inner membrane of the brood chamber; 3, brood chamber; 4, muscle attaching Nährboden to exoskeleton; 5, heart; 6, intestine; 7, Nährboden; 8, outer membrane in (A) or combined inner and outer membranes in (B), bounded on the outside by the carapace; 9, ovary; 10, oviduct; 11, anus; 12, tail spine. (After Butorina, 1968.)

membranes of the brood chamber may remain intact as described by Butorina (1968) for *P. pediculus* (Section 2.5). Rossi (1980) concluded that after birth the brood chamber is preserved as a canal at the posterior end of the brood chamber in *B. longimanus*. On the other hand, Zaffagnini (1964) observed for this species that the brood chamber ruptures and collapses during birth and eggs of the next generation are laid into a new brood chamber that begins to expand immediately. In either case, the next batch of parthenogenic eggs has already begun development by the time the mature brood has been released.

Associated with the brood chamber in marine and a few freshwater cladocerans is a placental organ, the "Nährboden", which appears to supply nutrients to embryos. This structure occurs in three evolutionary lines, namely, in all Onychopoda (*Bythotrephes*, *Evadne*, *Podon*, *Polyphemus*, etc.), in some Anomopoda (*Moina*) and in some Ctenopoda (*Penilia*). The Nährboden extends laterally and medially into the broodchamber as a ventral ridge of glandular tissue (Figure 4B). Cytological evidence supports the hypothesis that the Nährboden is the source of nutritive material in the brood pouch, but additional verification with modern histochemical and biochemical studies is needed. During the

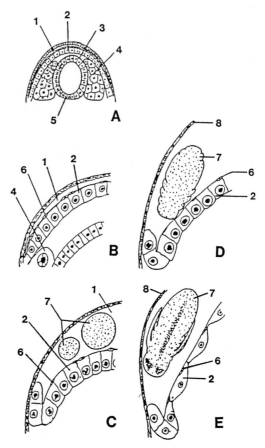

Figure 5   Schematic drawings of transverse sections through the brood chamber of a parthenogenic female of *Polyphemus pediculus*. (A) and (B): before ovulation; (C): after ovulation; (D) and (E): at successive stages during later development. 1, Brood chamber; 2, Nährboden; 3, dorsal blood sinus; 4, ovary; 5, intestine; 6, thin hypodermis covering the Nährboden; 7, parthenogenic embryos; 8, carapace. Membranes bounding the expanded brood chamber, as indicated in Figure 4, were not noted in this drawing. (After Patt 1947.)

reproductive cycle the Nährboden increases in size as a result of cell enlargement. These cells change in shape from cuboidal (Figure 5A, B) to columnar (Figure 5C). This enlargement coincides in *Polyphemus pediculus* with the period from ovulation to the first cleavage of the oocytes in the brood chamber. As the Nährboden expands, the cells at its margin are pushed to the lateral edges of the brood chamber, where they turn dorsally (Figure 5C) and appear to fuse with the hypodermis (Butorina,

1968), from which they were originally derived (Weismann, 1877). The Nährboden attains maximum volume after embryonic cleavage begins and maintains maximum volume until appendage buds appear. At this stage the medial cells begin to diminish in size, revert to a cuboidal shape (Figure 5D), and eventually shrink to a thin sheet (Figure 5E). In contrast, the marginal cells maintain their large volume throughout development. When the Nährboden is expanded, mitochondria and osmiophilic bodies associated with, and apparently arising from, the Golgi apparatus begin to aggregate around the periphery of a vesicle located at the distal or brood chamber side of the Nährboden cells (Patt, 1947). This vesicle, presumably a secretory structure, is firmly attached to the chitinous lining of the brood chamber. The absence of pores visible by light microscopy in the lining of the brood chamber led Patt (1947) to conclude that the lining may be a barrier to the passage of large organic molecules. Nonetheless, the secretory products of the Nährboden appear to move into the brood chamber because the appearance and staining properties of the fluid in the brood chamber are identical to the osmiophilic granules in the Nährboden cells (Kiernan, 1980). The osmiophilic material is presumably a lipid-containing, nourishing material. Before ovulation the fluid in the brood chamber does not stain with osmium tetroxide, indicating the absence of lipids. As development proceeds, the fluid becomes progressively more osmiophilic, changing from a light golden brown to black, indicative of unsaturated hydrophilic lipids (Kiernan, 1980). Throughout the period of development of embryos in the brood chamber, the embryos and extra-embryonic fluid in the brood chambers of *P. pediculus* are maintained in constant motion by the rhythmic contraction of paired muscles (Figure 4) attached to the Nährboden (Claus, 1877; Butorina, 1968). When the embryos reach the stage when release is imminent, the fluid no longer stains with osmium tetroxide (Patt, 1947).

The hypothesized nutritional role of the Nährboden is supported also by repeated failures to culture the nearly yolk-less eggs of *Moina*, *Polyphemus*, *Bythotrephes* and *Leptodora* outside the brood chamber. In contrast, yolk-filled eggs of *Daphnia* develop normally outside the brood chamber in isotonic Ringer's solution (Ramult, 1914) or in sterile pond water (Obreshkove and Fraser, 1940; Patt, 1947).

Additional evidence that the Nährboden plays a nutritional role is its expansion during development of the parthenogenic embryos (Claus, 1877; Kuttner, 1911; Jorgensen, 1933; Patt, 1947), but not in gamogenic females where the Nährboden remains small and undeveloped, with an intracellular appearance similar to that of pre-reproductive females (Patt, 1947). The yolk-rich eggs of gamogenic females develop apparently without exogenous nourishment from the Nährboden.

The brood chamber in *Penilia* is heart-shaped in cross-section with a dorsal apex and a ventral depression occupied by the intestine (Sudler, 1899). No other information is available on the Nährboden in *Penilia*.

2.2. Origin of Germ Cells

In the Polyphemidae, as represented by *Polyphemus pediculus*, the primordial germ cell or genital anlage appears first during the fourth cell division, at which time all but one cell in the eight-cell blastula divide meridionally (Kühn, 1913). This cell in the lower half of the blastula divides equatorially, creating two cells, one of which gives rise to the endoderm (the upper cell) and the other to the germ cell (the lower cell). The latter carries the remains of one of the undeveloped oocytes or nurse cells which had been incorporated earlier into the egg. At the next (fifth) division the germ cell does not divide, with the result that a 31-cell embryo is created. At the sixth division the germ cell divides for the first time thus creating two (left and right) germ cells at the 62-cell stage. Neither the cells of the genital anlage nor the endoderm divide at the next (seventh) division with the result that a 118-cell blastula is formed with four endodermal cells and two germ cells. These six cells are located at the vegetal pole of the embryo and surrounded by a horseshoe-shaped group of six mesodermal cells (see Figure 50, Kühn, 1913).

2.3. Oogenesis in Parthenogenic Females

The ovaries of female cladocerans exhibiting parthenogenic reproduction undergo gradual enlargement followed by a rapid diminishment at the time the eggs are transferred from the ovaries into the brood pouch. During oogenesis four oocytes typically become aligned in a row parallel to the long axis of ovaries as, for example, in *P. pediculus* (Claus, 1877). Ultrastructural studies in *Daphnia* reveal that these tetrads are derived from a single oogonium and are joined by intercellular bridges resulting from incomplete cytoplasmic division (Zaffagnini, 1987). This pattern of tetrads is absent in marine cladocerans. Gieskes (1970) found no indication of tetrads in *Evadne* spp. In *Penilia* clusters of four oocytes are scattered among larger clusters of oocytes but linear tetrads are absent (Sudler, 1899). During maturation of each set of four oogonia in *P. pediculus*, three oocytes are lost or engulfed by the single surviving oocyte. Intracellular debris from the disintegration of the engulfed cells has been described by Kühn (1913) for *P. pediculus* and by Sudler (1899) for *Penilia*.

In the Polyphemidae the small parthenogenic eggs obtain nourishment during development from material secreted into the brood chamber by the Nährboden (Rossi, 1980), thus vitellogenesis is virtually absent. Eggs of *Penilia* also contain little yolk and receive nourishment from a fluid secreted into the brood chamber (Sudler, 1899).

2.4. Embryonic Development of Subitaneous Eggs

In the absence of published embryological studies of marine podonids, studies of the freshwater *Polyphemus pediculus* by Kühn (1913) must suffice as an introduction to the early embryology of the order Onychopoda.

In *P. pediculus* the embryonic animal pole is marked by a single polar body, in contrast to the vegetal pole that is distinguished by remnants of three engulfed nurse cells. In *Daphnia* the single polar body that forms during oogenesis is diploid, and results from the extrusion of one set of chromosomes produced by ameiotic division. During this division, pairing of homologous chromosomes is absent, thus indicating that recombination and a reduction in chromosome number does not occur (Zaffagnini, 1987). Endomeiosis and genetic recombination has been described (Bacci *et al.*, 1961) for one strain of *Daphnia pulex* during the maturation of parthenogenic eggs, but it has not been confirmed by karyological or population studies in other daphnids.

Cleavage in *P. pediculus* is holoblastic, determinate, asynchronous and nearly radial (Kühn, 1913; Ivanova-Kazas, 1977). Total cleavage in *P. pediculus* is possible because of the reduction in yolk associated with the maternal nourishment of the embryos. Similar patterns may be found in all other Nährboden-bearing species. Cleavages in *P. pediculus* are meridional followed by an equatorial third division that creates four cells at the vegetal pole and four slightly larger cells at the animal pole. Mesoderm is derived ultimately from three of the four cells at the vegetal pole, while the endoderm and the germinal anlage arise from the fourth cell as described above (Section 2.2). Primary ectoderm develops subsequently from the cells of the animal pole, while secondary ectoderm emerges from the three cells at the vegetal pole that gave rise earlier to the mesoderm. After the 16-cell stage asynchronous divisions between the cells in the vegetal and animal hemispheres and between the cells within these halves result in embryos of 31, 62, 118, 236 and 460 cells after the fifth, sixth, seventh, eighth and ninth divisions respectively.

At the fifth division the embryo is a hollow, spherical blastula with 31 cells. After the sixth division (62-cell stage), six mesodermal cells are arranged at the vegetal pole in a crescent surrounding two germ cells; the crescent is filled at its open end by the four cells of the endodermal anlage (Ivanova-Kazas, 1977). This pattern persists through the 236-cell stage.

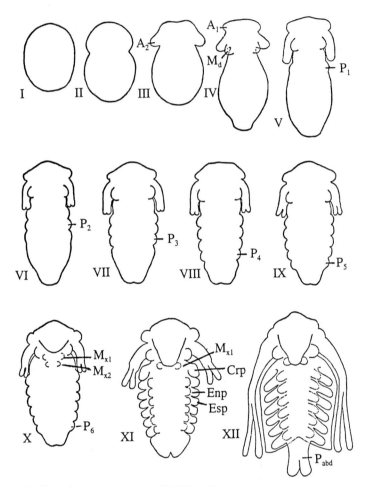

Figure 6 Developmental stages (I–XII) of embryos from parthenogenic eggs in the brood chamber of *Penilia avirostris* Dana. A_1, First antenna; A_2, second antenna; Crp, carapace; Enp, endopodite; Esp, exopodite; M_d, mandible; M_{x1}, first maxilla; M_{x2}, second maxilla, P_1–P_6, thoracic appendages; P_{abd}, postabdomen. (After Della Croce and Bettanin, 1965.)

At the ninth division gastrulation begins; initially the eight endodermal and 12 mesodermal cells move by ingression into the blastocoel, followed by the eight germ cells (Kühn, 1913); the eight endodermal cells subsequently divide, giving rise to a 460-cell gastrula. The ectoderm closes over the surface vacated by the ingression of cells and leaves behind only a slight trace of a blastoporal indentation in *P. pediculus* (see Kühn, 1913, Figure 75).

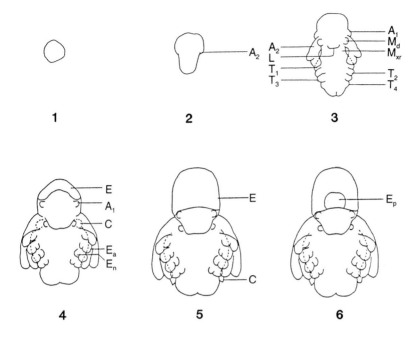

Figure 7 Developmental stages (1–6) of embryos from parthenogenic eggs in the brood chamber of *Evadne nordmanni*. A_1, First antennae; A_2, second antennae; C, carapace; E, eye; E_a, exopodite; E_n, endopodite; E_p, eye pigment; M_d, mandible; M_{xr}, maxillary and mandibular region; T_1–T_4, thoracic segments. (After Platt and Yamamura, 1986.)

Developmental patterns following gastrulation have been described for *Penilia avirostris* (Sudler, 1899; Della Croce and Bettanin, 1965), for *Evadne nordmanni* (Gieskes, 1970; Platt and Yamamura, 1986) and for *Podon leuckarti* (Gieskes, 1970). For *P. avirostris*, Della Croce and Bettanin (1965) arbitrarily divided the developmental process into 12 stages differentiated by changes in the shape and progressive appearance of the appendages (Figure 6). Platt and Yamamura (1986) followed a similar approach for *E. nordmanni* but delineated only six stages (Figure 7). Both systems are useful guides for the determination of developmental stages for use in physiological or ecological studies.

Absolute durations of overall embryonic development were determined (Fofonoff, 1994) for *Pleopis polyphemoides* at 10, 14, 18 and 22°C and for *Evadne nordmanni* at 14 and 18°C in laboratory cultures fed unialgal cultures of *Gymnodinium sanguineum* (an athecate dinoflagellate). Development times for *P. polyphemoides* reared on natural assemblages of phytoplankton containing gymnodinioid dinoflagellates were similar to those fed on cultured *G. sanguineum* (Fofonoff, 1994). *P. polyphemoides* neonates took 12 d from birth (and embryo deposition) to first embryo release at 10°C, but only 3 d at 22°C. *E. nordmanni*'s mean development times (6.1–7.6 d at 14°C and 3.6 d at 18°C) did not differ significantly from those of *P. polyphemoides* (5.0–6.1 d at 14°C and 3.9–4.3 d at 18°C). In some cases late developmental stages of *E. nordmanni* were greatly prolonged, probably because of stress or injury. These data were pooled and fitted to a second-degree logarithmic polynomial (Bottrell, 1975a) with the equation:

$$\log D = 0.847(\log T)^2 - 3.609 \log T + 3.796$$

where *D* is development time (days) and *T* is temperature (°C). Although 22°C was the highest water temperature used in these experiments, the equation predicts a 53-h development time at 28°C. This result is consistent with field estimates of development times of 2 d at 27–28°C for *Pseudevadne tergestina* (Onbé, 1974; Bryan, 1979).

Embryo development times of *Pleopis polyphemoides* and *Evadne nordmanni* are in the mid to upper range of cladocerans reared at similar temperatures (Figure 8). Because podonids lack a non-ovigerous juvenile stage, embryo development time is equivalent to generation time. This acceleration of the life cycle results in shorter generation times for *P. polyphemus* and *E. nordmanni* than those of any other known cladoceran (Figure 9) at equivalent temperatures. Only at temperatures above 28°C does one species, *Moina brachiata* (Maier, 1993), achieve a generation time as short as that estimated for *Pseudevadne tergestina* (Bryan, 1979). In comparison, among non-cladoceran zooplankton (Figure 10), only rotifers and appendicularians achieve generation times as short as *P. polyphemoides* and *E. nordmanni*.

A major factor in achieving these short generation times is the rapid growth of the oligolecithal eggs of marine cladocerans that results from the nourishment by the Nährboden. Despite their small size, eggs of marine cladocerans develop into neonates that are relatively larger than those of freshwater cladocerans (Butorina, 1968). The rapid growth during development is manifested by a volume increase from egg to neonate of five- to sixfold in *Polyphemoides pediculus* and more than tenfold in *Evadne anonyx*, as compared to a roughly twofold increase in *Daphnia* and *Bosmina* spp. (Patt, 1947; Rivier, 1968). Neonates of marine cladocerans are typically one-half to two-thirds their eventual length as adults (Hrbåcková, 1974). For example, *Penilia* neonates are 370–450 μm in length (Steuer, 1933) and *E. nordmanni* neonates are 320–460 μm (Bainbridge, 1958); these lengths are approximately 45–70%, respectively, the length of typical adults of these marine species (Smirnov and Timms, 1983). Fofonoff (1994) found similar size relationships between neonates

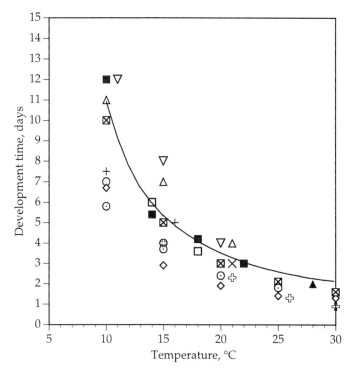

Figure 8 Embryo development times (egg to birth) of *Pleopis polyphemoides* (■: Fofonoff, 1994), *Evadne nordmanni* (□: Fofonoff, 1994) and *Pseudevadne tergestina* (▲: estimated from field data of Bryan, 1979) in comparison with those of selected cladoceran species. Solid line is fitted curve for *P. polyphemoides* and *E. nordmanni*. Key to data for non-podonid cladocerans: ×, *Penilia avirostris* (Paffenhöfer and Orcutt, 1986); ▽, *Sida crystallina* (Bottrell, 1975a); o, *Chydorus sphaericus* (Bottrell, 1975a); △, *Eurycerus lamellatus* (Bottrell, 1975a); ◇, *Bosmina (Eubosmina) longirostris* (Allan, 1977); +, *Bosmina longispina* (Kankaala and Wulff, 1981); ⊠, *Daphnia ambigua* (Allan, 1977); ⊙, *Daphnia obtusa* (Maier, 1993); ⊹, *Moina brachiata* (Maier, 1993).

and adults in *E. nordmanni* and *P. polyphemoides* in Narragansett Bay. In contrast, *Daphnia* neonates are typically less than one-half as long as adults. For example *D. schødleri* neonates are only one-third the length of the first adult instars (Lei and Clifford, 1974).

2.5. Release of Neonates

Near the end of development the inner membrane enveloping the brood chamber is pushed outward by enlarging embryos until it touches the

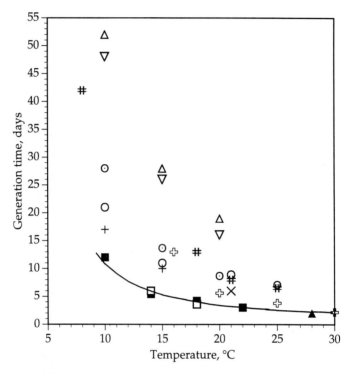

Figure 9 Generation times (birth to birth) of *Pleopis polyphemoides* (■: Fofonoff, 1994), *Evadne nordmanni* (□: Fofonoff, 1994) and *Pseudevadne tergestina* (▲: estimated from field data of Bryan, 1979) compared to selected cladoceran species. Solid line is fitted curve for *P. polyphemoides* and *E. nordmanni*. Note that in podonids, generation time is equal to embryo develop-ment time, while other cladocerans have a non-ovigerous juvenile stage. Key to data for non-podonid cladocerans: ×, *Penilia avirostris* (Paffenhöfer and Orcutt, 1986); ▽, *Sida crystallina* (Bottrell, 1975b); ○, *Chydorus sphaericus* (Bottrell, 1975b); △, *Eurycercus lamellatus* (Bottrell, 1975b); +, *Bosmina (Eubosmina) longispina* (Kankaala and Wulff, 1981); ⌗, *Daphnia magna* (MacArthur and Baillie, 1929); ☉, *Daphnia obtusa* (Maier, 1993); ⌗, *Moina brachiata* (Maier, 1993.)

hypodermis underlying the exoskeleton. At the points of contact short channels are formed. Ruptures in the membrane of the brood chamber at these channels appear to be the escape route for some podonids, for example, the Caspian species *Evadne anonyx* (Rivier, 1968). After the neonates are released from the old brood chamber in *E. anonyx*, the chamber contracts and a new brood of eggs is deposited into it. In other *Evadne* spp. a brood chamber with the next brood of embryos appears before the mature neonates are released from the old brood chamber

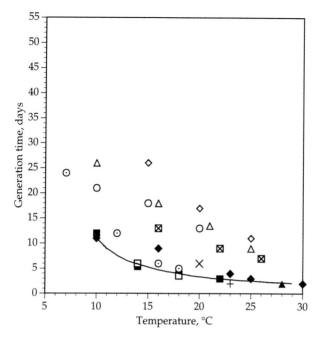

Figure 10 Generation times (birth to birth) of *Pleopis polyphemoides* (■: Fofonoff, 1994), *Evadne nordmanni* (□: Fofonoff, 1994) and *Pseudevadne tergestina* (▲: estimated from field data of Bryan, 1979) compared with selected, estuarine zooplankton species. Solid line is fitted curve for *P. polyphemoides* and *E. nordmanni*. Key to data for non-podonid zooplankton: (Cladocera) ×, *Penilia avirostris* (Paffenhöfer and Orcutt, 1986); (Copepoda, Calanoida) ⊠, *Acartia tonsa* (Heinle, 1966); ○, *Acartia hudsonica* (Landry, 1978); △, *Eurytemora affinis* (Heinle and Flemer, 1975); (Copepoda, Cyclopoida) ◇, *Oithona colcarva* (Lonsdale, 1981); (Rotifera) +, *Synchaeta cecilia* (Ardnt *et al.* 1990); ◆, *Brachionus plicatilis* (Hirayama and Kusano, 1972); (Appendicularia) ⊙, *Oikopleura dioica* (Paffenhöfer, 1975).

(Onbé, personal observations). Gieskes (1970) also found that the old brood chamber in *Evadne*, *Pleopis* and *Podon* spp. disintegrated after neonate release and a new chamber was formed to receive the next brood. This sequence of events is similar to that described for the freshwater *Bythotrephes longimanus* by Ischreyt (1930) and Zaffagnini (1964). One consequence of forming a new chamber for each brood is that the new brood may appear before the old brood is released, thus accelerating the rate of population growth (Gieskes, 1970).

In either case, the mature embryos escape into the space between the exoskeleton and the hypodermis, which by this time has separated from the exoskeleton in preparation for ecdysis. Shortly after the mature embryos enter the space enclosed only by the exoskeleton, the parent molts, liberating free-swimming neonates. This results in a significant reduction in the volume and a change in shape of the female as the carapace surrounding the brood chamber deflates (Bainbridge, 1958).

Release of the neonates in the freshwater *Polyphemus pediculus* and *Bythotrephes longimanus* appears not to be synchronized with molting of the parent. For example, in *B. longimanus* molting occurs 10–20 min after the release of the young and after a new brood of eggs has been deposited in the brood chamber (Zaffagnini, 1964). Release of the young in *P. pediculus* is accomplished through two fixed openings in the brood chamber that allow the embryos to move into the small, intermediary "dorsal sac" (Butorina, 1968). This sac is located at the posterior edge of the carapace dorsal to the abdomen. From this chamber the embryos are liberated through a single transverse slit, which lies dorsal to the rectum. Normally the anal lobes and the extended caudal process ("tail") cover this slit, which extends across the width of the organism. During birth in *P. pediculus* the caudal process swings ventrally to expose the slit, through which the neonates are released into the environment, head first and one at a time (Butorina, 1968). When giving birth *P. pediculus* females change from their normal swimming position (body horizontal, legs down) to a vertical position (head up), maintaining their position in the water column by movements of the antennae.

The developmental cycle and the release of neonates of podonids has a distinct diel cycle. Bryan (1979) found that mature embryos were rarely seen in populations of *Pseudevadne tergestina* sampled near the mouth of Chesapeake Bay during the day. After 20.00 hours the proportion of females with well-developed embryos increased; after 01.00 hours the release of neonates began and continued until pre-dawn, at which time virtually no females with mature embryos remained in the population. Similar nocturnal maturation and release has been described for populations of *E. nordmanni* and *P. tergestina* in the Inland Sea of Japan (Onbé, 1974), for *P. tergestina* in the Gulf of Mexico shelf and in the Inland Sea of Japan (Mullin and Onbé, 1992) and for *Podonevadne* spp. and *E. anonyx* in the Caspian Sea (Rivier, 1969). Using infrared videomicroscopy, Onbé (unpublished) has observed molting behavior and simultaneous neonate release occurring exclusively at night in total darkness for *P. tergestina* and *E. nordmanni*. In the Caspian Sea the release of neonates at night occurred at all stations, at all times of the year, and at all water temperatures (Rivier, 1969). Perotti (1988) made similar observations for *E. nordmanni* in coastal waters off Argentina. By contrast, *Penilia avirostris* from the Inland Sea of Japan and the Gulf of Mexico shelf has been shown to contain mature embryos at any time of day, but release tends to be maximal at night just before dawn (Mullin and Onbé, 1992).

3. GAMOGENESIS

3.1. Sexual Dimorphism, Sex Determination and Sex Ratios

3.1.1. Sexual Dimorphism

The sex of a marine cladoceran may be recognized on the basis of body size, body shape and other secondary sexual characteristics (Figure 11). Among the podonids, males (Figure 11A) are distinguished by their less voluminous shape, often appearing triangular in outline, by the presence of the paired testes and vas deferens, visible through the transparent body wall, and by the presence of a pair of slender, cylindrical copulatory organs (Figure 12A), terminally bifurcated, which arise from the trunk just posterior to the last pair of thoracopods (Lilljeborg, 1901). In male podonids, the head and eye are often relatively larger and more pigmented than those of the female; Rivier (1968) reported that the head and eye, measured along the longitudinal axis of the body, in males of the Caspian Sea *Podon camptonyx* is 41–42% of the body length, whereas in females it comprises only 33–36%. Less obvious in male podonids are the three setae on the terminal segment of the endopod of the first thoracopod. Of these setae, illustrated in Figure 12B, one is very stout and hook-shaped, presumably for copulation, and two are elongated, ventral and falciform (Onbé, 1978a). Males of *Pleopis polyphemoides* swim faster than females and with more rapid vertical excursions (Fofonoff, personal observations).

Podonid females (Figure 11B–D) reproducing by gamogenesis are readily recognized by the presence of one or two gamogenic embryos ("resting eggs") in the brood chamber and by a copulatory pore or vagina, which leads from a small funnel on the postero-dorsal side of the carapace through a narrow tube to the brood chamber.

In general, males of marine cladocerans are large relative to females. Males in marine cladocerans are seldom less than 80–90% of female length. In *Evadne nordmanni* and *Pleopis polyphemoides* the mean length of gamogenic females is greater than that of parthenogenic females, but this is not a reliable method to distinguish gamogenic and parthenogenic females because there is considerable overlap in size ranges (Bainbridge, 1958; Onbé, 1978a; Fofonoff, 1994). Parthenogenic females may equal gamogenic females in body length; in particular, parthenogenic females bearing their second or later broods and nearly all parthenogenic females bearing late-stage embryos may be as large as most gamogenic females.

Sexual dimorphism in *Penilia* is also apparent. The most conspicuous differences are the presence in males (Figure 13A) of an elongate pair of copulatory organs, an antennule that is much longer than in the females

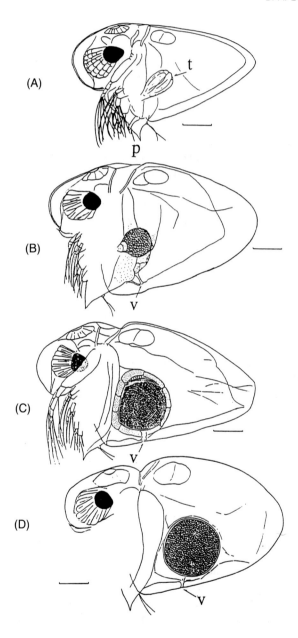

Figure 11   Pseudevadne tergestina (A) male and (B–D) gamogenic females with resting embryos at successive stages of development. p, penis; t, testes; v, vagina. Scale bar: 100 μm. (After Onbé, 1978a.)

(A) (B)

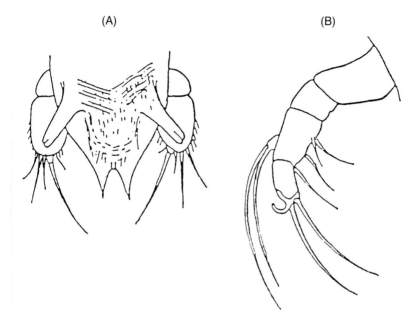

Figure 12 Evadne nordmanni male. (A) Paired penes projecting from the ventral side of the thorax posterior to the last (fourth) pair of thoracopods. (B) Endopodite of the first thoracopod. (After Lilljeborg, 1901; and Dolgopolskaya, 1958.)

(Figure 13B) and a rostrum that is not pointed as in the females. The paired copulatory organs arise from the sides of the male thorax posterior to the last (sixth) pair of thoracopods (Dolgopolskaya, 1958). The antennules of males are longer than those of females and extend nearly the full length of the carapace. Other differences include the generally shorter mean length of males; for example, 0.8 mm in males compared with 0.6–1.2 mm in females (Smirnov and Timms, 1983), the relatively small head of the male and the presence of small hooks on the distal endopodite segments of the first thoracopods (Steuer, 1933; Della Croce and Gaino, 1970).

3.1.2. *Sex Ratios and Sex Determination*

Sex ratios of individual broods and of populations vary in marine cladocerans. Individual broods of podonids contain both males and females (Bainbridge, 1958; Fofonoff, 1994) in contrast to *Daphnia* where single broods may contain all females, or all males or a mixture (Barker

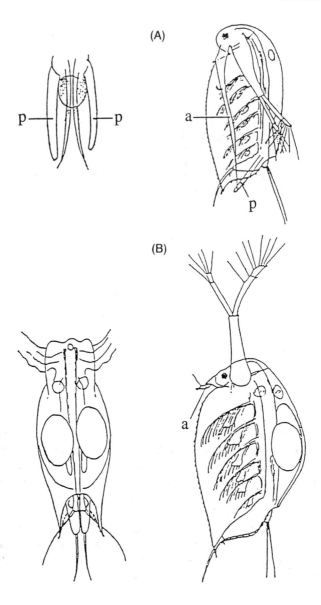

Figure 13 Penilia avirostris. (A) Ventral and lateral view of male. (B) Dorsal and lateral view of female with resting embryos. a, Antennule; p, penis. (After Steuer, 1933.)

and Hebert, 1986). Because parthenogenic females constitute a major proportion of the individuals in populations of marine podonids, population sex ratios favor females; Bainbridge (1958) found at most 26% males in populations of *Evadne nordmanni*. However, if only sexual individuals in the population are compared, the ratio of males to females is typically closer to 1 : 1 (Fofonoff, unpublished data).

Gamogenic females and males of marine species usually appear when population density is high and become most numerous just before the population disappears from the plankton (Onbé, 1978b). However, the shift from parthenogenic to gamogenic reproduction is never complete in field populations: the maximum recorded percentages of gamogenic individuals in a population ranges from less than 10% for *Pleopis polyphemoides* in the Inland Sea of Japan (Onbé, 1978b) and in Chesapeake Bay (Bosch and Taylor, 1973a) to 46% in autumnal populations in Narragansett Bay (Fofonoff, 1994), from 20 to 23% for *Pseudevadne tergestina* in the Inland Sea of Japan (Onbé, 1978a, b), from 50% for *Podon leuckarti* in the North Sea (Gieskes, 1971a) to 86% in the Bering Strait (Onbé *et al.*, 1996) and from 25% in Narragansett Bay for *Evadne nordmanni* (Fofonoff, 1994) to 60% in the Clyde Sea (Bainbridge, 1958). Low incidences of gamogenesis may result simply from the failure of parthenogenic females to be transformed to gamogenic females, as is apparently the case for *P. polyphemoides* in Narragansett Bay (Fofonoff, 1994). In the Caspian Sea males are rare or absent among the endemic Onychopoda (Mordukhai-Boltovskoi, 1967b), but they appear frequently in other populations of the same species in shallow lakes, perhaps as a result in those lakes of more extreme temperature regimes, including freezing.

Species may respond differently to identical environmental cues. In the Caspian Sea and Narragansett Bay gamogenic reproduction occurs in some species but not in others at the same site at the same time (Rivier, 1968; Fofonoff, 1994).

Although the mechanism of sex determination in cladocerans is unknown (Frey, 1965), environmental conditions appear to be influential, at least for freshwater species (Frey, 1982; Zaffagnini, 1984). When resources are favorable for population growth, parthenogenic reproduction predominates; when resources are less favorable, gamogenic reproduction predominates. The response to environmental cues, however, is complex (Hobæk and Larsson, 1990; Berner *et al.*, 1991) and has not been elucidated completely for any marine or freshwater cladoceran. Endogenous physiological or genetic factors also play a role in sex determination as evidenced by the presence of gynandromorphs in freshwater cladocerans (Frey, 1965). Barker and Hebert (1986) have postulated that sex ratio in *Daphnia magna* is controlled by the interaction of environmental cues with endogenous factors, not yet identified.

In the Daphniidae, short photoperiods appear essential for inducing males and gamogenic females (Stross, 1987; Hobæk and Larsson, 1990; Berner *et al.*, 1991). Rivier (1968) observed that gamogenic females and males of some Caspian species appear simultaneously at widely separated sites with otherwise different ambient conditions, for example, temperatures of 12–16°C versus 22–24°C. However, photoperiodic controls alone are insufficient to explain gamogenic reproduction because at the same sites at the same time some species will, and others will not, exhibit gamogenic reproduction, as noted by Rivier (1968) in the Caspian Sea and by Fofonoff (1994) in Narragansett Bay.

In other areas, gamogenic reproduction is correlated with temperature and not with photoperiod (Figure 14). In Chesapeake Bay gamogenic forms of *P. polyphemoides* appear during both spring and fall when water temperatures are similar but photoperiods are dissimilar. Bosch and Taylor (1973a) found gamogenic females and males in April and May when day lengths (sunrise to sunset) were 13–15 h, and in November and December when day length was less than 10 h. Temperatures during these two periods of gamogenic reproduction were between 11 and 13°C.

The onset of sexual reproduction in podonids generally follows a sharp decrease (Figure 15) in parthenogenic brood size (Bainbridge, 1958; Onbé, 1974; Corni and Gardenghi, 1974, 1975; Fofonoff, 1994), which is consistent with the hypothesis that food limitation may trigger sexual reproduction. Parthenogenic females carrying only amorphous material in the brood pouch frequently co-occur with gamogenic females carrying gamogenic eggs (Bainbridge, 1958), suggesting a link between prenatal mortality and gamogenic egg production. In Narragansett Bay, on the other hand, high rates of prenatal mortality did not always correspond with a high frequency of sexual reproduction (Fofonoff, 1994).

Temperature has also been implicated in the termination of sexual reproduction because gamogenic females do not appear above certain temperatures, for example, no gamogenic *Evadne nordmanni* are found in Narragansett Bay at temperatures above 19°C (Fofonoff, 1994). In *Pleopis polyphemoides* in Chesapeake Bay gamogenic females disappeared during the summer at temperatures exceeding 23°C (Bosch and Taylor, 1973a), while in Narragansett Bay they were found throughout the summer season, but always at temperatures less than 22.5°C (Fofonoff, 1994).

In Narragansett Bay, the timing of the onset and intensity of gamogenic reproduction varied among species, but did not seem to be clearly associated either with temperature or day length (Fofonoff, 1994). In *Pleopis polyphemoides* and *Evadne nordmanni*, males and females with gamogenic eggs occurred both during the spring–summer and fall. In two years of a field study, both dates and temperatures of first occurrence of sexual individuals sometimes differed considerably between years.

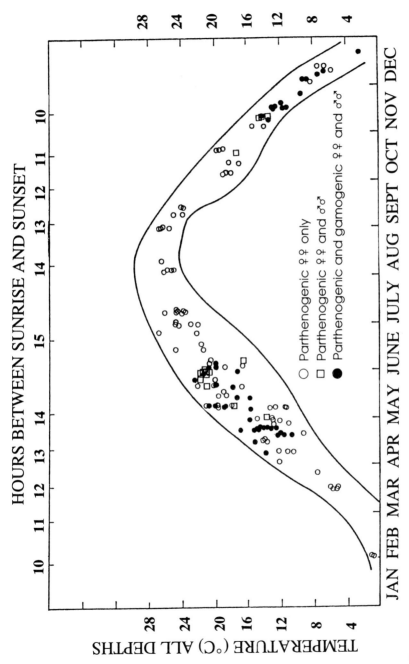

Figure 14 Seasonal temperatures and occurrence of gamogenic forms of *Pleopis polyphemoides* in Chesapeake Bay. The temperature envelope encloses the highest and lowest temperatures found within the water column at the time of sampling. Photoperiod is given as the time between sunrise and sunset, and does not take twilight into account. (After Bosch and Taylor, 1973a.)

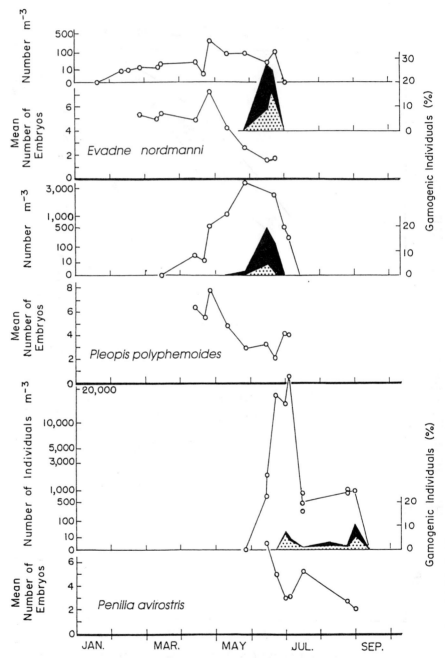

Figure 15 Percentages of gamogenic individuals in relation to changes in population density for three species of marine cladocerans. Solid areas: females with gamogenic embryos ("resting eggs"); dotted areas: males. (After Onbé 1974.)

Gamogenic *E. nordmanni* first occurred at 14°C on 6 June 1988 but at 10°C on 8 May 1989. During the fall, sexual individuals of *E. nordmanni* appeared at 6°C on 5 December 1988, but at 14°C on 15 October 1989. *P. polyphemoides* showed less variability in the onset of sexual reproduction. Sexual reproduction occurred on 6 June and 13 June 1988 and 1989, at 14 and 16.5°C, respectively. In the fall, gamogenic individuals were first collected on 3 October 1988 and 12 October 1989, at 18 and 13.5°C, respectively (Fofonoff, 1994).

Neither temperature nor photoperiod was a critical factor for gamogenic reproduction for the freshwater species, *Moina macrocopa*, in laboratory studies. D'Abramo (1980) demonstrated that gamogenesis in *Moina* can be suppressed in axenic cultures by feeding a high density of nutritious or non-nutritious particles. His results showed that the induction of gamogenesis in *Moina* depends on the number of particles ingested and not on chemical factors in the medium, as postulated by Banta and Brown (1939). The apparent success of earlier investigators in inducing gamogenesis in *Moina* by crowding, excretory metabolites, starvation, low temperatures and other "depressing" factors may be explained by the effect of these factors on ingestion rates.

Despite many experimental studies, the mechanism(s) of environmental sex determination in cladocerans is unclear. Banta and Brown (1939) inferred that environmental factors might act on the segregation of sex chromosomes because the critical period for induction of males by crowding in *Moina* precedes by 4 h the release of the eggs into the brood chamber. This period precedes by more than 3 h the maturation divisions of the oocyte. There is, however, no direct evidence for sex chromosomes in cladocerans (Mortimer, 1936), and allozyme loci do not segregate in male and female offspring of heterozygous females as expected if loci for sex determination were located on sex chromosomes (Young, 1983).

An alternative mechanism of environmental sex determination has been suggested by Ruvinsky *et al.* (1978). They found structural differences between the nuclei of somatic cells of male and female *D. pulex*, which they interpreted as differences in gene activity. These differences between males and females with apparently identical sets of chromosomes led them to conclude that environmental factors act directly on genes and not on chromosomal segregation. Barker and Hebert (1986) observed intraspecific variation in response to photoperiodic induction of gamogenesis in *Daphnia*. Their findings support the hypothesis that sex determination in *Daphnia* is controlled by polygenes. Clearly, more genetic and physiological studies are required before the mechanisms by which environmental cues influence gamogenesis in marine or freshwater cladocerans are elucidated.

3.2. Reproductive Anatomy of Males and Gamogenic Females

The paired testes of marine podonids are located in the hemocoel on both sides of the intestine (Figure 11A). The testes are lobed, compact and lack a distinct lumen in contrast to the extended cord-like testes of many other cladocerans (Zaffagnini, 1987). Each testis in podonids discharges through a vas deferens to the penes located just behind the last pair of thoracic limbs (Onbé, 1978a). In *Evadne* the spermatozoa pass through a sperm sac, described by Claus (1877) as a small bladder lying medial to the testes on the dorsal side of the intestine.

The paired testes of *Penilia* are elongate and connect to a medial seminal vesicle or sperm sac (Della Croce and Gaino, 1970), beyond which the system bifurcates (Figure 16) and leads to two elongated penes. These extend parallel to the post-abdomen and protrude beyond the edge of the carapace (Della Croce and Gaino, 1970).

Claus (1877) cites Leydig's (1860) observation that the testes of *Polyphemus pediculus* exhibit peristaltic movements but neither Claus nor Leydig demonstrated the presence of muscular tissue in the testicular epithelium of *Evadne* or *Polyphemus*. In *Polyphemus* the penes, very short mounds or papillae lying under the fourth thoracopod, can be extended by contraction of a muscle attached at their bases (Butorina, 1968).

The anatomy of the reproductive organs of parthenogenic and gamogenic females of marine podonids differ in two primary respects. These differences involve the presence of a copulatory aperture in gamogenic females and striking differences in the structure and function of the brood chamber. In mature gamogenic females a vagina or genital aperture is located on the postero-dorsal edge of an exoskeletal hump (Figure 11B). SEM observations reveal that in *Pseudevadne tergestina* and *Evadne nordmanni* this aperture is ellipsoid in outline with a long axis of 5–6 μm (Onbé and Hayashida, unpublished data). This opening is not present in parthenogenic females.

In contrast to *Polyphemus pediculus*, there is no dorsal (genital) sac in marine cladocerans that could serve as a seminal receptacle for incoming spermatozoa. The genital sac in *P. pediculus* is located at the posterior edge of the carapace dorsal to the abdomen, and opens to the outside through a transverse slit that serves as the vagina (Butorina, 1968).

The histology of the brood chambers in gamogenic and parthenogenic females is also distinct. Whereas the cells of the brood chamber in parthenogenic females are transparent and amorphous in outline, those in gamogenic females are cuboidal (Kuttner, 1911) or hexagonal (Rivier, 1968) and apparently contribute to the formation of the outer chitinous layer of the gamogenic egg. In *Evadne*, but not in *Podon*, the cells of the

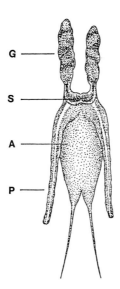

Figure 16 Penilia avirostris male reproductive organs and postabdomen. G, Gonads; S, seminal vesicle; P, penes; A, postabdomen. (After Della Croce and Gaino, 1970.)

brood sac in gamogenic females are easily recognized by the presence of dense granules (Kuttner, 1911).

In *Evadne anonyx* (Figure 17), *Podonevadne camptonyx* and other Caspian podonids the tissue of the brood sac develops independent of the egg in the space enclosed by the mother's domed exoskeleton. Only after the egg, nourished by a strand of tissue connected to the intestine, is fully developed does it move into hypodermal tissue, which then forms a brood chamber around it (Rivier, 1968).

The ovaries of gamogenic females in marine cladocerans appear similar in gross anatomical features to those of parthenogenic females, but no detailed studies of the internal anatomy of gamogenic marine cladocerans have been published. Gieskes (1970) was able to discern a slender strand, perhaps the germarium, connecting the left and right ovaries in *Podon leuckarti*.

Butorina's (1968, 1969) elegant studies of *Polyphemus pediculus* and the Caspian *P. exiguus* should be consulted as a guide to future studies. Butorina (1969) found that the ovaries of gamogenic females of *P. pediculus* are larger in all dimensions than in parthenogenic females. Posteriorly each ovary merges with a broad oviduct, each of which opens into a genital sac. These paired genital sacs, absent in the marine podonids, appear only in gamogenic females and are located on the dorsal side where

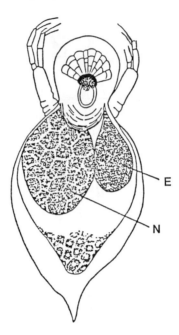

Figure 17   *Evadne anonyx* with developing gamogenic egg (E) and independently developing hypodermal tissue (N). The latter forms the membranes of the brood chamber after the egg moves into it. (After Rivier, 1968, and personal communication.)

the brood pouch and the caudal process (telson) meet. The genital sacs are formed from the unpaired dorsal chamber, present only in parthenogenic females (Section 2.5). Like the dorsal chamber, the genital sacs are connected through two openings on their dorsal anterior edge to the oviducts and through two additional openings on their dorsal posterior edge to the brood chamber. In gamogenic females each genital sac also opens to the outside through genital apertures or copulatory pores, located laterally at the posteroventral edge of the sac. These paired genital pores lie above the telson, which must be flexed ventrally to permit copulation and also for release of the gamogenic embryos to occur. In this respect *Polyphemus* differs from the marine podonids, which have only a single copulatory pore located in a more accessible position on the postero-dorsal side of the brood pouch.

Individual *Polyphemus pediculus* are capable of transforming from parthenogenic to gamogenic females (Makrushin, 1973). This transformation requires that the dorsal sac be divided internally and that the single transverse, ventral slit, through which parthenogenic neonates are released into the environment, be closed medially to form two separate and lateral genital pores. As a consequence an ovum in gamogenic females passes through the left or right oviduct to the respective genital sac, and thence into the brood chamber. (In parthenogenic females ova pass through the undivided dorsal sac.) After fertilization and maturation, the gamogenic embryo returns to the genital sac and is released through a genital pore which becomes enlarged by a rupture in the exoskeleton that allows the gamogenic embryo to be released. (In parthenogenic females mature embryos move from the brood chamber into the undivided dorsal sac and then to the outside through an existing transverse slit in the dorsal sac.)

Although the anatomical differences between parthenogenic and gamogenic marine podonids are not as great as they are for *Polyphemus*, Kuttner (1911) concluded that a transition from parthenogenic females to gamogenic females did not occur because of the requisite anatomical alterations: addition of a vagina and restructuring of the brood chamber. Nonetheless, there is increasing circumstantial evidence that parthenogenic females of podonids can transform themselves into gamogenic females. Rivier (1968) illustrated a female *Evadne anonyx* about to release neonates, but also carrying a developing gamogenic egg in the ovary. Embryonic and juvenile stages of gamogenic *E. nordmanni* females do not appear in field populations; mature embryos can be identified as either parthenogenic females or males, not as gamogenic females. When gamogenic females appear they are relatively large individuals about to produce diapausing embryos. Based on size relationships of parthenogenic and gamogenic females, transition from parthenogenesis to gamogenesis also appears likely in *Podon leuckarti* (Fofonoff, 1994). Fofonoff (1994) also observed that parthenogenic *Pleopis polyphemoides* females, collected during periods when gamogenesis was occurring in Narragansett Bay, produced resting embryos after bearing one or more parthenogenic broods in the laboratory cultures. These observations are consistent with the hypothesis that gamogenic females are derived from mature, parthenogenic females (Bainbridge, 1958).

3.3. Gametogenesis in Males and Females

Spermatogenesis in the Branchiopoda, including representatives of each of the three families of Onychopoda (Podonidae, Polyphemidae, Cercopagidae), has been described by Wingstrand (1978). For the Podonidae, Wingstrand examined the ultrastructure of the testes and spermatozoa of *Evadne nordmanni* and *Podon leuckarti*. In *E. nordmanni* the germinal and nutritive calls cannot be differentiated until the spermatids reach a moderate size (20 μm long by 15 μm wide). At that size each spermatid is partially enclosed in a cuplike "nutritive" cell. The spermatids are held on the inner sides of the nutritive cell, which attach to the testicular basement membrane on their outer side. The cytoplasm of each spermatid is relatively devoid of organelles, except for a concentration of mitochondria, ribosomes and endoplasmic reticulum in the region adjacent to the nutritive cells. As spermatogenesis proceeds, pseudopodia emerge from the area not enclosed by the nutritive cell and stretch across the testes, where they contact epithelial cells on the opposite wall of the testes. At maturity each *Evadne* spermatozoa is 70 μm long by 40–50 μm wide. Spermatogenesis in *Podon* differs from *Evadne* primarily in the absence of nutritive cells (Wingstrand, 1978). In *Podon*, spermatids and nutritive cells are scattered throughout the testes, where the spermatids are in contact with one or more nutritive cells and each other.

Jorgensen (1933) could not locate a germarium in gamogenic females of *E. nordmanni*, but Cheng (1947) saw a tube filled with cells attached to the posterior end of the ovary which he interpreted as the germarium. This interpretation is not consistent with the placement of the germarium at the anterior end of the ovary in parthenogenic females of this species (Jorgensen, 1933).

Oogenesis of gamogenic eggs begins, as in parthenogenic eggs (Section 2.3), with the appearance in the ovary of four large cells (tetrad) in a cluster in *Penilia* or aligned in a row in Podonidae. A gamogenic egg results invariably from the third cell from the anterior end of the ovary (Jorgensen, 1933; Cheng, 1947).

Claus (1877) thought that the three small cells that did not become the oocyte gave rise to the nurse cells in the brood chamber. This conclusion has been refuted by Jorgensen (1933) who reported that non-oocytes in the tetrad "disappeared" before the egg passed into the brood chamber. Cheng (1947) suggested that these cells may be absorbed by the enlarging oocyte. Rivier (1968) described a different fate for two of the non-egg cells in the Caspian *Evadne anonyx* and *Podonevadne angusta*; she observed that the two proximal cells contributed to a strand of tissue that served as an anchor and conduit for nourishment for the egg, while the terminal cell was incorporated eventually into the yolk of the enlarging egg.

One or two diapausing embryos may be produced by podonids. Gieskes (1970) never observed more than one gamogenic egg in the ovary of *Podon leuckarti* females, but one or two such eggs per female occur in other marine Podonidae, for example, *Evadne nordmanni* (Lilljeborg, 1901; Jorgensen, 1933; Cheng, 1947). Two fertile eggs evidently result from the simultaneous release and fertilization of an egg from each ovary (Jorgensen, 1933; Rivier, 1968).

Using neutral red stain, Jorgensen (1933) found glycogen present in the tetrad at the initial stages of oogenesis with a predominance of lipids in the developing egg as revealed by osmium tetroxide staining.

3.4. Male Gametes

Crustacean spermatozoa in general, and those of the Onychopoda in particular, lack many of the basic components of a typical metazoan spermatozoa except for a nucleus and the ability to fertilize eggs (Wingstrand, 1978). Spermatozoa of the Polyphemidae are 30–80 μm, ovoid, non-flagellated cells with a smooth surface, marginal vesicles and a cytoplasm filled with a dense mixture of filaments and tubules. Those of the podonids *Evadne nordmanni* and *Podon leuckarti* are filled with many, irregularly shaped mitochondria 2–6 μm in size with numerous cristae, extensive systems of smooth endoplasmic reticulum, a distinct zone of marginal vesicles, fiber bundles and an irregularly shaped, centrally located (*Podon*) or variably placed (*Evadne*) nucleus. The fiber bundles are unique organelles of unknown function in the podonids. They are composed of 80 Å microtubules scattered in the cytoplasm. The number of microtubules and the configuration of each bundle is species-specific. In *Evadne*, 10–50 microtubules per bundle form complex cross-sectional patterns that resemble "Y"s and "H"s; in *Podon* the cross-sectional patterns are plainer because there are fewer (1–10) microtubules per bundle. *Evadne* and *Podon* spermatozoa may be distinguished with light microscopy by their size: for example, those of *E. nordmanni* are large (70 μm long and 40–50 μm thick) compared to the smaller sperms (30 μm long and 29 μm thick) of *Podon leuckarti* (Wingstrand, 1978).

Spermatozoa of *Penilia* have not been described but may be similar to those of the freshwater Ctenopoda whose sperm have diameters of 20–80 μm (Wingstrand, 1978).

3.5. Reproductive Behavior including Mating

3.5.1. *Reproductive Behavior*

There are no published studies of mating behavior in marine cladocerans, but Butorina's (1968, 1986) studies of the swimming and mating of the freshwater *Polyphemus pediculus* provide a tantalizing introduction to the range of behaviors that are exhibited by at least one onychopod. Two primary categories of behavior relevant to reproduction characterize *Polyphemus*. One involves parthenogenic females and their parthenogenic young which tend to aggregate in dense patches in the upper 5 cm of lakes during the day with only a few parthenogenic females found below 25 cm. These daytime surface aggregations coincide with periods of intense feeding by these visual predators. At night, parthenogenic females and their young disperse horizontally and cease feeding. *Pleopis polyphemoides* exhibits similar behavior in Chesapeake Bay, swimming near the surface during the day and sinking passively at night (Bosch and Taylor, 1973b). Wickstead (1961) found similar vertical distributions of *Pseudevadne tergestina* and *Penilia avirostris* in the South China Sea.

In contrast, gamogenic *Polyphemus* females and males aggregate in dense patches at the surface only for a short period near dawn, apparently for the purpose of mating. During the rest of the day they disperse vertically and horizontally and are not found in the surface waters occupied by the parthenogenic females and their young. Sexual swarms have been reported also in *Daphnia magna* populations where males and unmated gamogenic females, both rare outside the swarms, become densely aggregated within them (Young, 1978).

The distribution patterns of *P. pediculus* females are correlated with the kinds of behavior exhibited in the laboratory. Butorina (1986) observed parthenogenic females aggregating at the top of shallow, lighted containers as a result of a positive phototaxis and exhibiting horizontal movements that alternate with spinning in one place (Figure 18A). These movements are characteristic of feeding individuals that alternate long-distance attacks on prey, including their own young, with shorter zig-zag movements or hovering in one place. Likewise, males and gamogenic females when observed separately move in the same feeding patterns as described for parthenogenic females. When males and gamogenic females are mixed the behavior changes (Figure 18B). The linear velocity of both sexes increases with males swimming faster than gamogenic females. The individual path lengths of both sexes become short and turning frequency increases. The result is a dense aggregation in which the probability of encounters between males and females is increased. In laboratory vessels mating swarms were observed at the bottom of the vessel whereas in lakes they occur at the surface; this discrepancy is not explained by Butorina (1968), but may be the result of high artificial light intensities, to which sexual females and males normally exhibit negative phototaxis. In confinement of the laboratory vessels, aggregation and mating behavior were observed but not at the low light intensities that occur during the dawn aggregations of sexual females and males in lakes.

3.5.2. *Mating*

Mating in marine podonids has not been described, but Weismann (1880), Zacharias (1884) and Butorina (1968) have described mating in

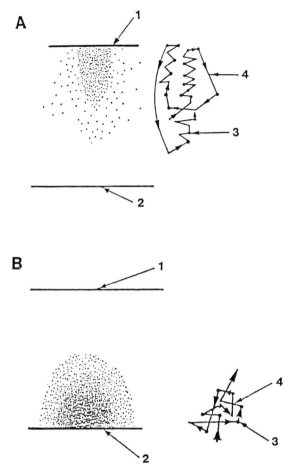

Figure 18 *Polyphemus pediculus* aggregations (left) and movement patterns of individuals (right) in shallow dishes lit from above. On the left, each dot represents the position of an individual in a shallow observation chamber. On the right, each dot represents successive positions of an individual in the horizontal plane. (A) Parthenogenic females only. (B) Gamogenic females and males. 1, Air–water interface; 2, bottom of Petri dish; 3, individual spinning in one place; 4, path of individual moving to next position. (After Butorina, 1986.)

Polyphemus pediculus. In this species males use their large swimming antennae to explore the ventral side of encountered individuals. Some encounters result in a male following a female and ultimately mating. Mating takes two primary forms. In young primiparous females the males mount a female on the dorsal side. This position permits them to continue swimming with their heads pointed in the same direction. The male grasps the young female's brood chamber with his first pair of trunk limbs, which appear modified for grasping (Section 3.1). His second and third pair of trunk limbs are also used to hold the female and to maneuver himself into a position for mating. When the male is in position near the female's genital apertures his fourth pair of trunk limbs is spread apart thereby exposing the penes. At this time the caudal process of the male turns ventrally, becoming nearly perpendicular to the long axis of the male's body. By flexing the caudal process at this angle, the male creates a favorable angle for insertion of the penes into the female's genital apertures. After insertion spermatozoa can move directly from the penes to the mouth of the oviducts. The male does not move for 15–30 s, during which time spermatozoa are transferred to the female. After copulation the partners disengage and swim away.

By contrast, mating proceeds differently between males and older females that have previously reproduced parthenogenically. The enlarged brood chamber of these females prevents the males from mounting on the dorsal side. Consequently, a male approaches a larger female from the ventral side, with his head pointed in the opposite direction to the female. In this position coordinated swimming movements are impossible, but the antennae of both partners continue to move during copulation. Subsequent movements of the male are identical to those described above, except that the engagement is lengthened to about 40 s. Often mating in this position ends with the female grasping the male with her second and third pair of limbs and cannibalizing him.

4. GAMOGENIC EGGS AND EMBRYOS

4.1. Fertilization

Pseudopodia and ameboid movements of spermatozoa have been observed in *Polyphemus* (Leydig, 1860; Zacharias, 1884) and *Evadne* (Claus, 1877). Movements of pseudopodia occur for only a few minutes in exsected sperm. The observed pseudopodia may be artifacts resulting from the premature liberation of developing spermatozoa (Wingstrand, 1978). Consistent with this hypothesis are Claus's (1877, Taf, Figure 16) drawings of pseudopodia on exsected "sperm cells" that closely resemble the immature, intratesticular spermatids. Whatever the nature of the observed pseudopodia, clear vesicles that line the outer margin of spermatozoa are potential sources of the membranes (Wingstrand, 1978). In mature sperm, pseudopodia could play a vital role in the initial contact and final positioning of the sperm on the surface of the egg prior to fertilization.

Careful *in situ* observations of mature sperm during transfer and fertilization will be required to establish the existence and function of pseudopodia before and during fertilization.

Only 10–20 large spermatozoa are found in mature testes of male *Evadne* or *Podon* (Wingstrand, 1978). These few spermatozoa suffice apparently because the distances between the egg and the point of release of spermatozoa are short and only one or two eggs are fertilized per female. Moreover, if gamogenic females reproduce only once, the storage of sperm would be unnecessary. Indeed, seminal receptacles are absent (Rossi, 1980).

Fertilization in marine cladocerans probably occurs before the gamogenic egg moves from the ovary into the brood chamber, as is the case in *Daphnia* (Ojima, 1958; Zaffagnini, 1987). Rivier (1968) reported for the Caspian *E. anonyx* and *Podonevadne angusta* that the third cell in the germarium begins to enlarge *and* divide before it leaves the ovary. For a cladoceran spermatozoon this presents a logistical problem, perhaps solved by the postulated ameboid movements, which would allow the spermatozoon to move from the genital aperture across the brood chamber to the egg in the ovary. In *Polyphemus* Butorina (1968) concluded that spermatozoa are released near the openings of the oviducts in the genital sacs with fertilization taking place in the oviduct or in the ovary.

For marine cladocerans the prevailing assumption is that gamogenic zygotes are produced from two haploid gametes. This may generally be so, but the existence of obligate parthenogenicity (thelytoky) has not been ruled out. In *Daphnia* diapausing embryos may be produced gamogenic-ally or parthenogenically. Karyological studies of four *Daphnia* species have shown that viable, diapausing embryos can be produced from unfertilized eggs (the so-called "pseudosexual eggs") in a manner identical to the production of subitaneous (parthenogenic) eggs, namely, without a pairing of homologous chromosomes and without a reduction divi-sion (Zaffagnini, 1987). Even when males are present, females may produce diapausing embryos without gamogenesis. This superfluous male phenomenon in field populations of *Daphnia* has been detected by deviations from expected Hardy–Weinberg equilibria and by the presence of sibling species and apomictic complexes (Hebert, 1987). No comparable investigations have been undertaken to determine if thelytoky or super-fluous males occur in populations of marine cladocerans.

4.2. Development of Gamogenic Eggs before Diapause

The early developmental pattern of gamogenic eggs in cladocerans appears similar in marine and freshwater species. The cell that will become a gamogenic egg arises initially as one of four cells in a linear tetrad in the ovary, as is the case for the development of parthenogenic eggs (see Section 2.3). During differentiation of the gamogenic egg three of these cells are either absorbed by the egg or otherwise lost in marine (Cheng, 1947; Della Croce and Bettanin, 1969) and freshwater taxa (Allen and Banta, 1929).

The sequence of development for gamogenic eggs (Figure 19) has been divided into discrete stages in *Penilia avirostris* (Della Croce and Bettanin, 1969). Initially (stage I) four similar oocytes form a linear quartet in the ovary. These quartets are differentiated from those forming parthenogenic eggs by their pronounced color and finely granular cytoplasm. During stage II the third oocyte from the anterior end becomes elongated and yolk appears in that cell only. This cell has a mean length of 145 μm at stage I and 341 μm at stage II. At stage III the second cell from the anterior end has disappeared, probably by absorption. Of the three remaining cells, the elongated middle cell continues to exhibit yolk formation, but the cytoplasm remains indistinguishable from that of the adjacent cells. By stage IV the anteriormost cell has disappeared and the margin of the expanded ovum becomes folded, conforming to the shape of the gut against which it is closely adpressed. This intimate association of the egg and the intestine may facilitate the transfer of nutrients from the intestine. The ovum develops rapidly in size during stage IV–VI. During this enlargement, folds appear and become progressively larger on two sides of the egg; three of the folds in contact with the intestine on the inner edge become larger and more rounded than those on the outer side (stage VI). The folds on the side opposite the intestine begin as flat protrusions in stage IV and subsequently expand in size and number in later stages. Curiously, the pattern on the side opposite the intestine appears initially to mimic that on the side in contact with the intestine.

By stage V the third small cell of the original tetrad has disappeared. Enlargement of the oocyte continues through stage VI until the oocyte reaches a maximum length of about 500 μm. After stage VI condensation of the ovum to a final length of less than 360 μm occurs quickly; Della Croce and Bettanin (1969) saw no intermediate stages between stages VI and VII. The condensed ovum in stage VII retains a concavity on the side toward the intestine, which may (Onbé, 1985) or may not (Della Croce and Bettanin, 1969) be retained in released eggs.

Because gamogenic and parthenogenic eggs of all cladocerans produce morphologically equivalent parthenogenic females, the developmental patterns of their embryos should be identical. This assumption has been confirmed only for *Daphnia pulex* by Baldass (1941). The only significant difference he found related to the differentiation of the germ cells. Differentiation begins earlier and is more prolonged in the development

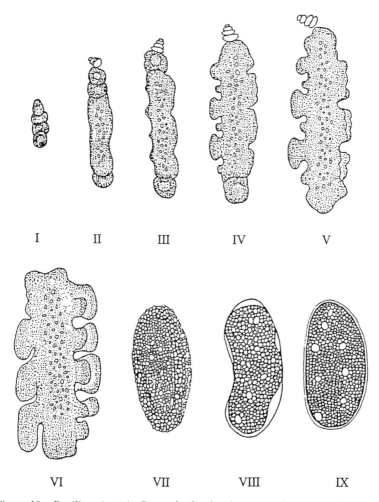

Figure 19 *Penilia avirostris.* Stages in the development of a gamogenic egg from four original oocytes. The primary oocyte in stages III–VI conforms to the folds of the intestine on the left side of the drawing (see Section 4.2). The cells of the germarium are visible in stages II–V. (After Della Croce and Bettanin, 1969.)

of gamogenic than in parthenogenic eggs. In particular, the final differentiation of the germ cells in gamogenic embryos of *D. pulex* is delayed until after diapause, which occurs between the tenth and eleventh cleavage divisions, whereas germ cells are differentiated before the eleventh division in parthenogenic embryos (Baldass, 1941). The developmental stage or stages at which diapause begins in podonids and *Penilia* is unknown.

At the earliest stages of gamogenic embryo development, when the tetrad of ovarian cells is forming in the ovary, extraembryonic cells simultaneously begin to appear in the brood chamber in podonids. In *Evadne nordmanni* these are commonly called "nurse cells" and originate from the same dorsal ridge where the Nährboden is formed during parthenogenic development (see Section 2.1). These nurse cells, whose exact function is unknown, differ from the cells of the Nährboden in parthenogenic females by their larger size, polyhedral rather than cuboidal shape, smaller number, and cytology; nurse cells in gamogenic females have a central nucleus and granular cytoplasm containing lipid droplets (Jorgensen, 1933) whereas those of the Nährboden in parthenogenic females have large secretion vacuoles (Patt, 1947). The early development of the nurse cells in gamogenic females resembles that of cells in the ovum, despite their separate origins. When the ovum has expanded in size to the width of the intestine, the nurse cells partially fill the brood chamber (Jorgensen, 1933). At this stage the embryo is covered by a thin membrane which attaches to the thorax near the site of the former ovary. Eventually, the nurse cells fill the brood chamber and assume polyhedral shapes as they press against the developing embryo, the walls of the chamber and the other nurse cells. As the embryo grows further into the brood chamber it becomes enveloped by the nurse cells which diminish in number and size as the embryo grows. When the embryo is fully formed only a few isolated nurse cells remain in the brood chamber. This pattern of development contrasts with that in cladocerans without nurse cells in the brood chamber, where gamogenic embryos do not grow after being deposited.

Gamogenic females with more than one egg have not been seen in *Pleopis polyphemoides* (Onbé, 1974; Corni and Gardenghi, 1975; Fofonoff, 1994) and they are not common in *Evadne nordmanni*. Two eggs are typically accommodated in the brood chamber if one egg is displaced dorsally, away from the trunk (Jorgensen, 1933). In females with two gamogenic embryos neither embryo attains the size of a single embryo, although a specimen having two gamogenic embryos, each as large as a typical single one, has recently been found in an *E. nordmanni* from the Bering Straits (Onbé *et al.*, unpublished observation). When two embryos are present, one or both may abort. Cheng (1947) interpreted irregular masses of cytoplasm in the brood chamber of *E. nordmanni* females bearing a single gamogenic egg as the residues of a degenerate, second egg.

The Caspian *E. anonyx* exhibits a strikingly different developmental pattern which illustrates the plasticity of this group (Rivier, 1968). The extraembryonic tissue and embryos of *E. anonyx* develop independently in the brood chamber and come into physical contact only late in development (Rivier, 1968). During their independent growth the egg and extraembryonic, hypodermal tissues are connected to the trunk of the mother by separate strands of tissue through which nourishment is presumably obtained (Figure 17). Not until the egg nears its final dimensions and detaches from its connection to the trunk does the egg move into the hypodermal tissue, which proceeds to form membranes around the egg. During this process the gamogenic egg is reduced slightly in diameter as the chitinous membrane envelops it (Rivier, 1968; personal communication).

At maturity, resting embryos of *Penilia* are ovoid in outline. Ecotypic variation occurs in egg size of *Penilia* (Onbé, 1972). The smallest resting embryos produced by this species are reported from the Inland Sea of Japan where the eggs are on average 180 μm (140–200 μm) wide, 250 μm (210–290 μm) long and 100 μm thick (Onbé, 1972, 1985). Resting embryos from the Black Sea are somewhat larger with mean dimensions of 210 μm wide by 327 μm long (Braiko, 1966). Populations in the western Atlantic Ocean produce the largest resting embryos; 279 μm wide by 356 μm long (Della Croce and Bettanin, 1969).

4.3. Release, Diapause and Hatching of Gamogenic Embryos

When the development of one or two resting embryos is complete, the female may molt and cast off the embryo(s) ("resting eggs") or may cease feeding, die and retain the embryo(s) in her brood chamber (Rammner, 1930; Rivier, 1968; Onbé, 1972). The production of more than one brood of gamogenic eggs by a gamogenic female has not been established. Embryos sink to the bottom (Braiko, 1966; Rivier, 1968) and are found in sediments, as described originally by Purasjoki (1945) and by Onbé (1972, 1973, 1974). Onbé (1991) provides a thorough review of the distribution, abundance, morphology, development and hatching of resting embryos.

Ephippia—transformed elements of the carapace which enclose the resting embryos of freshwater anomopods—are absent in marine cladocerans. Outer egg membranes that are chitinous and sometimes calcified, provide adequate protection for the resting embryos of the marine taxa as they do in freshwater Ctenopoda, Onychopoda and Haplopoda. In *Penilia* the outer membrane is calcified thus increasing the weight of the egg and ensuring that it sinks after release from the female and perhaps providing protection from predators in the sediments (Braiko, 1966). Braiko (1966) and Onbé (1985) have speculated that resting embryos may survive passage through the digestive systems of planktivores and detritivores. In addition, resting embryos that sink may be an important adaptation to avoid transport by strong oceanic currents to distant, unfavorable habitats.

Resting embryos of *Polyphemus pediculus* are released near the surface just before dawn by females in dense aggregations (Butorina, 1986). Marine species exhibit a similar periodicity in the release of parthenogenic neonates (see Section 2.5) but no information is available on the timing of the release of resting embryos.

In the Inland Sea of Japan during periods of intense gamogenic reproduction, resting embryos in the sediments attain densities of $7.0–14.0 \times 10^4$ eggs m^{-2} for *Penilia* and $1.5–1.8 \times 10^4$ eggs m^{-2} for *Evadne* and *Podon*, but densities of less than 1.0×10^4 eggs m^{-2} are typical at other times. After isolation from sediments by flotation in a sugar solution (Onbé, 1978c), the resting embryos of most marine species are sufficiently distinct to permit their identification. Onbé (1974, 1985) has described resting embryos from the Inland Sea of Japan. From size, shape, surface ornamentation, relative transparency and color in transmitted and reflected light, he was able to differentiate the resting embryos of five species: *Penilla avirostris*, *Pseudevadne tergestina*, *Evadne nordmanni*, *Podon leuckarti* and *Pleopis polyphemoides*. Onbé and Grice (unpublished) identified the resting embryo of *P. intermedius* in Buzzards Bay, Massachusetts (Onbé, 1991).

The podonid resting embryos are mostly spherical, opaque and darkly colored with maximum diameters of 150–230 μm. Resting embryos of *Penilia* are also opaque, dark in transmitted light and gray in reflected light, but differ from podonid embryos in shape. They are flattened to a thickness of about 100 μm (Figure 20B), concave on the one side and ovoid in outline (Onbé, 1985).

Hatching of resting embryos of marine cladocerans (Onbé, 1974; Iwasaki *et al.*, 1977; Takami and Iwasaki, 1978) is governed by salinity, as well as by temperature and duration of the dark incubation (photorefractory) period, as is the case in freshwater species (Schwartz and Hebert, 1987; Stross, 1987). In general, optimal hatching conditions in the laboratory correspond to those that prevail at the time the species normally appears in nature. For example, resting embryos of podonids and *Penilia* from the Inland Sea of Japan exhibit maximal hatching in the laboratory when salinity (> 19‰) and temperature (> 12°C) are similar to those in spring when these embryos normally hatch. Maximum hatching rates for *Evadne* spp. (60%) and *Podon* spp. (80–90%) were obtained at 19‰ and 15°C by Iwasaki *et al.* (1977). For *Penilia* maximum hatching of 80% was attained at 25‰ and 18°C. Light intensities also affect hatching: at 1000 lux the rate of hatching is greater than at 5000 lux or in continual darkness (Iwasaki *et al.*, 1977).

All resting embryos produced in the laboratory by captive females can

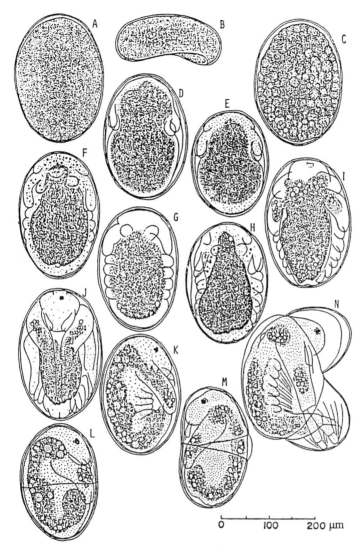

Figure 20 Development to hatching of the resting embryo of *Penilia avirostris*. (A) Embryo, dorsal view; (B) embryo, lateral view; (C) embryo showing first sign of development; (D) formation of second antennae; (E) formation of the first antennae; (F) formation of the mandibles and thoracic legs 1–3; (G, H) formation of the fourth and fifth thoracic legs, respectively; (I) thoracic leg formation complete, postabdomen becomes bifurcated, and eye appears; (J) eye becomes pigmented, morphogenesis nearly complete; (K) embryo turns sideways, the characteristic position before hatching; (L) dehiscent line becomes visible along the equator of the outer membrane; (M) rupture of the outer membrane and swelling of the inner membrane; (N) embryo just before hatching, enveloped by a swollen inner membrane. (After Onbé, 1978b.)

be hatched (Onbé, 1974) but hatchability begins to diminish immediately after their release, whether or not the embryos are removed from the sediments. Embryos recently isolated from natural muds often show hatching rates less than 50% and the rate declines rapidly. One day of storage at 5°C reduced the hatching rate of resting *Penilia* embryos to 40%; after 17 or 66 days at 5°C less than 10% of the embryos hatched, yet some resting embryos of *Penilia* remained viable for as long as 13 months of storage at room temperature (Onbé, 1974).

From the above observations we can draw few conclusions about the environmental conditions that govern the viability and hatchability of resting embryos of marine cladocerans. Carefully designed experiments using uniform populations of resting embryos are needed to show the role of temperature, salinity, photoperiod and other environmental factors in the hatching process.

Kankaala (1983) studied hatching of resting embryos and the population dynamics of *Bosmina (Eubosmina) longispina* in the Baltic. This species hatches at times and places not conducive for growth of adult populations. Light or photoperiodic responses, presumably inherited from freshwater ancestors, are thought to be non-adaptive because temperature and other environmental factors in the Baltic follow different and/or less predictable cycles than in lakes.

Development and hatching stages have been described for *Pseudevadne tergestina* (Onbé, 1974) and *Penilia* (Onbé, 1974, 1978b). In *Penilia* (Figure 20) the outer membrane ruptures first, followed by continued swelling and rupture of the inner membrane. A similar process has been described for *Daphnia laevis* (Wood and Banta, 1937). When the volume increases sufficiently to permit the embryo to move its antennae freely, the inner membrane ruptures, liberating the neonate. Braiko (1966) reported that the outer membrane separates from the inner membrane of *Penilia* from the Black Sea causing the embryo to float to the surface. In contrast, Onbé (1974) reported that resting embryos of *Penilia* from the Inland Sea of Japan do not float. These differences in the buoyancy of *Penilia* embryos appear unrelated to ambient water density. In the Black Sea *Penilia* embryos float at salinities of 17–18‰ (Mordukhai-Boltovskoi and Rivier, 1971b) whereas they sink in the Inland Sea of Japan at salinities of 25–34‰ (Onbé, 1977).

5. ANNUAL CYCLES AND POPULATION DYNAMICS

Annual cycles of abundance have been recorded for many populations of marine cladocerans. Most of the work cited in Table 4 are reports of single or multi-year studies that correlate population densities through one or more seasonal cycles with selected environmental factors, usually temperature and salinity. Typical populations are multivoltine (more than one generation per year), monacmic or diacmic (one or two periods of great abundance per year), and monocyclic or dicyclic (one or two annual changes in mode of reproduction, i.e. from parthenogenic to gamogenic) (Hutchinson, 1967). Populations are usually initiated by the hatching of overwintering embryos in spring followed by one or two annual periods of abundance. Periods of abundance are often attained rapidly, but sustained only briefly. Densities as high as $37\,000\,\text{m}^{-3}$ and $85\,000\,\text{m}^{-3}$ have been reported for *Pleopis polyphemoides* in the Inland Sea of Japan (Onbé, 1974) and in Narragansett Bay, Rhode Island (Smayda, unpublished data), while densities of $100\,000$ individuals m^{-3} have been reported for *Evadne nordmanni* in St Margaret's Bay, Nova Scotia (Platt, 1977). For *Penilia avirostris*, densities of over $50\,000$ and $150\,000$ individuals m^{-3} have been recorded, respectively, in Kingston Harbor, Jamaica (Grahame, 1976) and in the Inland Sea of Japan (Onbé, 1974). These densities are attained as a result of high rates of embryonic and postembryonic growth combined with parthenogenic reproduction accelerated by pædogenesis, as discussed in Section 5.2.1. Population maxima are typically preceded or accompanied by the onset of gamogenic reproduction and a sharp reduction in parthenogenic reproduction. After producing resting embryos, the population may disappear but parthenogenic reproduction may continue for several months, especially during the summer or, exceptionally, throughout the year (Della Croce and Venugopal, 1973).

5.1. Distribution and Abundance

Marine cladocerans exhibit distinct patterns of abundance in relation to temperature and salinity. Each of the seven species that occur in Chesapeake Bay has a unique range (Figure 21). The salinity–temperature ranges of *Pleopis schmackeri*, the only marine species that does not occur in Chesapeake Bay, are 21–27°C and 19–34‰ (Onbé, 1983; Rocha, 1985; Kim and Onbé, 1989b; Onbé and Ikeda, 1995) in the northwestern Pacific and Southern Atlantic Oceans; these ranges overlap those of *Podon intermedius*, *Pseudevadne tergestina* and *Penilia avirostris* in Chesapeake Bay at high salinities, but the patterns are unique for *Pleopis schmackeri* at salinities below 25‰ and temperatures above 21°C. Thus each marine species appears to occupy a unique niche in relation to salinity and temperature. Patterns of abundance in one locale, however, are not necessarily representative of patterns elsewhere. Each species might be

Table 4 Sources of information and locations of studies on abundance cycles of oceanic and neritic cladocera.

Locality	Latitude	Longitude	Species*	Reference
White Sea	65°N	34°E	En. Pl.	Makrushin (1984)
Gulf of Finland	60°N	25°E	Bl. En. Pi. Pl. Pp.	Vuorinen and Ranta (1987)
Baltic Sea	60°N	20°E	Bl.	Purasjoki (1958)
Baltic Sea	58°N	17°E	Bl. En. Pi. Pp.	Ackefors (1965)
Baltic Sea	58°N	17°E	Bl.	Kankaala (1983)
Kiel Fjord	54°N	10°E	En. Pl.	Poggensee and Lenz (1981)
Skagerrak	54°N	11°E	En. Pi. Pl. Pp.	Eriksson (1974)
North Sea	56°N	2°W	En. Pp. Pi.	Jorgensen (1933)
North Sea	51°N	4°E	Pl. Pi.	Gieskes (1971b)
Clyde Sea (Millport)	56°N	4°E	En.	Bainbridge (1958)
St Margaret's Bay, Canada	44°N	64°E	En. Es. Pi. Pl. Pp.	Platt (1977)
North Atlantic Ocean	41°–65°N	70°W–10°E	En. Es. Pi. Pl. Pp.	Gieskes (1971a)
Gulf of Lions	42°N	3°E	En. Es. Pa. Pi. Pp. Pt.	Thiriot (1968, 1971, 1972–73); Thiriot and Vives (1969)
Gulf of Trieste	45°N	13°E	En. Es. Pa. Pt.	Specchi and Fonda (1974); Specchi *et al.* (1974)
Adriatic Sea	44°N	13°E	En. Es. Pa. Pi. Pp. Pt.	Corni (1971); Corni and Cattani (1979, 1980); Corni and Gardenghi (1974, 1975)
Bay of Naples	40°N	14°E	Pa.	Della Croce and Bettanin (1964–65)

Location				Reference
Northwest Atlantic Ocean	36–44°N	65–75°W	Pa.	Colton (1985)
Woods Hole	41°N	70°W	En. Pi. Pp. Pt	Fish (1926)
Narragansett Bay	41°N	71°W	En. Pa. Pi. Pl. Pp.	Fofonoff (1994)
Chesapeake Bay	37°N	76°W	En. Pa. Pi. Pl. Pp. Pt.	Bryan and Grant (1974); Bryan (1977)
Chesapeake Bay	37°N	76°W	Pp.	Bosch and Taylor (1973a)
St Andrew's Bay	30°N	86°W	Pa. Pp. Pt.	Hopkins (1966)
Monterey Bay	36°N	121°W	En. Es. Pp. Pt.	Baker (1938)
Toyama Bay	37°N	137°E	En. Es. Pa. Pl. Pp. Ps. Pt.	Onbé and Ikeda (1995)
Onagawa Bay	38°N	141°E	En. Pa. Pl. Pp. Pt.	Uye (1982)
Inland Sea of Japan	34°N	133°E	En. Pa. Pl. Pp. Pt.	Onbé (1974, 1985)
Northwestern Pacific Ocean	10–40°N	110–150°E	Ps.	Kim and Onbé (1989b)
Northwestern Pacific Ocean	0–43°N	100–163°E	Pa.	Kim and Onbé (1995)
Singapore Strait	1°N	104°E	Pa. Pt.	Wickstead (1961)
Hooghly Estuary (W. Bengal)	22°N	88°E	Pt.	Sarkar and Choudhury (1987)
Sierra Leone (Freetown)	8°N	13°W	Pt.	Bainbridge (1960)
Santos (Brazil)	24°S	46°W	Es. Pa. Pi. Pt.	Rocha (1982)
Agulhas Bank (South Africa)	35°S	21°E	Pa.	Angelino and Della Croce (1975); Della Croce and Venugopal (1973)
Argentine Sea	41°S	59°W	En. Pi.	Ramirez and Perez Seijas (1985)

* Bl., *Bosmina (Eubosmina) longispina.* Pa., *Penilia avirostris.* Pp., *Pleopis polyphemoides.*
En., *Evadne nordmanni.* Pi., *Podon intermedius.* Ps., *Pleopis schmackeri.*
Es., *Evadne spinifera.* Pl., *Podon leuckarti.* Pt., *Pseudevadne tergestina.*

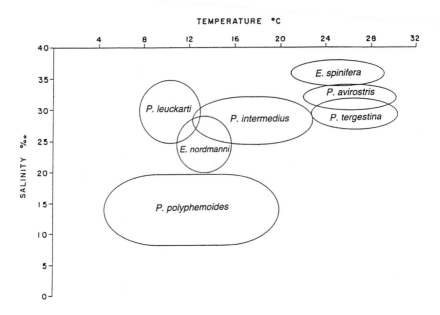

Figure 21 Optimal ranges of temperature and salinity for seven species of cladocerans in Chesapeake Bay: *Evadne nordmanni, Evadne spinifera, Podon intermedius, Podon leuckarti, Pseudevadne tergestina, Pleopis polyphemoides* and *Penilia avirostis.* (After Bryan and Grant, 1979.)

adapted to local regimes of temperature and possibly salinity. *Pleopis polyphemoides* appears at a temperature range of 7–20°C in the west coast of Sweden (Eriksson, 1974), whereas in the Inland Sea of Japan it is most abundant at temperatures of 12–27°C (Onbé, 1974). Likewise, in the western Mediterranean, *Penilia avirostris* (Thiriot, 1972–73) is abundant at salinities (< 35‰) and temperatures (14–21°C) that are lower than those correlated with periods of its abundance in Chesapeake Bay.

No single pattern of seasonal abundance is exhibited and exceptions exist even for species exhibiting otherwise consistent seasonal cycles. A common trend is the latitudinal shift in timing and duration of occurrence, probably related to shifts in each species' preferred temperature and salinity range. At the same time, factors other than temperature, such as hydrography, phytoplankton cycles and seasonal variations in the abundance of predators, may influence annual cycles.

Evadne nordmanni, predominately a species of cool water, often shows a single prolonged period of abundance during the summer in high latitudes (North Atlantic: Gieskes, 1970), but occurs only briefly during the spring near the southern edge of its range, for example, in Chesapeake

Bay (Bryan, 1977) and at 30°N in the northwestern Pacific (Kim and Onbé, unpublished). Separate spring and fall periods of abundance occur in some temperate regions (Kiel Fjord: Poggensee and Lenz, 1981; Mediterranean: Corni and Gardenghi, 1974; Narragansett Bay: Fofonoff, 1994). The timing of the cycle may shift regionally, with the peak occurring in August–September on the continental shelf of the western North Atlantic Ocean, at least one month earlier (July) in the western North Sea, and two months earlier (June) on the eastern edges of the North Sea (Gieskes, 1971a).

Podon leuckarti exhibits a single summer period of abundance in most years in the White Sea (Makhrushin, 1984), in the North Sea and off Iceland (Gieskes, 1971a), but near the southern edge of its range in Narragansett Bay (Fofonoff, 1994), Chesapeake Bay (Bryan, 1977) and the Inland Sea of Japan (Onbé, 1974, 1985), it occurs only for a few weeks or months in spring. Double seasons are rarely reported for *P. leuckarti*, but in Onagawa Bay, Japan, this species has two distinct peaks separated by 6 months (Uye, 1982). In the North Atlantic, this relatively cool-water species peaks in abundance before the warmer-water species *P. intermedius* (Gieskes, 1970, 1971b). Similarly, *E. nordmanni* precedes *E. spinifera* in the North Atlantic (Gieskes, 1970, 1971a), but the temporal separation of this congeneric pair is not as distinct as for the two *Podon* species because *E. nordmanni* persists through late summer, thereby overlapping with the later bloom of *E. spinifera*. In the Gulf of Lions, *P. intermedius* is never dominant compared to the other five cladoceran species which occur there, but is present for the entire warm season of 5–6 months (Thiriot, 1972–73).

A similar mixture of broad latitudinal patterns with local variability is seen in the warm-water species *Evadne spinifera* and *Pseudevadne tergestina*. *E. spinifera* has a single population maximum in the North Sea (Gieskes, 1970) but two distinct periods of abundance in the Adriatic Sea (Gulf of Trieste) (Specchi and Fonda, 1974; Specchi *et al.*, 1974) or an extended period of abundance of 4–6 months in the western Mediterranean (Gulf of Lions) (Thiriot, 1968, 1972–73). Similarly, *P. tergestina* exhibits extended periods of abundance from June to September–October Mediterranean (Thiriot, 1972–73; Specchi *et al.*, 1974) and the Inland Sea of Japan (Onbé, 1974, 1985). In the North Pacific, the maximum population size is not achieved until late summer (Uye, 1982).

Populations of *Penilia avirostris* are often associated with the warmer water of late summer or autumn in the northern hemisphere (Thiriot, 1972–73; Specchi and Fonda, 1974; Uye, 1982; Colton, 1985) and with particle-rich intrusions which result from upwellings on continental shelves where densities of 1–2×10^4 *Penilia* m^{-3} have been recorded (Paffenhöfer *et al.*, 1984; Yoo and Kim, 1987). Extended periods of abundance spanning five months or more have been recorded in the Bay of Naples (Della Croce and Bettanin, 1964–65) and in the Inland Sea of Japan (Onbé, 1974, 1985), while on the Agulhas Bank (35°S, 21°E) parthenogenic females bearing eggs occur throughout the year (Della Croce and Venugopal, 1973).

   Pleopis polyphemoides is a summer species in the Baltic (Ackefors, 1965; Vuorinen and Ranta, 1987), Mediterranean (Theriot, 1972–1973), North Pacific (Uye, 1982) and Nova Scotia (Platt 1977), but has separate spring to early summer and fall seasons in Narragansett Bay (Fofonoff, 1994) and Chesapeake Bay (Bosch and Taylor, 1973a). In the Inland Sea of Japan this species usually occurs in spring to early summer, but sometimes reappears in the late fall to early winter (Onbé *et al.*, 1977; Onbé, 1985). At the southern limit of its range in the western Atlantic in, St Andrew's Bay, Florida, *P. polyphemoides* is a winter to spring species (Hopkins, 1966).

5.2. Population Dynamics

5.2.1. *Pædogenesis*

The typical parthenogenic life cycle without pedogenesis is exhibited by *Penilia avirostris* and most freshwater cladocerans. In these taxa, neonates are reproductively immature and molt typically three times before the first brood of eggs is produced (Pavlova, 1959a). Subsequently, as each new generation leaves the brood chamber, the eggs of the next generation are deposited in a vacant chamber.

   The reproductive cycle of marine podonids differs from this cycle because it is accelerated by pædogenesis, namely, the development of parthenogenic eggs beginning in the brood chambers of embryos *before* they are released from their parent. In *Evadne nordmanni* cleavage may begin in the ovary (Kuttner, 1911), a phenomenon expressed more commonly in the Podonidae of the Azov and Caspian Seas (Rivier, personal communication), where water temperatures and development rates are higher than in the North Atlantic, the source of Kuttner's (1911) specimens. Mature eggs or early second-generation embryos are transferred to the brood chamber while the parent, still in its own parent's brood chamber, is only at the late embryonic stage when the eye pigment first appears. By the time the neonates are released from the brood pouch they carry individuals of the next generation in their brood pouches as late blastula stage embryos.

   Pædogenesis results in high population growth rates and the rapid attainment of high population densities. In Nova Scotia (44°N) *Evadne nordmanni* the maximum instantaneous population growth rate is

$0.245 \, d^{-1}$, which is equivalent to a doubling time of $2.8 \, d$ (Platt and Yamamura, 1986). In Narragansett Bay (41°N) *E. nordmanni* and *Pleopis polyphemoides* exhibited maximum population growth rates of $0.70 \, d^{-1}$, yielding a doubling time of 1 d or less. An instantaneous maximum growth rate of $0.424 \, d^{-1}$, or a doubling time of $1.63 \, d$, was extrapolated from data for *Penilia avirostris* populations in the Inland Sea of Japan and the Gulf of Mexico shelf (Mullin and Onbé, 1992). These are extraordinary growth rates for planktonic metazoans.

The presence of embryos in the brood pouch of even the smallest females in field samples (Bainbridge, 1958; Fofonoff, 1994; Onbé, unpublished), except in populations of recently hatched resting embryos (Gieskes, 1970, Figure 5; Onbé, 1974, Figure 60; Onbé, unpublished data) indicates that pædogenesis occurs widely in podonids. The apparent exceptions reported for *Pleopis schmackeri* in coastal waters off Brazil (Rocha, 1985) and in the western Pacific (Onbé, 1983) may be explained by the rapid rate of embryonic development in the warm water (21°C and 25–28.5°C, respectively) coupled with the nocturnal release of neonates (Onbé, 1974; Bryan, 1979; Perotti, 1988). Under these conditions mature embryos bearing pædogenic embryos will rarely be seen in specimens collected during the day. Females collected during the day will carry only early-stage embryos without pedogenic embryos of their own. Kim and Onbé (1989a) have confirmed the presence of pædogenic embryos in *P. schmackeri*.

In addition to pedogenesis, reproduction may be accelerated in some species by the release of a new brood into a newly formed brood chamber before the preceding brood is released, as reported for *Evadne nordmanni* (Gieskes, 1970, Figure 5; Onbé, 1974, Figure 60; Onbé, unpublished) and for *Pseudevadne tergestina* (Onbé, 1974, Figure 61; Onbé, unpublished). In these species individual females simultaneously bear early- and late-stage embryos in separate brood chambers.

Further acceleration of brood production has been reported for *Podon leuckarti* (Gieskes, 1970, figure 15) and for *P. intermedius* (Le Tourneau, 1961). In these species more than one brood of different ages may be found in the same brood chamber. Gieskes describes *P. leuckarti* females carrying three broods in a single brood chamber; the broods, each of 5–8 embryos, were recognizable because of their distinctly different stages of development ranging from blastulae to mature embryos.

5.2.2. Brood Size

Brood size depends on many factors. In theory, there should be more eggs per brood chamber in marine than in freshwater species because the eggs of the former are oligolecithal and therefore smaller than eggs supplied with yolk (Claus, 1877). Likewise, there should be a direct relationship between body size and egg number with larger females carrying more eggs. The latter was confirmed by Della Croce and Venugopal (1973), who found that the number of eggs or embryos of *Penilia avirostris* increased from 2 to 12 per female as body length increased from 0.5 to 1.0 mm. Cheng (1947) documented an increase in mean fecundity as a positive function of body size in *Evadne nordmanni*. For three size classes (300–400, 400–500 and 500-600 μm) the mean was 2.7, 3.4 and 4.2 eggs per brood, respectively. Similarly for four body size classes (0.8–0.9, 0.9–1.0, 1.0–1.1 and 1.1–1.2 mm) of *Podon intermedius*, Cheng (1947) found that the mean egg number per brood was 2.7, 2.6, 3.0 and 4.4 eggs per brood, respectively. This relationship between body size and brood size is not always strong. In Narragansett Bay correlations of egg number and standard body length were significant ($p < 0.05$) on only 16 of 34 sampling dates of field populations of *E.nordmanni*, and on only 10 of 30 sampling dates for *Pleopis polyphemoides*. Maximum r^2 values were 0.44 for *E. nordmanni* and 0.49 for *P. polyphemoides* (Fofonoff, 1994).

Brood sizes of marine cladocerans are generally highest during the initial phases of population growth and decrease rapidly as populations increase (Bainbridge, 1958; Della Croce and Bettanin, 1965; Onbé, 1974; Platt and Yamamura, 1986; Fofonoff, 1994). Mean brood sizes of *Evadne nordmanni* during the spring maximum abundances ranged from 9 to 12 embryos per female in Nova Scotia and Narragansett Bay, to 6–7 embryos per female in the Clyde Sea, Scotland (Bainbridge, 1958), the Mediterranean (Corni and Gardenghi, 1974), the Inland Sea of Japan (Onbé, 1974) and the Baltic (Poggensee and Lenz, 1981). Individual *E. nordmanni* may carry as many as 26 embryos during the spring maximum (Fofonoff, 1994). In *Polyphemus* and *Bythotrephes*, brood size is largest for the first spring generation (Mordukhai-Boltovskoi and Rivier, 1987). In *Pleopis polyphemoides*, mean brood sizes are 7–10 embryos per female, with individuals bearing up to 19 embryos (Onbé, 1974; Corni and Gardenghi, 1975; Fofonoff, 1994). During declining phases of podonid populations females may carry only one or two embryos (Bainbridge, 1958; Onbé, 1974; Platt and Yamamura, 1985b). Similar seasonal patterns occur in some populations of *Penilia avirostris* in Narragansett Bay, the Bay of Naples and the Inland Sea of Japan, where in early spring or summer females carry 8–10 embryos. In late fall, only 2–4 embryos are carried (Figure 15) (Della Croce and Bettanin, 1965; Della Croce, 1966; Onbé, 1974). Other *P. avirostris* populations (Beaufort, North Carolina; Gulf of Trieste) show no clear seasonal trends in brood size (Della Croce, 1966; Specchi and Fonda, 1974).

5.2.3. *Birth Rates and Life Tables*

Birth rates of *Pleopis polyphemoides* and *Evadne nordmanni* in Narragansett Bay were calculated by Paloheimo's (1974) method using brood sizes in field populations and development times of embryos in laboratory cultures (Fofonoff, 1994). During the late spring to summer seasons, birth rates of both species were lowest when populations first appeared in the plankton, and showed an upward trend, peaking in July at $0.8-1.0\,d^{-1}$ as temperature increased. Maximum birth rates generally occurred at or shortly before the summer population collapse, while maximum population growth rates ($0.64-0.70\,d^{-1}$) occurred in June, during the early phases of population growth. During the fall, both birth rates and growth rates of each species were initially high ($0.5-1.0\,d^{-1}$) and declined as temperature decreased and the proportion of sexual individuals increased. Birth rates were affected more by temperature and its influence on development time than by fluctuations in brood size (Fofonoff, 1994). Development times were obtained, however, in Fofonoff's study only at high food concentrations in laboratory cultures; if development times are prolonged in field populations by low ambient food quantity or quality, birth rates in field populations would be less than those estimated.

Complete life tables for marine cladocerans have not been constructed, thus preventing a rigorous analysis of their population dynamics. While there have been many studies tracking field abundance and a few that have quantified one or more reproductive parameters (Bainbridge, 1958, Della Croce and Bettanin, 1965; Onbé, 1974; Platt and Yamamura, 1986; Mullin and Onbé, 1992), only Fofonoff (1994) has identified both birth and death rates based on combined field and laboratory data. Nonetheless, culture methods for podonids and *Penilia* are still imperfect and estimates of longevity and lifetime fecundity cannot be made. Additional work is needed to determine all relevant parameters under a full range of ambient conditions.

5.2.4. *Prenatal mortality*

Prenatal mortality of embryos in the brood chamber in both freshwater and marine cladocerans conceivably allows adjustment of reproductive rates to the prevailing environmental conditions. In *Penilia avirostris* under stress, all embryos in the brood pouch may die simultaneously (Fofonoff, unpublished observations) but in podonids, embryo mortality is more selective so that fewer embryos are found in late-stage broods than earlier. Since embryos apparently absorb nutrients from fluids in the brood chamber, surviving podonid embryos may benefit from the death of their siblings. The result will be fewer but larger offspring (Bainbridge, 1958) and an immediate reduction in the population growth rate. Yet, if environmental conditions change and become favorable for population growth, the population retains its potential for immediate and rapid increase. To the extent that egg number is positively correlated with body size (Section 5.2.2 and Cheng, 1947; Bainbridge, 1958) this potential can be realized by the greater number of eggs that can be accommodated in the brood chamber of the large survivors.

Platt and Yamamura (1986) found a seasonal and progressive increase in "relative" prenatal mortality for *Evadne nordmanni* in St Margaret's Bay, Nova Scotia from the first appearance of this species in May to its decline in October. Lacking laboratory data on development times, they determined relative prenatal mortality per generation from field populations by assuming that the frequency of occurrence of embryos at each of six arbitrarily defined developmental stages was proportional to the time taken to complete each stage. They estimated that relative prenatal mortality was near zero in early June but reached values sufficiently high to result in a decrease in population size in September and October. In contrast, prenatal mortality of *E. nordmanni* and *Pleopis polyphemoides* in Narragansett Bay did not increase during either of the two annual growth seasons. Instead, sharp spikes of prenatal mortality were followed by periods of decreased embryo mortality and increasing brood size (Fofonoff, 1994) (Figures 22 and 23). The low correlations of prenatal mortality with demographic patterns in Narragansett Bay populations may be, in part, because faster development times offset decreased brood size at higher temperatures and, in part, because postnatal mortality during decline of population size greatly exceeded prenatal mortality (Fofonoff, 1994). The apparently greater importance of prenatal mortality in Nova Scotian waters compared to Narragansett Bay may reflect differences in temperature, phytoplankton and predators. Overall, the capacity of marine cladocerans to alter population growth rates through a change in brood size as a function of environmental conditions prevailing within a day or two prior to birth gives these organisms unusually flexible control over population growth rates.

5.2.5. *Temperature and Salinity*

Effects of salinity, temperature and other environmental factors on fecundity have not been studied experimentally, but some field studies indicate an effect of salinity and temperature on body size, which affects fecundity as noted above. In *Evadne spinifera* and *Pseudevadne tergestina* body size is positively correlated with salinity, and negatively correlated

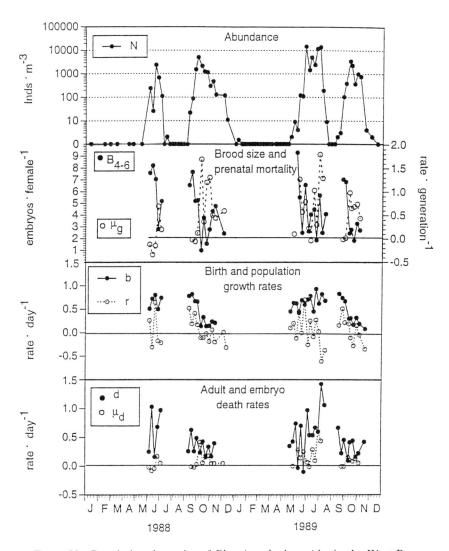

Figure 22 Population dynamics of *Pleopis polyphemoides* in the West Passage of Narragansett Bay. N, Abundance m^{-3}; B$_{4-6}$, numbers of stages 4–6 embryos per brood; b, birth rate day^{-1}; d, death rate per day; r, population growth rate per day; μ_d, absolute prenatal mortality (embryo death rate or mortality per day); μ_g = the relative prenatal mortality (mortality per generation)/generation time. (After Fofonoff, 1994.)

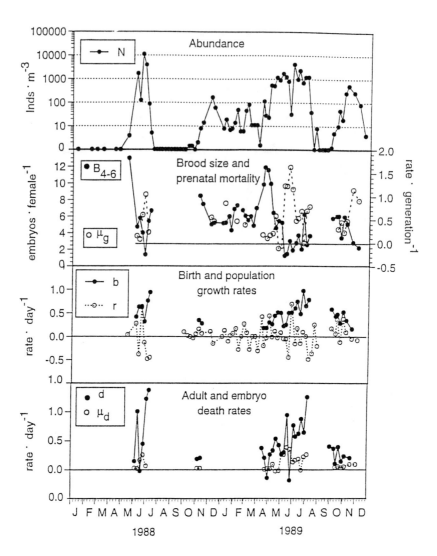

Figure 23 Population dynamics of *Evadne nordmanni* in the West Passage of Narragansett Bay. N, Abundance m^{-3}; B$_{4-6}$, numbers of stages 4–6 embryos per brood; b, birth rate per day; d, death rate per day; r, population growth rate per day; μ_d, absolute prenatal mortality (embryo death rate or mortality per day); μ_g = the relative prenatal mortality (mortality per generation)/generation time. (After Fofonoff, 1994.)

and scanning laser densitometry (Schulz and Yurista, 1995) are techniques that are also yielding useful information.

Optimal salinities for cultivating marine cladocerans need to be determined. As with other zooplankters, the optimal salinity for cultivation may prove to be different from that of their natural environments. For example, ephippia of the chydorid *Alona taraporevalae* found in sediments in an estuary at 26‰ will hatch in laboratory vessels at salinities of 1.5–18‰, with minimal incubation duration (5 d) at 6–9‰ (Shirgur and Naik, 1977). Similarly, the estuarine *Diaphanosoma celebenesis*[2] can be reared in cultures at salinities from 0.1 to 38‰, but maximal net reproduction (total offspring per female) is obtained at 5‰ (Segawa and Yang, 1987).

Attempts to culture the ctenopods *Penilia avirostris* (Gore, 1980) and the brackish water *Disphanosoma celebensis* (Segawa and Yang, 1987, 1990) yielded mixed results. Given the similarity of limbs and mouthparts (Meurice and Monoyer, 1984) of *Penilia* to those of other ctenopods, most investigators have assumed that small food particles are preferred. Takami and Iwasaki (1978) succeeded in establishing *P. avirostris* in cultures which produced more than one generation by daily additions of *Isochrysis galbana* (4.5 μm cell diameter) at 15×10^4 cells per animal (equivalent to 60–375×10^3 cells ml^{-1}). At 20°C, 25‰ and 1000 lux, *Penilia* increased rapidly in numbers after a lag phase of 30–35 d. The intrinsic growth rates (r) following the lag phase were approximately 0.1 d^{-1}. A mixed diet, containing *Isochrysis* and *Chlamydomonas*, gave a shorter lag phase and more neonates per adult but lower final population size (Takami *et al.*, 1978). Although growth rates were at the low end of the range observed for marine cladocerans during bloom conditions, this work demonstrates that laboratory cultivation of *Penilia* is possible.

Autotrophic flagellates were not the only food available in the experiments of Takami and Iwasakis (1978) and Takami *et al.* (1978). High concentrations of bacteria in the cultures were reduced but not eliminated by weekly changes of the medium. It is probable that the bacteria or bacterivorous microflagellates (Pavlova, 1959b) were at least a supplementary and perhaps a primary source of food in these experiments. We know that *Penilia* captures heterotrophic microflagellates (2–5 μm) at clearance rates in excess of 1.0 ml animal^{-1} h^{-1} (Turner *et al.*, 1988). Diatoms such as *Thalassiosira pseudonana* and *T. weissflogii* are also eaten at substantial rates (Turner *et al.*, 1988) and are a potential food.

---

[2] *Diaphanosoma volzi* Chiang, 1956 (= *D. aspinosum*) and *D. celebensis* (Stingelin, 1900) are distinct species but easily confused (Korovchinsky, 1989). Segawa and Yang (1987) originally identified their material as *D. aspinosum*, but later changed their identification to *D. celebensis* (Segawa and Yang, 1990).

Paffenhöfer and Orcutt (1986) cultured *Penilia* on *Isochrysis galbana* in a series of growth and feeding experiments. They observed growth and reproduction only at relatively low algal concentrations of 0.63 and 2.1×10^3 cells ml^{-1} (6 and 20 μg C ml^{-1}); multiple broods in some females occurred only at the *lower* of these two concentrations. At algal concentrations of 6.3 and 21×10^3 cells ml^{-1}, no growth or reproduction occurred. While these results demonstrate the ability of *Penilia* to reproduce marginally at low concentrations of one algal species, they do not illuminate either the quantitative or qualitative contributions of other food items for growth and reproduction under natural conditions, for example, in eutrophic harbors (Wong *et al.*, 1992).

Further progress in the cultivation of podonids will require improved culture conditions and diets that will sustain parthenogenic reproduction. A wide array of potential food organisms, including microflagellates, dinoflagellates, diatoms (Guillard, 1988), rotifers, ciliates (Stoecker and Egloff, 1987) and other elements of microzooplankton have been cultivated in recent years, facilitating the search for an improved diet. Additives to culture media should also be investigated. A chelating agent, such as EDTA, is added routinely at low concentrations to natural sea water medium used to culture ciliates (Stoecker *et al.*, 1981) and a rotifer (Egloff, 1988). The basis for beneficial results is unknown but may involve reduction or stabilization of heavy metal concentrations (Stoecker *et al.*, 1986).

6.1.2. *Physiology of Development*

Developmental physiology of marine cladocerans has received scant attention despite its unique characteristics. The entire developmental process deserves careful study to elucidate the biochemical and physiological mechanisms, especially the role of the Nährboden as the probable source of nutrients for developing embryos and possibly its involvement in the resorption and redistribution of nutrients from aborted eggs and embryos.

To date, the nutritive value of dissolved organic matter to adult marine cladocerans has been investigated only by Pavlova (1959b), who found that *Penilia* adults could not be sustained on organic broths. This is not unexpected given the impermeability of typical crustacean exoskeletons to organic solutes (Anderson and Stephens, 1969), yet the rapid growth of oligolecithal eggs and embryos in the brood chambers of marine cladocerans cannot occur unless nutrients are obtained from an exogenous source. A better understanding of the process will depend on biochemical characterizations of the fluid in the brood chamber as well as new studies of the histology and permeability of the exoskeleton of embryonic marine cladocerans.

The control of growth and development undoubtedly has similar elements in both freshwater and marine cladocerans. In that regard, the demonstrations that juvenile hormone affects embryogenesis in *Daphnia magna* (Templeton and Laufer, 1983), that high concentrations of high-quality foods are necessary for maximal growth, reproduction and survival of *D. magna* (Goulden *et al.*, 1982), that vitamin B_{12} is required for normal reproduction in *D. pulex* (Keating, 1985) and that neurosecretory activity is correlated with molting and development (van den Bosch de Aguilar, 1969) may provide additional leads for future research.

6.2. Biochemical and Genetic Analyses of Populations

Two primary explanations have been advanced to explain the relatively few species of cladocerans in the oceans. Historically, the ancestors of the podonids may have invaded the sea from freshwater origins relatively recently. It has been suggested that the Onychopoda ("Polyphemoidea") colonized the sea from the mesosaline basins of the Ponto-Caspian regions when those basins became connected to the Mediterranean through the Dardanelles during the Riss–Würm interglacial (Hutchinson, 1957, 1967), approximately 100 000 years ago. The invasion of *Penilia* and other cladocerans via estuaries may be more recent (Lochhead, 1954; Egborge, 1987). In either case, sufficient time may have elapsed for greater evolutionary divergence of either *Penilia* or the oceanic podonids from their ancestral stocks than has occurred.

Another explanation for the relative scarcity of cladoceran species in the sea may be the extent to which parthenogenesis has enhanced the rate of gene flow between distant populations. Because continuous production of multiple generations of parthenogenic females facilitates wide dispersion in oceanic surface currents, successful immigrants will carry genotypes relatively unchanged from their point of origin, thus reducing the probability that distant populations will evolve as new species. The apparent lack of speciation based on phenotypic data supports this interpretation. However, Frey (1987) argues that similar conclusions about the chydorid cladocerans are not valid. Moreover, we know in *Daphnia* that genetic diversification as revealed by allozymes is much greater than shown by classical taxonomic techniques (Hebert, 1987).

Molecular characterizations of widely dispersed populations will be required to reveal the extent to which long-distance transport of cladocerans by ocean currents or other means influences the reproductive dynamics of local populations. For example, Makrushin (1981) has hypothesized that the mixing of indigenous populations of *Evadne nordmanni* and *Podon leuckarti* in the White Sea with other individuals carried by currents from the North Atlantic Ocean and Barents Sea would explain the asynchronous pattern of gamogenic reproduction that he observed. A direct test of his hypothesis with respect to mixing could be made if specimens from the North Atlantic, Barents and White Seas were characterized by biochemical (allozymes) or molecular (mitochondrial or nuclear DNA) techniques.

Future studies using molecular and biochemical techniques to identify distinct cladoceran populations will be necessary to determine the extent to which certain populations of cladocerans are associated with specific oceanic water masses. The patterns of abundance (Figure 21) with respect to temperature and salinity in Chesapeake Bay (Bryan and Grant, 1979) and in oceanic water masses (Gieskes, 1971a) need further investigation to determine the factors that control these patterns. These studies could be initiated wherever cladocerans are seasonally abundant, but sites that have been recently invaded by a marine species have a special attraction because the critical factors controlling distribution and abundance may be easier to identify in populations on the edge of their distribution range. For example, in this century *Penilia* has appeared suddenly in the well-studied water of the Bay of Naples, Italy (Cattley and Harding, 1949), Great Harbor, Woods Hole, USA (Lochhead, 1954) and George's Bank, USA and Canada (Colton, 1985). Colton framed the problem succinctly for the population of *Penilia* on the Georges Bank. The key questions are whether this species occurs on the Georges Bank, relative to adjacent areas, primarily because sea surface temperatures are high enough to trigger gamogenic reproduction or primarily because bottom water temperatures are high enough to trigger the development of resting embryos. The striking differences between temperatures on and off the Bank makes this a promising site for future work.

ACKNOWLEDGEMENTS

We are grateful to Norberto Della Croce for providing original prints of Figures 6 and 19 and to Michael Kryzytski for translations of some of the major Russian publications on Caspian and other cladocerans. Funds for travel to libraries and for translations were generously provided by Oberlin College. Many persons at the libraries of the Marine Biological Laboratory (Woods Hole), the Marine Biological Association of the United Kingdom (Plymouth), University of Rhode Island and Oberlin College were extraordinarily patient in our quest to find literature from many obscure sources. Paul Fofonoff wishes to thank Professors Theodore Smayda, Edward Durbin and Saran Twombly, and fellow graduate students Cindy Heil, David Borkman and Bob Campbell of the Graduate School of Oceanography, University of Rhode Island, for assistance and constructive criticism during his doctoral research on cladocerans in Narragansett Bay. Finally, D. Rae Barnhisel, Dorothy B. Berner, David G. Frey, Irina Rivier and Fulvio Zaffagnini read various drafts of the manuscript and each made important contributions and suggestions for its improvement.

REFERENCES

Ackefors, H. (1965). On the zooplankton fauna at Askö (The Baltic–Sweden). *Ophelia* **2**, 269–280.

Aladin, N. V. (1979). Hypoosmotic regulation in the marine cladoceran *Penilia avirostris*. *Journal of Evolutionary Biochemistry and Physiology* [English translation of *Zhurnal Evolyutsionnoi Biokhimii i Fiziologii*] **15**, 503–504.

Aladin, N. V. (1982). Salinity adaptations and osmoregulation abilities of the Cladocera. 1. Forms from open seas and oceans. *Zoologicheskii Zhurnal* **61**, 341–351.

Aladin, N. V. (1983). Amphiosmotic regulation in the euryhaline cladocera, *Moina mongolica*. *Hydrobiological Journal* [English translation of *Gidrobiologicheskii Zhurnal*] **19**, 76–81.

Aladin, N. V. (1995). The conservation ecology of the Podonidae from the Caspian and Aral Seas. *Hydrobiologia* **307**, 85–97.

Allan, J. D. (1976). Life history patterns in zooplankton. *American Naturalist* **110**, 165–180.

Allan, J. D. (1977). An analysis of seasonal dynamics of a mixed population of *Daphnia*, and the associated cladoceran community. *Freshwater Biology* **7**, 505–512.

Allen, E. and Banta, A. M. (1929). Growth and maturation in the parthenogenetic and sexual eggs of *Moina macrocopa*. *Journal of Morphology and Physiology* **48**, 123–151.

Alonso, M. (1985). *Daphnia (Ctenodaphnia) mediterranea*: a new species of hyperhaline waters, long confused with *D. (C.) dolichocephala* Sars, 1895. *Hydrobiologia* **128**, 217–228.

Anderson, J. W. and Stephens, G. C. (1969). Uptake of organic material by aquatic invertebrates. VI. Role of epiflora in apparent uptake of glycine by marine crustacea. *Marine Biology* **4**, 243–249.

Angelino, M. I. and Della Croce, N. (1975). Osservazioni sul ciclo biologico di *Penilia avirostris* Dana in acque sud-africane: Banco Agulhas e Laguna di Knysna. *Cahiers de Biologie Marine* **16**, 551–558.

Ardnt, H., Schröder, C. and Schese, W. (1990). Rotifers of the genus *Synchaeta* —an important component of the zooplankton in the coastal waters of the southern Baltic. *Limnologica* **21**, 233–235.

Bacci, G., Cognetti, G. and Vaccari, A. M. (1961). Endomeiosis and sex determination in *Daphnia pulex*. *Experientia* **17**, 505–506.

Bainbridge, V. (1958). Some observations on *Evadne nordmanni* Lovén. *Journal of the Marine Biological Association of the United Kingdom* **37**, 349–370.

Bainbridge, V. (1960). The plankton of inshore waters off Freetown, Sierra Leone. *Fishery Publications, London* **13**, 1–48.

Baker, H. M. (1938). Studies on the Cladocera of Monterey Bay. *Proceedings of the California Academy of Sciences* **23**, 311–365.

Baldass, F. V. (1941). Entwicklung von *Daphnia pulex*. *Zoologische Jahrbücher, Abteilung für Anatomie und Ontogenie der Tiere* **64**, 1–60.

Banta, A. M. and Brown, L. A. (1939). Control of male and sexual-egg production. *In* "Studies on the Physiology, Genetics, and Evolution of some Cladocera" (A. M. Banta, ed.), Paper No. 39, pp. 106–130. Carnegie Institution of Washington, Washington, DC, USA.

Barker, D. M. and Hebert, P. D. N. (1986). Secondary sex ratio of the cyclic parthenogen *Daphnia magna* (Crustacea: Cladocera) in the Canadian Arctic. *Canadian Journal of Zoology* **64**, 1137–1143.

Barlow, J. P. (1955). Physical and biological processes determining the distribution of zooplankton in a tidal estuary. *Biological Bulletin, Woods Hole* **109**, 211–225.

Beadle, L. C. (1943). An ecological survey of some inland saline waters of Algeria. *Journal of the Linnean Society of London Zoology* **41**, 218–242.

Bell, G. (1982). *The Masterpiece of Nature: The Evolution and Genetics of Sexuality*. University of California Press, Berkeley, USA, 635 pp.

Berg, D. J. and Garton, D. W. (1994). Genetic differentiation in North American and European populations of the cladoceran *Bythotrephes*. *Limnology and Oceanography* **39**, 1503–1516.

Bernard, M. F. (1955). Étude préliminaire quantitative de la répartition saison-nière du zooplancton de la Baie d'Alger. *Bulletin de l'Institut Océanographique, Monaco* **1065**, 1–27.

Berner, D. B., Nguyen, L., Nguy, S. and Burton, S. (1991). Photoperiod and temperature as inducers of gamogenesis in a dicyclic population of *Scapholeberis armata* Herrick (Crustacea: Cladocera: Daphniidae). *Hydrobiologia* **225**, 269–280.

Bosch, H. F. (1970). Ecology of *Podon polyphemoides* (Crustacea, Branchiopoda) in Chesapeake Bay. *Technical Reports Chesapeake Bay Institute of Johns Hopkins University* **66**, 1–77.

Bosch, H. F. and Taylor, W. R. (1968). Marine cladocerans in the Chesapeake Bay Estuary. *Crustaceana* **15**, 161–164.

Bosch, H. F. and Taylor, W. R. (1973a). Distribution of the cladoceran *Podon polyphemoides* in Chesapeake Bay. *Marine Biology* **19**, 161–171.

Bosch, H. F. and Taylor, W. R. (1973b). Diurnal vertical migration of an estuarine Cladoceran, *Podon polyphemoides* in Chesapeake Bay. *Marine Biology* **19**, 172–181.

Bottrell, H. H. (1975a). The relationship between temperature and duration of egg development in some epiphytic Cladocera and Copepoda from the River Thames, Reading, with a discussion of temperature functions. *Oecologia* **18**, 63–84.

Bottrell, H. H. (1975b). Generation time, length of life, instar duration and frequency of moulting, and their relationship to temperature in eight species of cladocerans from the River Thames, Reading. *Oecologia* **19**, 129–140.

Braiko, V. D. (1966). Biology of the winter eggs of *Penilia avirostris*. *Doklady Biological Sciences* [English translation of *Doklady Akademii Nauk SSSR*] **170**, 681–683.

Brooks, J. L. (1959). Cladocera. *In* "Fresh-water Biology" (W. T. Edmondson, ed.), pp. 587–656. John Wiley, New York, USA.

Bryan, B. B. (1977). The ecology of the marine Cladocera of lower Chesapeake Bay. PhD Dissertation, University of Virginia, Charlottesville, USA, 101 pp.

Bryan, B. B. (1979). The diurnal reproductive cycle of *Evadne tergestina* Claus (Cladocera, Podonidae) in Chesapeake Bay, USA. *Crustaceana* **36**, 229–236.

Bryan, B. B. and Grant, G. C. (1974). The occurrence of *Podon intermedius* (Crustacea, Cladocera) in Chesapeake Bay, a new distributional record. *Chesapeake Science* **15**, 120–121.

Bryan, B. B. and Grant, G. C. (1979). Parthenogenesis and the distribution of the Cladocera. *Bulletin of the Biological Society of Washington* **3**, 54–59.

Bur, M. T., Klarer, D. M. and Kreiger, K. A. (1986). First records of a European cladoceran, *Bythotrephes cederstroemi*, in Lakes Erie and Huron. *Journal of Great Lakes Research* **12**, 144–146.

Butorina, L. G. (1965). Observations on the feeding of *Polyphemus pediculus* and the function of its appendages in food capture. *Trudy Instituta Biologii Vnutrennikh Vod* **8**, 44–55.

Butorina, L. G. (1968). The reproductive organs of *Polyphemus pediculus*. *Trudy Instituta Biologii Vnutrennikh Vod* **17**, 41–57.

Butorina, L. G. (1969). On the morphology of *Polyphemis exiguus* Sars. *Trudy Instituta Biologii Vnutrennikh Vod* **19**, 137–157.

Butorina, L. G. (1986). On the problem of aggregations of planktonic crustaceans *Polyphemus pediculus* (L.), Cladocera. *Archiv für Hydrobiologie* **105**, 355–386.

Calver, M. C. (1984). A review of ecological applications of immunological techniques for diet analysis. *Australian Journal of Ecology* **9**, 61–74.

Cattley, J. G. and Harding, J. P. (1949). *Penilia*, a cladoceran normally found off tropical and sub-tropical coasts, recorded in North Sea plankton. *Nature, London* **164**, 238–239.

Cheng, C. (1947). On the fertility of marine Cladocera with a note on the formation of the resting egg in *Evadne nordmanni* Lovén and *Podon intermedius* Lilljeborg. *Journal of the Marine Biological Association of the United Kingdom* **26**, 551–561.

Claus, C. (1877). Zur Kenntniss des Baues und der Organisation der Polyphemiden. *Denkschriften Kaiserlichen Akademie der Wissenschaften, Wien, Mathematisch-Naturwissenschaftlichen Classe* **37**, 137–160.

Colton, J. B. Jr (1985). Eastward extension of the distribution of the marine cladoceran *Penilia avirostris* in the Northwest Atlantic: a case of ecesis? *Journal of the Northwest Atlantic Fishery Science* **6**, 141–148.

Corni, M. G. (1971). Fluttuazioni stagionali dei Cladoceri nelle acque de Fano (Adriatico) durante il 1969. *Bollettino di Pesca Piscicoltura e Idrobiologia* **26**, 113–123.

Corni, M. G. and Cattani, O. (1979). Aspetti biologici ed ecologici di *Penilia avirostris* Dana (Cladocera, Sididae) nel plancton di Fano. *Nova Thalassia* (Suppl.) **3**, 89–112.

Corni, M. G. and Cattani, O. (1980). Cicli eterogonici ed aspetti morfofisiologici del maschio in *Podon polyphemoides* Leuckart (Cladocera, Podonidae). *Nova Thalassia* **4**, 43–61.

Corni, M. G. and Gardenghi, G. (1974). Osservazioni anatomiche e biometriche su *Evadne nordmanni* Lovén (Crustacea, Phyllopoda) del medio Adriatico. *Bolletino di Pesca Piscicicoltura e Idrobiologia* **29**, 157–166.

Corni, M. G. and Gardenghi, G. (1975). Osservazioni sull'anatomia e sulla biologia della femmina di *Podon polyphemoides* Leuckart (Crustacea, Phyllopoda). *Bolletino di Pesca Piscicicoltura e Idrobiologia* **30**, 225–234.

D'Abramo, L. R. (1980). Ingestion rate decrease as the stimulus for sexuality in populations of *Moina macrocopa*. *Limnology and Oceanography* **25**, 422–429.

Dejdar, E. (1931). Bau und Funktion des Sog. "Haftorgans" bei marinen Cladoceren. *Zeitschrift für Morphologie und Oekologie der Tiere* **21**, 617–628.

Della Croce, N. (1966). Observations on the marine cladoceran *Penilia avirostris* in northwestern Atlantic waters. *US Bureau of Sport Fisheries and Wildlife, Technical Papers* **3**, 1–13.

Della Croce, N. (1974). Cladocera. *In* "Fiches d'Identification du Zooplancton" Vol. 143, pp. 1–4. Conseil International pour L'Exploration de la Mer. Copenhagen, Denmark.

Della Croce, N. and Angelino, M. (1987). Marine Cladocera in the Gulf of Mexico and the Caribbean Sea. *Cahiers de Biologie Marine* **28**, 263–268.

Della Croce, N. and Bettanin, S. (1964–65). Osservazioni sul ciclo biologico di *Penilia avirostris* Dana del Golfo di Napoli. *Bollettino dei Musei e degli Istituti Biologici dell' Universita di Genova* **33**, 49–68.

Della Croce, N. and Bettanin, S. (1965). Sviluppo embrionale della forma partenogenetica di *Penilia avirostris* Dana. *Cahiers de Biologie Marine* **6**, 269–275.

Della Croce, N. and Bettanin, S. (1969). Formazione delle uova durevoli in *Penilia avirostris* Dana. *Cahiers de Biologie Marine* **10**, 95–102.

Della Croce, N. and Gaino, E. (1970). Osservazioni sulla biologia del maschio di *Penilia avirostris* Dana. *Cahiers de Biologie Marine* **11**, 361–365.

Della Croce, N. and Venugopal, P. (1972). Distribution of marine cladocerans in the Indian Ocean. *Marine Biology* **15**, 132–138.

Della Croce, N. and Venugopal, P. (1973). *Penilia avirostris* Dana in the Indian Ocean (Cladocera). *Internationale Revue der Gesamten Hydrobiologie* **58**, 713–721.

Di Caporiacco, L. (1938). Cladocières marins recueillis pendant les croisères du Prince Albert de Monaco. *Bulletin de l'Institut Océanographique, Monaco* **740**, 1–12.

Dolgopolskaya, M. A. (1958). The marine Cladocera of the Black Sea. *Trudy Sevastopol'skoi Biologicheskoi Stantsii* **10**, 27–75.

Edmondson, W. T. (1960). Reproductive rates of rotifers in natural populations. *Memorie dell'Istituto Italiano di Idrobiologia* **12**, 21–77.

Egborge, A. B. M. (1987). Salinity and distribution of Cladocera in Warri River, Nigeria. *Hydrobiologia* **145**, 159–167.

Egloff, D. A. (1988). Food and growth of the marine microzooplankter, *Synchaeta cecilia* (Rotifera). *Hydrobiologia* **157**, 129–141.

Eriksson, S. (1974). The occurrence of marine Cladocera on the west coast of Sweden. *Marine Biology* **26**, 319–327.

Feller, R. J., Taghon, G. L., Gallagher, E. D., Kenny, G. E. and Jumars, P. A. (1979). Immunological methods for food-web analysis in a soft-bottom benthic community. *Marine Biology* **54**, 61–74.

Fish, C J. (1926). Seasonal distribution of the plankton of the Woods Hole region. *Bulletin of the United States Bureau of Fisheries* **41**, 91–179.

Flössner, D. (1972). Krebstiere, Crustacea, Kiemen- und Blattfüsser,

Branchiopoda, Fischläuse, Branchiura. *Die Tierwelt Deutschlands und der Angrenzenden Meersteile* **60**, 1–501.

Fofonoff, P. W. (1994). Feeding and growth of the marine cladoceran *Podon polyphemoides*. In "Marine Cladocerans in Narragansett Bay", pp. 1–54. PhD Dissertation, University of Rhode Island, Kingston, USA.

Frey, D. G. (1965). Gynandromorphism in the chydorid cladocera. *Limnology and Oceanography* **10**, R103–R114.

Frey, D. G. (1982). Contrasting strategies of gamogenesis in northern and southern populations of Cladocera. *Ecology* **63**, 223–241.

Frey, D. G. (1987). The taxonomy and biogeography of the Cladocera. *Hydrobiologia* **145**, 5–17.

Frey, D. G. (1993). The penetration of Cladocerans into saline waters. *Hydrobiologia* **267**, 233–248.

Fryer, G. (1987a). Morphology and the classification of the so-called Caldocera. *Hydrobiologia* **145**, 19–28.

Fryer, G. (1987b). A new classification of the branchiopod Crustacea. *Zoological Journal of the Linnean Society* **91**, 357–383.

Gieskes, W. W. C. (1970). The Cladocera of the North Atlantic and the North Sea: biological and ecological studies. PhD Dissertation, McGill University (Marine Sciences Centre). Montreal, Canada, 204 pp.

Gieskes, W. W. C. (1971a). Ecology of the Cladocera of the North Atlantic and the North Sea, 1960–1967. *Netherlands Journal of Sea Research* **5**, 342–376.

Gieskes, W. W. C. (1971b). The succession of two *Podon* (Crustacea, Cladocera) species in the North Sea. *Netherlands Journal of Sea Research* **5**, 377–381.

Gieskes, W. W. C. (1971c). Removal of "*Podon*" *polyphemoides* from the Genus *Podon*. *Hydrobiologica* **38**, 61–66.

Gore, M. A. (1980). Feeding experiments on *Penilia avirostris* Dana (Cladocera: Crustacea). *Journal of Experimental Marine Biology and Ecology* **44**, 253–260.

Goulden, C. E. (1968). The systematics and evolution of the Moinidae. *Transactions of the American Philosophical Society* **58**, 3–101.

Goulden, C. E., Comotto, R. M., Hendrickson, J. A. Jr., Hornig, L. L. and Johnson, K. L. (1982). Procedures and recommendations for the culture and use of *Daphnia* in bioassay studies. In "Aquatic Toxicology and Hazard Assessment: Fifth Conference", ASTM Special Technical Report 766 (J. G. Pearson, R. B. Foster and W. E. Bishop, eds), pp. 139–160. American Society for Testing and Materials, Philadelphia, USA.

Grahame, J. (1976). Zooplankton of a tropical harbour: the numbers, composition and response to physical factors of zooplankton in Kingston Harbour, Jamaica, *Journal of Experimental Marine Biology and Ecology* **25**, 219–237.

Guillard, R. R. L. (1988). The Center for Culture of Marine Phytoplankton: history, structure, function, and future. *Journal of Protozoology* **35**, 255–256.

Hall, D. J. (1964). An experimental approach to the dynamics of a natural population of *Daphnia galeata mendotae*. *Ecology* **45**, 94–112.

Hebert, P. N. D. (1987). Genotypic characteristics of the Cladocera. *Hydrobiologia* **145**, 183–193.

Heinle, D. R. (1966). Production of a colanoid copepod, *Acartia tonsa*, in the Patuxent River estuary. *Chesapeake Science* **7**, 59–74.

Heinle, D. R. and Flemer, D. A. (1975). Carbon requirements of a population of the estuarine copepod *Eurytemora affinis*. *Marine Biology* **31**, 235–247.

Hirayama, K. and Kusano, T. (1972). Fundamental studies on the physiology of rotifer for its mass culture II. Influence of temperature on population growth of rotifer. *Bulletin of the Japanese Society of Scientific Fisheries* **38**, 1357–1363.

Hobæk, A. and Larsson, P. (1990). Sex determination in *Daphnia magna*. *Ecology* **71**, 2255–2268.

Hopkins, T. L. (1966). The plankton of the St Andrew Bay system, Florida. *Publications of the Institute of Marine Science, University of Texas* **11**, 12–64.

Hrbáčková, M. (1974). The size of primiparae and neonates of *Daphnia hyalina* Leydig (Crustacea: Cladocera) under natural and enriched food conditions. *Vestnik Ceskoslovenske Spolecnosti Zoologicke* **38**, 98–105.

Hutchinson, G. E. (1937a). A contribution to the limnology of arid regions. *Transactions of the Connecticut Academy of Arts and Sciences* **33**, 47–132.

Hutchinson, G. E. (1937b). Limnological studies in Indian Tibet. *Internationale Revue der Gesamten Hydrobiologie* **35**, 134–176.

Hutchinson, G. E. (1957). "A Treatise on Limnology. Vol. I: Geography, Physics, and Chemistry." John Wiley, New York, USA.

Hutchinson, G. E. (1967). "A Treatise on Limnology. Vol. II: Introduction to Lake Biology and the Limnoplankton." John Wiley, New York, USA.

Inoue, M. and Aoki, M. (1971). Reproduction of Cladocera. *Diaphanosoma* sp. cultured with seawater-acclimatized *Chlorella* as basic diet under different chlorinity. *Journal of the College of Marine Sciences and Technology, Tokai University* **5**, 1–8.

Ischreyt, G. (1930). Ueber Körperbau und Lebersweise des *Bythotrephes longimanus* Leydig. *Archiv für Hydrobiologie* **21**, 241–324.

Ivanova-Kazas, O. M. (1977). Analysis of the early development of Crustaceans. I. Principal variants of early development. *Soviet Journal of Marine Biology* [English translation of *Biologiya Morya*] **3**, 1–112.

Iwasaki, H., Takami, A. and Onbé, T. (1977). Studies on the cultivation of marine Cladocera-I. Factors affecting the hatch of resting eggs. *Bulletin of Plankton Society of Japan* **24**, 62–65.

Jagger, R. A., Kimmerer, W. J. and Jenkins, G. P. (1988). Food of the cladoceran *Podon intermedius* in a marine embayment. *Marine Ecology Progress Series* **43**, 245–250.

Johnsen, G. (1983). Egg age distribution, the direct way to cladoceran birth rates. *Oecologia* **60**, 234–236.

Jorgensen, O. M. (1933). On the marine Cladocera from the Northumbrian plankton. *Journal of the Marine Biological Association of the United Kingdom* **19**, 177–226.

Kankaala, P. (1983). Resting eggs, seasonal dynamics, and production of *Bosmina longispina maritima* (P. E. Müller) (Cladocera) in the northern Baltic proper. *Journal of Plankton Research* **5**, 53–69.

Kankaala, P. and Wulff, F. (1981). Experimental studies on temperature-dependent embryonic and postembryonic rates of *Bosmina longispina maritima* (Cladocera) in the Baltic. *Oikos* **36**, 137–146.

Keating, K. I. (1985). The influence of vitamin B_{12} deficiency on the reproduction of *Daphnia pulex* Leydig (Cladocera). *Journal of Crustacean Biology* **5**, 130–136.

Khlebovich, V. V. and Aladin, N. V. (1976). Hypotonic regulation in marine daphnids *Evadne nordmanii* [sic] and *Podon leuckarti*. *Journal of Evolutionary*

Biochemistry and Physiology [English translation of *Zhurnal Evolyutsionnoi Biokhimii i Fiziologii*] **12**, 528–529.

Kiernan, J. A. (1980). "Histological and Histochemical Methods." Pergamon Press, Oxford, UK. 344 pp.

Kim, S. W. (1989). "Studies on the ecology of marine cladocerans in the northwestern Pacific Ocean." PhD Thesis, Hiroshima University, Higashi-Hiroshima, Japan. 180 pp.

Kim, S. W. and Onbé, T. (1989a). Observations on the biology of the marine cladoceran *Podon schmackeri*. *Journal of Crustacean Biology* **9**, 54–59.

Kim, S. W. and Onbé T. (1989b). Distribution and zoogeography of the marine cladoceran *Podon schmackeri* in the northwestern Pacific. *Marine Biology* **102**, 203–210.

Kim, S. W. and Onbé, T. (1995). Distribution and zoogeography of the marine cladoceran *Penilia avirostris* in the northwestern Pacific. *Bulletin of Plankton Society of Japan* **42**, 19–28.

Kim, S. W., Onbé, T. and Yoon, Y. H. (1989). Feeding habits of marine cladocerans in the Inland Sea of Japan. *Marine Biology* **100**, 313–318.

Kim, S. W., Yoon, Y. H. and Onbé, T. (1993). Note on the prey items of marine cladocerans. *Journal of the Oceanological Society of Korea* **28**, 69–71.

Kleppel, G. S., Frazel, D., Pieper, E. and Holliday, D. V. (1988). Natural diets of zooplankton off southern California. *Marine Ecology Progress Series* **49**, 231–241.

Korovchinsky, N. M. (1989). Redescription of *Diaphanosoma celebensis* Stingelin, 1900 (Crustacea, Cladocera). *Hydrobiologia* **184**, 7–22.

Kühn, A. (1913). Die Sonderung der Keimesbezirke in der Entwicklung der Sommereier von *Polyphemus pediculus* de Geer. *Zoologische Jahrbücher, Abteilung für Anatomie und Ontogenie der Tiere* **35**, 243–340.

Kurashova, E. K., Tinenkhova, D. Kh. and Elizarenko, M. M. (1992). *Podon intermedius* (Cladocera, Podonidae) in the Caspian Sea. *Journal of Ichthyology* **32**, 27–29.

Kuttner, O. (1911). Mitteilungen über marine Cladoceren. *Sitzungsberichte Gesellschaft Naturforschender Freunde zu Berlin* **2**, 84–93.

Landry, M. R. (1978). Population dynamics and production of a planktonic marine copepod, *Acartia clausi*, in a small temperate lagoon on San Juan Island, Washington. *Internationale Revue der Gesamten Hydrobiologie* **63**, 77–119.

Lehman, J. T. (1991). Causes and consequences of cladoceran dynamics in Lake Michigan: implications of species invasion by *Bythotrephes*. *Journal of Great Lakes Research* **17**, 437–445.

Lei, C-H. and Clifford, H. F. (1974). Field and laboratory studies of *Daphnia schødleri* Sars from a winterkill lake of Alberta. *Canadian National Museum Yearbook, Ottawa* **9**, 1–53.

Le Tourneau, M. (1961). Contribution à l'étude des Cladocères du plancton du Golfe de Marseille. *Recueil des Travaux de la Station Marine d'Endoume* **22**, Fasc 36, 123–151.

Leydig, F. (1860). "Naturgeschichte der Daphniden (Crustaca Cladocera)." Verlag. H. Laupp'schen Buchhandlung, Tübingen, Germany, 252 pp.

Leider, U. (1983). Revision of the genus *Bosmina* Baird, 1845 (Crustacea, Cladocera). *Internationale Revue der Gesamten Hydrobiologie* **68**, 121–139.

Lilljeborg, W. (1901). Cladocera Sueciae oder Beiträge zur Kenntniss der in Schweden lebenden Krebsthiere von der Ordnung der Branchiopoden und der

Unterordnung der Cladoceren. *Nova Acta Regiae Societatis Scientiarum Upsaliensis (Seriei tertiae)* **19**, 1–701.

Lochhead, J. H. (1954). On the distribution of a marine cladoceran, *Penilia avirostris* Dana (Crustacea, Branchiopoda), with a note on its reported bioluminescence. *Biological Bulletin, Woods Hole* **107**, 92–105.

Longhurst, A. R. and Seibert, D. L. R. (1972). Oceanic distribution of *Evadne* in the eastern Pacific. *Crustaceana* **22**, 239–248.

Lonsdale, D. J. (1981). Influence of age-specific mortality on the life history traits of two estuarine copepods. *Marine Ecology Progress Series* **5**, 333–340.

MacArthur, J. W. and Baillie, W. H. T. (1929). Metabolic action and the duration of life. I. Influence of temperature on longevity in *Daphnia magna. Journal of Experimental Zoology* **53**, 221–242.

Maier, G. (1993). The life histories of two temporarily coexisting pond dwelling cladocerans. *Internationale Revue der Gesamten Hydrobiologie* **78**, 83–93.

Makrushin, A. V. (1973). Adaptations of *Polyphemus pediculus* (Cladocera, Polyphemidae) to short northern autumn. *Zoologicheskii Zhurnal* **52**, 1868–1870.

Makrushin, A. V. (1981). Hypothesis of the cause of incomplete correlation between generation cycles and dynamics of environmental conditions in Podonidae. *Soviet Journal of Marine Biology* [English translation of *Biologiya Morya*] **7**, 350–352.

Makrushin, A. V. (1984). On incomplete consistency between the cycle of Cladocera (Crustacea) *Podon leuckarti* and *Evadne nordmanni* generation and seasonal environmental changes. *Ekologiya Morya* **18**, 59–62.

Meurice, J.-C. (1982). Aspects morphologiques comparés des cladocères marins calyptomères et gymnomères en microscopie électronique à balayage. *Annales de la Societé Royale Zoologique de Belgique* **112**, 165–174.

Meurice, J.-C. and Dauby, P. (1983). Scanning electron microscope study and computer analysis of taxonomic distances of the marine Podonidae (Cladocera). *Journal of Plankton Research* **5**, 787–795.

Meurice, J.-C. and Goffinet, G. (1983). Ultrastructural evidence of the ion-transporting role of the adult and larval neck organ of the marine gymomeran Cladocera (Crustacea, Branchiopoda). *Cell and Tissue Research* **234**, 351–363.

Meurice, J.-C. and Goffinet, G. (1990). Étude préliminaire de l'organe nucal de *Penilia avirostris* cladocère marine calyptomère (Crustacea, Branchiopoda). *Bulletin de la Societé Royale de Liège* **59**, 83–88.

Meurice, J.-C. and Monoyer, Ph. (1984). Étude au microscope électronique à balayage de la morphologie des mandibules chez les Cladocères marins. *Hydrobiologia* **111**, 185–191.

Moore, J. E. (1952). The Entomostraca of southern Saskatchewan. *Canadian Journal of Zoology* **30**, 410–450.

Moraitou-Apostolopoulou, M., Verriopoulos, G. and Tsipoura, N. (1986). Dimensional differentiation between five planktonic organisms living in two areas characterized by different salinity conditions. *Archiv für Hydrobiologie* **105**, 459–469.

Mordukhai-Boltovskoi, Ph. D. (1965). Polyphemidae of the Pontocaspian Basin. *Hydrobiologia* **25**, 212–220.

Mordukhai-Boltovskoi, Ph. D. (1966). Distribution, biology and morphology of Polyphemidae in fresh and brackish waters of the Pontocaspian Basin. *Verhandlungen der Internationalen Vereinigung für Theoretische und Angewandte Limnologie* **16**, 1677–1683.

Mordukhai-Boltovskoi, Ph. D. (1967a). On the taxonomy of the genus *Cornigerius* (Cladocera, Polyphemidae). *Crustaceana* 12, 74–86.
Mordukhai-Boltovskoi, Ph. D. (1967b). On males and gamogenetic females of the Caspian Polyphemidae (Cladocera). *Crustaceana* 12, 113–123.
Mordukhai-Boltovskoi, Ph. D. (1968a). On the taxonomy of the Polyphemidae. *Crustaceana* 14, 197–209.
Mordukhai-Boltovskoi, Ph. D. (1968b). The order Cladocera. *In* "Atlas Bespozvonocknyh Kaspiiskogo Moria" (I. A. Birshtein *et al.*, eds), pp. 120–160. Izd. Pishchevaia promyshlennost, Moscow, Russia.
Mordukhai-Boltovskoi, Ph. D. (1978). A contribution to the taxonomy of marine Podonidae (Cladocera). *Zoologicheskii Zhurnal* 57, 523–529.
Mordukhai-Boltovskoi, Ph. D. and Rivier, I. K. (1971a). On some species and gamogenetic forms of Caspian Polyphemoidea (Cladocera). *Crustaceana* 20, 1–8.
Mordukhai-Boltovskoi, Ph. D. and Rivier, I. K. (1971b). A brief survey of the ecology and biology of the Caspian Polyphemoidea. *Marine Biology* 8, 160–169.
Mordukhai-Boltovskoi, Ph. D. and Rivier, I. K. (1987). Khishchnye vetvistousye Podonidae, Polyphemidae, Cercopagidae i Leptodoridae fauny mira. [Predatory Cladocera: Podonidae, Polyphemidae, Cercopagidae, and Leptodoridae of the World Fauna]. *Fauna SSSR* 148, 1–182.
Morey-Gaines, G. (1979). The ecological role of red tides in the Los Angeles–Long Beach Harbor food web. *In* "Toxic Dinoflagellate Bloom" (D. L. Taylor and H. H. Seliger, eds), pp. 315–320. Elsevier North Holland, New York, USA.
Mortimer, C. H. (1936). Experimentelle und cytologische Untersuchungen über den Generationswechsel der Cladoceren. *Zoologische Jahrbücher, Abteilung für Allegemeine Zoologie und Physiologie der Tiere* 56, 323–388.
Mullin, M. M. and Onbé, T. (1992). Diel reproduction and vertical distributions of the marine cladocerans, *Evadne tergestina* and *Penilia avirostris*, in contrasting coastal environments. *Journal of Plankton Research* 14, 41–59.
Nielsen, T. G. (1991). Contribution of zooplankton grazing to the decline of a *Ceratium* bloom. *Limnology and Oceanography* 36, 1091–1106.
Nival, S. and Ravera, S. (1979). Morphological study of the appendages of the marine cladoceran *Evadne spinifera* Müller by means of the scanning electron microscope. *Journal of Plankton Research* 1, 207–213.
Obreshkove, V. and Fraser, A. W. (1940). Growth and differentiation of *Daphnia magna* eggs *in vitro*. *Biological Bulletin, Woods Hole* 78, 428–436.
Ojima, Y. (1958). A cytological study of the development and maturation of the parthenogenetic and sexual eggs of *Daphnia pulex*. *Kwansei Gakuin University Annual Studies* 6, 123–176.
Onbé, T. (1972). Occurrence of the resting eggs of a marine cladoceran, *Penilia avirostris* Dana on the sea bottom. *Bulletin of the Japanese Society of Scientific Fisheries* 38, 305.
Onbé, T. (1973). Preliminary notes on the biology of the resting eggs of marine cladocera. *Bulletin of Plankton Society of Japan* 20, 74–77.
Onbé, T. (1974). Studies on the ecology of marine cladocera. *Journal of the Faculty of Fisheries and Animal Husbandry Hiroshima University* 13, 83–179. [In Japanese with English abstract.]
Onbé, T. (1977). The biology of marine cladocerans in a warm temperate water. *In* "Proceedings of the Symposium on Warm Water Zooplankton." Special Publication, National Institute of Oceanography, Goa, pp. 383–398.

Onbé, T. (1978a). Gamogenetic forms of *Evadne tergestina* Claus (Branchiopoda, Cladocera) of the Inland Sea of Japan. *Journal of the Faculty of Fisheries and Animal Husbandry, Hiroshima University* 17, 43–52.

Onbé, T. (1978b). The life cycle of marine cladocerans. *Bulletin of Plankton Society of Japan* 25, 41–54.

Onbé, T. (1978c). Sugar flotation method for sorting the resting eggs of marine cladocerans and copepods from sea-bottom sediment. *Bulletin of the Japanese Society of Scientific Fisheries* 44, 1411.

Onbé, T. (1983). Preliminary observations on the biology of a marine cladoceran *Pleopis ("Podon") schmackeri* Poppe. *Journal of the Faculty of Applied Biological Science, Hiroshima University* 22, 55–64.

Onbé, T. (1985). Seasonal fluctuations in the abundance of populations of marine cladocerans and their resting eggs in the Inland Sea of Japan. *Marine Biology* 87, 83–88.

Onbé, T. (1991). Some aspects of the biology of resting eggs of marine cladocerans. *In* "Crustacean Egg Production", Crustacean Issues 7 (A. Wenner and A. Kuris, eds), pp. 41–55, plates I–III. A. A. Balkema, Rotterdam, Netherlands.

Onbé, T. and Ikeda, T. (1995). Marine cladocerans in Toyama Bay, southern Japan Sea: seasonal occurrence and day–night vertical distributions. *Journal of Plankton Research* 17, 595–609.

Onbé, T., Mitsuda, T. and Murakami, Y. (1977). Some notes on the resting eggs of the marine cladoceran *Podon polyphemoides*. *Bulletin of Plankton Society of Japan* 24, 9–17.

Onbé, T., Tanimura, A., Fukuchi, M., Hattori, H., Sasaki, H. and Matsuda, O. (1996). Distribution of marine cladocerans in the northern Bering Sea and the Chukchi Sea. *In* "Proceedings of the NIPR Symposium on Polar Biology", No. 9, pp. 141–152 The National Institute of Polar Research, Tokyo, Japan.

Paffenhöfer, G.-A. (1975). On the biology of Appendicularia of the southeastern North Sea. *In* "Population Dynamics of Marine Organisms in Relation with Nutrient Cycling in Shallow Waters" (G. Persoone and E. Jaspers, eds), 10th European Marine Biology Symposium, Vol. 2, pp. 437–455. Universa Press, Wetteren, Belgium.

Paffenhöfer, G.-A. and Orcutt, J. D., Jr (1986). Feeding, growth and food conversion of the marine cladoceran *Penilia avirostris*. *Journal of Plankton Research* 8, 741–754.

Paffenhöfer, G.-A., Wester, B. T. and Nicholas, W. T. (1984). Zooplankton abundance in relation to state and type of intrusions onto the southeastern United States shelf during summer. *Journal of Marine Research* 42, 995–1017.

Paloheimo, J. E. (1974). Calculation of instantaneous birth rate. *Limnology and Oceanography* 19, 692–694.

Patt, D. I. (1947). Some cytological observations on the Nährboden of *Polyphemus pediculus* Linn. *Transactions of the American Microscopical Society* 66, 344–353.

Pavlova, E. V. (1959a). Development cycle and some data on the growth of *Penilia avirostris* Dana in the Sevastopol Bay. *Trudy Sevastopol'skoi Biologicheskoi Stantsii* 11, 54–62. [English translation No. 908, 1967, Fisheries Research Board of Canada.]

Pavlova, E. V. (1959b). On grazing by *Penilia avirostris*. *Trudy Sevastopol'skoi Biologicheskoi Stantsii* 11, 63–71. [English translation No. 967, 1968, Fisheries Research Board of Canada.]

Perotti, M. G. (1988). Sobre la presencia de un ritmo embriogenico en hembras partenogenetica de *Evadne*. *Anales Museo Historia Natural del Valparaiso* **19**, 15–20.

Platt, T. (1977). Population ecology of marine cladocera in St Margaret's Bay, Nova Scotia. *Fisheries and Marine Service, Canada. Technical Report* **698**, 1–142.

Platt, T. and Yamamura, N. (1986). Prenatal mortality in a marine cladoceran, *Evadne nordmanni*. *Marine Ecology Progress Series* **29**, 127–139.

Poggensee, E. and Lenz, J. (1981). On the population dynamics of two brackish-water Cladocera *Podon leuckarti* and *Evadne nordmanni* in Kiel Fjord. *Kieler Meeresforschungen*, Sonderheft **5**, 268–273.

Potts, W. T. W. (1959). The sodium fluxes in the muscle fibres of a marine and a freshwater Lamellibranch. *Journal of Experimental Biology* **36**, 676–689.

Potts, W. T. W. and Durning, C. T. (1980). Physiological evolution in the branchiopods. *Comparative Biochemistry and Physiology* **67B**, 475–484.

Purasjoki, K. J. (1945). Quantitative Untersuchungen über die Mikrofauna des Meeresbodens in der Umgebung der Zoologischen Station Tvärminne an der Südküste Finnlands. *Societas Scientiarum Fennica, Commentationes Biologicae* **9**, 1–24.

Purasjoki, K. J. (1958). Zur Biologie der Brackwasserkladozere *Bosmina coregoni maritima* (P. E. Müller). *Annales Zoologici Societatis Zoologicae-Botanicae Fennicae 'Vanamo'* **19**, 1–117.

Ramirez, F. C. and Perez Seijas, G. M. (1985). New data on the ecological distribution of cladocerans and first local observations on reproduction of *Evadne nordmanni* and *Podon intermedius* (Crustaces, Cladocera) in Argentina sea waters. *Physis A* **43**, 131–143.

Rammner, W. (1930). Phyllopoda. *Die Tierwelt der Nord- und Ostee* (Lief.) **18**, Teil Xa: 1–32.

Rammner, W. (1931). Mitteilungen über marine Cladoceren. *Biologisches Zentralblatt* **51**, 618–633.

Ramult, M. (1914). Untersuchungen über die Entwicklungsbedingungen der Summereier von *D. pulex* und anderen Cladoceren. *Bulletin International de l'Academie des Sciences de Craçovie* **1914**, 481–514.

Rasmussen, E. (1973). Systematics and ecology of the Islefjord marine fauna (Denmark). *Ophelia* **11**, 1–507.

Rivier, I. K. (1968). The reproduction of Caspian Podoninae. *Trudy Instituta Biologii Vnutrennikh Vod* **17**, 58–69.

Rivier, I. K. (1969). Diurnal reproduction cycles in Caspian Polyphemidae. *Trudy Instituta Biologii Vnutrennikh Vod* **19**, 128–136.

Rocha, C. E. F. da (1982). Distribution of marine cladocerans (Crustacea, Branchiopoda) off Santos, Brazil. *Boletin de Zoologia; Universidade de Sao Paulo* **7**, 155–169.

Rocha, C. E. F. da (1985). The occurrence of *Pleopis schmackeri* (Poppe) in the southern Atlantic and other marine cladocerans on the Brazilian coast. *Crustaceana* **49**, 202–204.

Rossi, F. (1980). Comparative observations on the female reproductive system and parthenogenetic oogenesis in Cladocera. *Bollettino di Zoologia* **47**, 21–38.

Ruvinsky, A. O., Cellarius, S. F. and Cellarius, Yu. G. (1978). The possible role of genome activity changes in the sex determination of *Daphnia pulex*. *Theoretical and Applied Genetics* **52**, 269–271.

Sarkar, S. K. and Choudhury, A. (1987). Occurrence of a cladoceran swarm in the lower stretch of Hooghly estuary, West Bengal, India. *Current Science, Bangalore* **56**, 431.

Sars, G. O. (1897). Pelagic Entomostraca of the Caspian Sea. *Annuaire du Musée zoolgique de l'Academie impériale des sciences de St Petersbourg* **2**, 1–73.

Schram, F. R. (1986). "Crustacea." Oxford University Press, Oxford, UK, 606 pp.

Schulz, K. L. and Yurista, P. M. (1995). Diet composition from allozyme analysis in the predatory cladoceran *Bythotrphes cederstromi. Limnology and Oceanography* **40**, 821–826.

Schwartz, S. S. and Hebert, P. D. N. (1987). Methods for the activation of the resting eggs of *Daphnia. Freshwater Biology* **17**, 373–379.

Segawa, S. and Yang, W. T. (1987). Reproduction of an estuarine *Diaphanosoma aspinosum* (Branchiopoda: Cladocera) under different salinities. *Bulletin of Plankton Society of Japan* **34**, 43–51.

Segawa, S. and Yang, W. T. (1990). Growth, moult, reproduction and filtering rate of an estuarine cladoceran, *Diaphansoma celebensis,* in laboratory culture. *Bulletin of Plankton Society of Japan* **37**, 145–155.

Sergeev, V. and Williams, W. D. (1983). *Daphniopsis pusilla* Serventy (Cladocera : Daphniidae), an important element in the fauna of Australian salt lakes. *Hydrobiologia* **100**, 293–300.

Sergeev, V. and Williams, W. D. (1985). *Daphniopsis australis* nov. sp. (Crustacea : Cladocera), a further daphniid in Australian salt lakes. *Hydrobiologia* **120**, 119–128.

Sherman, K. (1966). Seasonal and areal distribution of zooplankton in coastal waters of Gulf of Maine, 1964. *US Fish and Wildlife Service, Special Scientific Report—Fisheries* **530**, 1–11.

Sherman, K., Smith, W. G., Green, J. R., Cohen, E. B., Berman, M. S., Marti, K. A. and Goulet, J. R. (1987). *In* "Georges Bank" (R. H. Backus and D. W. Bourne, eds), pp. 268–282. Massachusetts Institute of Technology Press, Cambridge, MA, USA.

Shirgur, G. A. and Naik, A. A. (1977). Observations on morphology, taxonomy, ephippial hatching and laboratory culture of a new species of *Alona (Alona taraporevalae* Shirgur and Naik), a chydorid cladoceran from Back Bay, Bombay. *In* "Proceedings of the Symposium on Warm Water Zooplankton." Special Publication, National Institute of Oceanography, Goa, pp. 48–59.

Smirnov, N. N. and Timms, B. V. (1983). A revision of the Australian Cladocera (Crustacea). *Records of the Australian Museum* (Suppl.) **1**, 1–130.

Specchi, M. and Fonda, S. (1974). Alcune osservaziona sul ciclo biologico di *Penilia avirostris* dana nel Golfo di Trieste (Alto Adriatico). *Bollettino di Pesca Piscicoltura e Idrobiologia* **29**, 11–19.

Specchi, M., Dollinar, L. and Fonda-Umani, S. (1974). I Cladoceri del genere *Evadne* nel Golfo di Trieste: Notizie sul ciclo biologico di *Evadne nordmanni, Evadne tergestina* ed *Evadne spinifera. Bollettino di Pesca Piscicoltura e Idrobiologia* **29**, 107–122.

Spittler, P. and Schiller, H. (1984). The effect of salinity on the distribution and population growth of *Chydorus sphaericus* (Cladocera). *Limnologica* **15**, 507–512.

Steuer, A. (1933). Zur Fauna des Canal di Leme bei Rovigno. *Thalassia* **1**, 1–44.

Stoecker, D. K. and Egloff, D. A. (1987). Predation by *Acartia tonsa* Dana on ciliates and rotifers. *Journal of Experimental Marine Biology and Ecology* **110**, 53–68.

Stoecker, D., Guillard, R. R. L. and Kavee, R. M. (1981). Selective predation by *Favella ehrenbergii* (Tintinnia) on and among dinoflagellates. *Biological Bulletin, Woods Hole* **160**, 136–145.

Stoecker, D., Sunda, W. G. and Davis, L. H. (1986). Effects of copper and zinc on two planktonic ciliates. *Marine Biology* **92**, 21–29.

Stross, R. G. (1987). Photoperiodism and phased growth in *Daphnia* populations: coactions in perspective. *In* "Daphnia" (R. H. Peters and R. de Bernardi, eds), pp. 413–437. *Memorie dell'Istituto Italiano di Idrobiologia*, Vol. 45. Verbania, Pallanza, Italy.

Sudler, M. T. (1899). The development of *Penilia schmackeri* Richard. *Proceedings of the Boston Society of Natural History* **29**, 109–132.

Sukhanova, E. R. (1971). *Moina microphthalma* (Cladocera, Daphniidae) under ultrahaline conditions in the North Caucasus. *Zoologicheskii Zhurnal* **50**, 285–287.

Takami, A. (1991). Annual fluctuations in abundance of the resting eggs of the marine cladoceran, *Penilia avirostris*, in the bottom sediments of Atsumi Bay, central Japan. *Bulletin of Plankton Society of Japan* **38**, 25–31.

Takami, A. and Iwasaki, H. (1978). Cultivation of marine Cladocera, *Penilia avirostris* Dana. *Bulletin of the Japanese Society of Scientific Fisheries* **44**, 393.

Takami, A., Iwasaki, H. and Nagoshi, M. (1978). Studies on the cultivation of marine Cladocera. II. Cultivation of *Penilia avirostris*. *Journal of the Faculty of Fisheries, Prefectural University of Mie* **5**, 47–68.

Templeton, N. S. and Laufer, H. (1983). The effects of a juvenile hormone analog (Altosid ZR-515) on the reproduction and development of *Daphnia magna* (Crustacea: Cladocera). *International Journal of Invertebrate Reproduction* **6**, 99–110.

Threlkeld, S. T. (1979). Estimating cladoceran birth rates: the importance of egg mortality and the egg age distribution. *Limnology and Oceanography* **24**, 601–612.

Thiriot, A. (1968). Les cladocères de Méditerranée occidentale. I. Cycle et répartition des espèces du genre *Evadne* à Banyuls-sur-Mer (Golfe du Lion) 1967. *Vie et Milieu* **19B**, 361–394.

Thiriot, A. (1971). Les cladocères de Méditerranée occidentale. II. Cycle et répartition de *Podon intermedius* et *Penilia avirostris* á Banyuls-sur-Mer (Golfe du Lion) 1967. *Vie et Milieu* **22B**, 75–92.

Thiriot, A. (1972–73). Les cladocères de Méditerranée occidentale: III. Cycle et répartition à Banyuls-sur-Mer (Golfe Du Lion): synthèse des anneés 1965–1969. *Vie et Milieu* **23B**, 243–295.

Thiriot, A. and Vives, F. (1969). *Evadne nordmanni* Lovén en Méditerranée occidentale. *Vie et Milieu* **20B**, 145–157.

Trégouboff, G. (1963). La distribution verticale des Cladocères au large de Villefranche-sur-Mer. *Bulletin de l'Institut Océanographique, Monaco* **61**, 1–12.

Turner, J. T. and Graneli, E. (1992). Zooplankton feeding ecology: grazing during enclosure studies of phytoplankton blooms from the west coast of Sweden. *Journal of Experimental Marine Biology and Ecology* **157**, 19–31.

Turner, J. T., Tester, P. A. and Ferguson, R. L. (1988). The marine cladoceran *Penilia avirostris* and the "microbial loop" of pelagic food webs. *Limnology and Oceanography* **33**, 245–255.

Uye, S. (1982). Seasonal cycles in abundance of major holoplankton in the innermost part of Onagawa Bay, Northeast Japan. *Journal of the Faculty of Applied Biological Science, Hiroshima University* **21**, 1–10.

van den Bosch de Aguilar, P. (1969). Nouvelles données morphologiques et hypothèses sur le rôle du système neurosécréteur chez *Daphnia pulex* (Crustacea: Cladocera). *Annales de la Societé Royale Zoologique de Belgique* **99**, 27–44.

Vijverberg, J. (1989). Culture techniques for studies on the growth, development and reproduction of copepods and cladocerans under laboratory and *in situ* conditions: a review. *Freshwater Biology* **21**, 317–373.

Vuorinen, I. and Ranta, E. (1987). Dynamics of marine meso-zooplankton at Seili, Northern Baltic Sea, in 1967–1975. *Ophelia* **28**, 31–48.

Weismann, A. (1877). Beiträge zur Naturgeschichte der Daphnoiden. III. Die Abhändigkeit der Embryonal-Entwicklung vom Fruchtwasser der Mutter. *Zeitschrift für Wissenschaftliche Zoologie* **28**, 176–211.

Weismann, A. (1880). Beiträge zur Naturgeschichte der Daphnoiden. VI. Samen und Begattung der Daphnoiden. *Zeitschrift für Wissenschaftliche Zoologie* **33**, 55–110.

White, M. J. D. (1973). "Animal Cytology and Evolution", 3rd edn. The University Press, Cambridge, UK.

Wiborg, K. F. (1955). Zooplankton in relation to hydrography in the Norwegian Sea. *Report of the Norwegian Fisheries Investigations* **2**, 1–66.

Wickstead, J. H. (1961). A quantitative and qualitative study of some Indo-West-Pacific plankton. *Fishery Publications, London* **16**, 1–200.

Wingstrand, K. G. (1978). Comparative spermatology of the Crustacea Entomostraca. I. Subclass Branchiopoda. *Kongelige Danske Videnskabernes Selskab Biologiske Skrifter* **22**, 1–66.

Wong, C. K., Chan, A. L. C. and Tang, K. W. (1992). Natural ingestion rates and grazing impact of the marine cladoceran *Penilia avirostris* Dana in Tolo Harbour, Hong Kong. *Journal of Plankton Research* **14**, 1757–1765.

Wood, T. R. and Banta, A. M. (1937). Hatchability of *Daphnia* and *Moina* sexual eggs without drying. *Internationale Revue der Gesamten Hydrobiologie* **35**, 229–242.

Yoo, K. I. and Kim, S. W. (1987). Seasonal distribution of marine cladocerans in Chinhae Bay, Korea. *Journal of the Oceanological Society of Korea* **22**, 80–86.

Young, J. P. W. (1978). Sexual swarms in *Daphnia magna*, a cyclic parthenogen. *Freshwater Biology* **8**, 279–281.

Young, J. P. W. (1983). The population structure of cyclic parthenogens. *In* "Protein Polymorphism: Adaptive and Taxonomic Significance" (G. S. Oxford and D. Rollinson, eds), pp. 361–378. Academic Press, London, UK.

Young, R. T. (1924). "The Life of Devil's Lake, North Dakota." North Dakota Biological Station, Devil's Lake, North Dakota, USA.

Yurista, P. M. (1992). Embryonic and postembryonic development in *Bythotrephes cederstroemi*. *Canadian Journal of Fisheries and Aquatic Sciences* **49**, 1118–1125.

Zacharias, O. (1884). Über die amoeboiden Bewegungen der Spermien von *Polyphemus pediculus*. *Zeitschrift für Wissenschaftliche Zoologie* **41**, 252–258.

Zaffagnini, F. (1964). Il ciclo riproduttivo partenogenetico di *Bythotrephes longimanus* Leydig (Cladocera : Polyphemidae). *Rivista di Idrobiologia* **3**, 97–109.

Zaffagnini, F. (1984). Considerazioni sulla produzione di uova efippiali e sulla determinazione del sesso nei Dafnidi (Crostaci, Cladoceri). *Atti dell'Accademia delle Scienze dell'Istituto di Bologna* (Classe di Scienze Fisiche, Serie XIV) **I**, 205–226.

Zaffagnini, F. (1987). Reproduction in *Daphnia*. *In*: "Daphnia" (R. H. Peters and R. de Bernardi, eds), pp. 245–284. *Memorie dell'Istituto Italiano di Idrobiologia*, Vol. 45. Verbania, Pallanza, Italy.

Zozulya, S. S. and Mordukhai-Boltovskoi, Ph. D. (1977). Seasonal variability of *Bythotrephes longimanus* (Crustacea, Cladocera). *Doklady Akademii Nauk SSSR* **232**, 493–495.

The Role of Microscale Turbulence in the Feeding Ecology of Larval Fish

John F. Dower[1], Thomas J. Miller[2] and William C. Leggett[1]

[1] *Department of Biology, Queen's University, Kingston, Ontario K7L 3N6, Canada*
[2] *Chesapeake Biological Laboratory, Center for Environmental and Estuarine Studies, University of Maryland, Solomons, MD 20688-0038, USA*

ADVANCES IN MARINE BIOLOGY VOL. 31
ISBN 0-12-026131-6

1. INTRODUCTION

For 80 years the primary aim of fisheries oceanography has been to identify and to understand the processes that regulate interannual variability in recruitment to marine fish populations. Following the ground-breaking work of Johan Hjort in the early part of this century, most of the research within this field has been focused on the early life history stages of fish, particularly the egg and larval stages. It was Hjort (1914) who first proposed that year-to-year fluctuations in fish abundance resulted from variation in recruitment of young fish to the population. Prior to this, the commonly held belief had been that fluctuations in fish abundance resulted from interannual variations in the migrations and/or distributions of adult fish (Smith, 1994). Specifically, Hjort (1914) suggested that larval mortality and, ultimately, recruitment were regulated by the amount of food available during a "critical period" when larvae switched from endogenous to exogenous nutrition. Years when food was abundant resulted in strong year classes; years with low food availability resulted in mass starvation and poor recruitment. Hjort's critical period concept was later generalized by Cushing (1972, 1990) in his "match–mismatch hypothesis", which proposes that recruitment strength in a given year depends on the *overall* temporal synchrony between larval production and food availability rather than being confined specifically to the period during the shift to exogenous feeding.

These hypotheses share two features. Both focus on starvation as the primary mechanism of larval mortality, and both infer the primacy of abiotic (i.e. density-independent) factors as regulators of the interannual variation in food availability to larvae and, hence, recruitment. Although Hjort did not identify these abiotic factors explicitly, Cushing (1975) considered them to be a combination of radiance and wind strength. Together, these factors strongly influence the timing of the spring phytoplankton bloom and thus influence the degree of temporal overlap between larval production and spring zooplankton production. There have been many attempts to test both the critical period and match–mismatch hypotheses. Leggett and DeBlois (1994) reviewed the work to date and concluded that the available evidence does not support Hjort's "critical period" as an important general contributor to recruitment variation in marine fish populations. However, they found that much of the evidence available was *broadly* consistent with Cushing's "match–mismatch" prediction of a positive relationship between the overlap of the seasonal production/larval abundance cycles and larval survival. They note, however, that the precision of this coupling may be less important than previously assumed, and that other factors also influence the survival and recruitment of larvae.

One such factor is predation, which interacts with larval growth rate and stage duration to regulate mortality during the larval period (Bailey and Houde, 1989). Leggett and DeBlois (1994) noted that food-mediated changes in larval growth, condition and performance are likely to alter both the frequency and the intensity of predation on post-yolk-sac stages of most, if not all, marine fish larvae. Two current hypotheses bear on this link between food availability and the potential impact of predation on recruitment. Importantly, both hypotheses are based on the implicit assumption that food availability is an important regulator of larval growth rates, size at age and stage durations. The "bigger is better" hypothesis holds that larvae that are larger at any given age or developmental stage will be less susceptible to predation, and therefore should contribute preferentially to cohort survival (see Litvak and Leggett, 1992, for a discussion of the evolution of this hypothesis). Hence, favorable feeding conditions, which lead to size differences within and between cohorts, could directly influence the intensity and impact of predation. The "stage duration" hypothesis holds that larvae which grow fast will metamorphose to the juvenile stage sooner. Since mortality rates decline exponentially throughout the larval and juvenile stages (Houde, 1987), reduced stage durations within or between cohorts should translate into higher survival and enhanced recruitment. Chambers and Leggett (1987) estimated that the observed variability in age-at-metamorphosis in winter flounder (*Pseudopleuronectes americanus*) could result in up to 100-fold changes in the number of larvae surviving to the juvenile stage.

The concepts underlying the critical period, match–mismatch, bigger is better and stage duration hypotheses, have greatly influenced the development of research in fisheries oceanography (Leggett and DeBlois, 1994). In particular, much research has sought to document the fact that variations in the quantity/quality of food available to larvae are important determinants of recruitment, and that abiotic processes influence this availability. Much of the field and laboratory research stimulated by these hypotheses has *assumed* that food abundance *per se* directly influences feeding success and, hence, survival and recruitment. Consequently, these studies have commonly attempted to link variation in food abundance in areas occupied by larvae to growth and survival, and to assess the abiotic forces (e.g. aggregative and advective) that regulate food abundance. Conspicuously absent from these analyses, however, is a consideration of the physical factors that influence the dynamics of feeding interactions between larvae and their prey (Gerritsen and Strickler, 1977).

The broad and frequently uncritical acceptance of a direct causal link between food abundance, feeding success and survival has been sustained partly by the frequent observation that the minimum prey densities required to support larval growth and survival in the laboratory are commonly much higher than prey densities in the sea (reviewed by Leggett, 1986). MacKenzie *et al.* (1990) reviewed the available laboratory and field data on the relationship between prey abundance and larval feeding rates. They found that *in situ* ingestion rates were consistently higher at any given prey density than would be predicted from laboratory data and that, in fact, *in situ* rates were commonly at (or near) satiation and appeared to be largely independent of prey density.

Rothschild and Osborn (1988) had previously suggested that a pattern such as that described by MacKenzie *et al.* (1990) may result from problems of bias and failure to include key concepts. They suggested that the paradox revealed by MacKenzie *et al.* could be partially resolved by removing the bias that arises from considering ingestion as a simple function of the relative densities of larvae and their prey when, instead, it must also be a function of their relative motions. Rothschild and Osborn argued that, at the low prey concentrations typical of aquatic systems, planktonic predators must exhibit motion relative to their prey to ensure encounters. Thus, if laboratory and field systems differ in the relative motion induced between predator and prey as a consequence of purely physical forces in the system, one should expect to observe different ingestion rates, even if prey concentrations are equal. Rothschild and Osborn (1988) were the first to advance the idea that turbulence, a ubiquitous feature of natural aquatic systems, can generate relative motion between predators and prey in addition to the relative motion that results from their respective swimming. Moreover, Rothschild and Osborn noted that the absence of microscale turbulence in laboratory experiments had been largely ignored. This absence would lead to reduced rates of encounter relative to field systems even if prey densities were similar.

Together, the findings of Rothschild and Osborn (1988) and MacKenzie *et al.* (1990) have sparked a growing interest in the potential role of microscale turbulence as a regulator of feeding success and survival in zooplankton and larval fishes. In particular, the ideas that the *effective* prey concentrations in the sea are higher than indicated by measures of absolute abundance, and that microscale turbulence represents a possible mechanism to explain this discrepancy, have led to a fundamental change in the spatiotemporal scale (large $\rightarrow$ small) and approach (correlational $\rightarrow$ mechanistic) of recent investigations into the availability of food to larval fishes, and the influence of food on the dynamics of these populations.

In this paper, we review the development and application of what we term the "turbulence theory" specifically as it relates to the feeding ecology of larval fish. We begin with a brief review of the physics of oceanic turbulence, and then trace the development of the original Rothschild and Osborn (1988) model. Next, we consider more recent attempts to model the effects of microscale turbulence on encounter rates. We also review both empirical and field studies designed to test various elements of the turbulence theory. Finally, we identify unresolved questions and suggest some specific lines of investigation designed to further enhance our understanding of small-scale physical–biological coupling. Throughout the paper we reference parallel research on the importance of microscale turbulence to the feeding ecology of zooplankton and protozoans.

2. TURBULENCE IN THE OCEAN

Mixing processes in the ocean are responsible for the transfer of kinetic energy from the largest to the smallest scales. The spatial scale of these processes spans ten orders of magnitude from ocean basin scales of $\times 10^7$ m down to the viscous scales of $\times 10^{-3}$ m at which energy is dissipated as heat by molecular viscosity (Denman and Gargett, 1995). However, our focus on the interactions between individual larval fish and their prey restricts our interest primarily to millimetre-scale turbulent motions.

Turbulence is ubiquitous in the world's oceans (Yamazaki and Osborn, 1988; Denman and Gargett, 1995). Furthermore, the spatial scales over which turbulence occurs (i.e. kilometres $\rightarrow$ millimetres) suggest it is the dominant influence of water movement on the local environment and biology of aquatic organisms (Mann and Lazier, 1991). Turbulent motions are produced when local buoyancy and/or shear forces generate instabilities in local pressure and density fields (Price et al., 1987). In the surface mixed layer of the open ocean common forcing mechanisms that generate turbulence include wind mixing, convective heat loss (leading to buoyancy instabilities), current shear across density interfaces and the breaking of surface waves. In shallow coastal waters, bottom friction generated by tidal flow and other currents is also a major source of turbulence. Once created, turbulence extracts energy from larger scales and transfers it to ever smaller scales (Landahl and Mollo-Christensen, 1986; Kundu, 1990). The rate at which this turbulent kinetic energy (TKE) is removed is referred to as the "turbulent dissipation rate" and is denoted by ε, with units W kg^{-4}. At large scales (tens of kilometres $\rightarrow$ metres), eddies frequently exhibit distinct orientations and direction, and are responsible for the majority of turbulent transport. At submetre scales, however, turbulence becomes homogeneous and isotropic (Townsend, 1976). One further characteristic of small-scale turbulence is that it is governed entirely by ε and the kinematic viscosity of water. Hence, microscale turbulence is independent of the generating source.

One of the difficulties in incorporating the effects of turbulence into studies of plankton feeding ecology has been, and continues to be, the fact that direct measurements of ε are difficult to make in the field (Gargett, 1989). Until recently, direct measurements were few in number and required specialized instrumentation available to only a very few physical oceanographers. However, the advent of techniques employing high-resolution conductivity–temperature–depth probes (CTDs) to detect overturning events and for calculating associated vertical scales and potential energy (Galbraith and Kelley, 1996) suggest that field measurements of ε may become more common in the near future. In the interim, for situations in which wind forcing is the main source of turbulence, the empirical relationship between wind speed and turbulent dissipation demonstrated by Oakey and Elliott (1982) remains a useful predictor of ε. These authors show that ε is roughly proportional to the cube of the wind speed:

$$\varepsilon = (5.82 \times 10^{-9})W^3/z \tag{1}$$

and can be approximated for any depth in the mixed layer. Using a data set comprised of about 800 literature-derived TKE profiles, MacKenzie and Leggett (1993) evaluated the explanatory power of this relationship and found that the simple model where W is wind speed in $\mathrm{m\,s^{-1}}$ and z is depth in metres accounted for 58% of the variation in TKE as a function of wind speed and depth, and that 68% of the observations fell within a factor of 5 of model predictions. MacKenzie and Leggett (1993) also showed that the model produced estimates of ε accurate to within a factor of 5 even in situations where wind mixing was not the sole source of turbulence as, for example, in upwelling zones, near surface wave-breaking zones and in shelf waters characterized by high current shear.

The cut-off point for turbulent motions is generally taken to be that length scale at which molecular viscosity damps out any remaining TKE faster than it can be supplied by turbulence at larger scales (Denman and Gargett, 1995). This scale is known as the Kolmogorov scale, η, and can be defined as $\eta = (v^3/\varepsilon)^{1/4}$ (Lazier and Mann, 1989), where v is the kinematic viscosity of the fluid $(\mathrm{m^2\,s^{-1}})$ and ε is the dissipation rate $(\mathrm{W\,kg^{-4}})$. The Kolmogorov scale is also proportional to the length scale of the smallest sustainable turbulent eddies; Lazier and Mann (1989) suggest that such eddies are typically an order of magnitude larger than η. Therefore, taking a typical value for v as 10^{-6} $(\mathrm{m^2\,s^{-1}})$ (Pond and Pickard, 1983) and values of ε between 10^{-6} and $10^{-9}\,\mathrm{W\,kg^{-1}}$ (Oakey and Elliott, 1980; Lazier and Mann, 1989; MacKenzie and Leggett, 1993) yields typical values of η in the range of 1–5 mm. Consequently, the smallest turbulent eddy diametres are likely to be in the order of 1–5 cm. For reference, wind speeds between 0.5 and 5.0 m s^{-1} ($\sim$ 1–10 knots) generate values of ε similar to those used above.

Biologically, the Kolmogorov scale also represents a boundary because movements of organisms of size $\gg \eta$ (e.g. juvenile/adult fish and larger invertebrate zooplankton) are unlikely to be affected by motions at the scale of η, whereas the movements of organisms with length scales $\approx \eta$ (notably crustacean zooplankton and larval fish) are likely to be *directly* affected by such small-scale motions. Consequently, it was the latter group of organisms that Rothschild and Osborn (1988) proposed were most likely to benefit from increased encounter rates resulting from microscale turbulence. Interestingly, while it had long been assumed that organisms of size $\ll \eta$ (e.g. protozoans, bacteria and phytoplankton) were un-affected by turbulent motion, as they inhabit a hydrodynamic regime in which viscous forces dominate over inertial forces, recent theoretical and laboratory work has suggested that turbulent effects may also be manifested at sub-Kolmogorov scales (Hill *et al.*, 1992).

Under classical super-Kolmogorov conditions an estimate of ε can be used to calculate ω (ms^{-1}), the turbulent velocity in the water column, as:

$$\omega^2 = 3.6(\varepsilon r)^{2/3} \qquad (2)$$

where r is a characteristic length scale (m). Hill *et al.* (1992) have shown that Equation 2 is valid considerably below η. However, there has been much debate in the literature regarding the correct definition of r. Most often r has been defined as the average prey separation distance (Sundby and Fossum, 1990; MacKenzie and Leggett, 1991; MacKenzie *et al.*, 1994; Sundby *et al.*, 1994). In contrast, Muelbert *et al.* (1994) suggested that r may be defined as the Kolmogorov scale, whereas Davis *et al.* (1991) assumed it to equal the eddy separation distance. Most recently, several authors have suggested that r should be defined as the larval fish reactive distance (Evans, 1989; Mackenzie *et al.*, 1994; Denman and Gargett, 1995; Kiørboe and MacKenzie, 1995; Kiørboe and Saiz, 1995). We will consider the consequences of using these different definitions of r in Section 4.

3. ENCOUNTER RATES AND FEEDING IN LARVAL FISH

3.1. The Encounter Process

Key to establishing the role of turbulence in the feeding ecology of larval fish is a clear understanding of the actual encounter and feeding processes. To begin, we emphasize that *encountering* prey and *ingesting* prey are separate events, and that increased encounter rates do not necessarily result in increased ingestion rates. Simply stated, a predator does not necessarily ingest every prey that it encounters.

For larval fish, the probability of successfully ingesting prey is actually the multiplicative probability of *at least* three separate events: prey encounter, prey attack and prey capture (Holling, 1959). We can write this relationship as:

$$p(Ingestion) = p(Encounter) \cdot p(Attack) \cdot p(Capture) \qquad (3)$$

Furthermore, each of these stages is influenced by many physical and behavioral processes. Encounter is the most likely to be *directly* affected by microscale turbulence, however, because it occurs on the same spatial scale. We therefore focus on encounter in this section. The latter two terms in the above equation will be considered in the next section as they relate to attempts to model the effect of turbulence on larval feeding ecology. Note that we use "ingestion" to refer to the successful placement of food in the mouth, and use the more general term "feeding" to refer to the entire suite of larval behaviors involved in obtaining food.

The basic components determining the encounter rate between predator and prey are the absolute density of prey and the relative velocity difference between predator and prey. Relative velocity is important because two particles with exactly the same velocity can never encounter each other. The simplest means of expressing this relationship is as:

$$Z = DA \qquad (4)$$

where Z is the encounter rate between predator and prey, D is the number of prey per unit length scanned by the predator and A is the relative velocity at which the predator is moving.

3.2. The Gerritsen–Strickler Model

Gerritsen and Strickler (1977) first attempted to adapt this basic formulation to accurately represent encounter rates between planktonic predators and prey. Indeed, it was this model that Rothschild and Osborn (1988) later elaborated upon when considering the effects of microscale turbulence. In their model, Gerritsen and Strickler rewrote the density and velocity terms from Equation 4 as:

$$D = \pi R^2 N \qquad (5)$$

and

$$A = \frac{(u^2 + 3v^2)}{3v} \qquad (6)$$

where D is still the number of prey per unit length, R is the predator's contact radius in mm (i.e. the maximum distance at which the predator

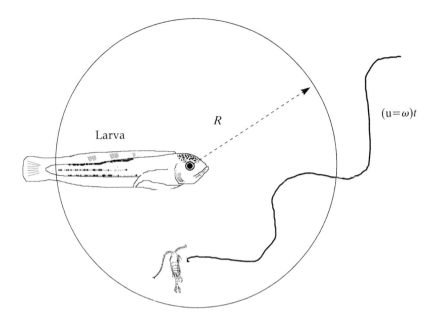

R

Larva

$(u=\omega)t$

Figure 1 Schematic of the factors necessary for encounter. A larval fish is shown frozen in time. The larvae can see a distance R, which encompasses the area shown in this figure as a circle (note that this representation is highly simplified, as the actual search geometry is 3-dimensional and is likely not to be spherical). A prey organism is initially outside the reactive volume. However, during some period of time, t, it is moved (either under its own locomotion, by turbulence or by a combination of the two) so that it enters the reactive volume, hence leading to an encounter.

can perceive prey), N is the prey concentration, and u and v are the velocities of the prey and predator, respectively (m s^{-1}). Note that this formulation assumes that predator velocity, v, is greater than prey velocity, u, as is generally the case for larval fish pursuing prey. For the less common situation in which predator velocity is $\leq$ prey velocity (as is the case for many predatory gelatinous zooplankton), the u and v of Equation 5 are simply swapped. The Gerritsen and Strickler (1977) formulation assumes that there is a random distribution and random movement of predators and prey, and that R is unaffected by either predator or prey swimming speed. It also excludes behavioral flexibility of either predator or prey (we reconsider these assumptions in Section 4).

Inspection of these equations reveals that a predator can increase its encounter rate with prey either by increasing its swimming velocity, v, or by increasing the contact radius, R (see also Figure 1). Gerritsen and Strickler (1977) point out that, of these two, the contact radius of the predator has the greatest effect on encounter rate as it enters as a squared term. This has important implications since our understanding of the perceptive abilities of larval fish is changing rapidly. All of the terms in Equations 5 and 6 exhibit strong size-dependencies (Miller *et al.*, 1988) and we will explore the implications of this later.

The original Gerritsen and Strickler (1977) model also explored the impact of predator strategy on encounter rates. Most planktonic predators can be broadly categorized as either cruising predators or ambush predators (Greene, 1985). Planktonic cruise predators swim almost constantly and capture their prey as they swim. Ambush predators, although they do display active capture behaviors, remain motionless for much of the time and only search for prey while stationary. Amongst the invertebrate zooplankton, various herbivorous copepods, many scyphozoan medusae, and ctenophores such as *Mnemiopsis maccradyi* (Gerritsen and Strickler, 1977; Larson, 1988) are considered cruise predators. Ambush predators include chaetognaths (Greene, 1985) and many carnivorous copepods (Landry *et al.*, 1985; Jonsson and Tiselius, 1990; Tiselius and Jonsson, 1990). Many larval fish (e.g. herring, *Clupea harengus*, and northern anchovy, *Engraulis mordax*) fall into the category of cruise predators (Rosenthal and Hempel, 1970; MacKenzie and Kiørboe, 1995) although some, such as the "pause–travel predators" white crappie (*Pomoxis annularis*) and cod (*Gadus morhua*) can reasonably be considered as ambush predators (Browman and O'Brien, 1992a; MacKenzie and Kiørboe, 1995).

Based largely on bioenergetics arguments, Gerritsen and Strickler (1977) suggested that ambush predation is the most efficient strategy for capturing fast-moving prey, while cruising is optimal for predation on slow-moving prey. The basis for this assertion is that increased predator swimming velocity necessitates an increased energy cost to overcome the stronger hydrodynamic drag. To benefit from an increased cruising speed, any increase in encounter rate experienced by a predator must be sufficient to offset the increased energetic cost of faster swimming. Gerritsen and Strickler found that the energetic benefit of increased encounter rate outweighed the associated increased energetic cost only when prey were moving very slowly. However, they also suggested that as the swimming efficiency (defined as a combination of hydrodynamic and propulsion efficiencies) of a predator increases, so too does its optimal cruising velocity and, hence, its encounter rate with prey. Thus, increased swimming efficiency permits a predator to cruise effectively at higher velocities. One result of this analysis was that Gerritsen and Strickler (1977) predicted that larval fish, which are more efficient swimmers than most crustacean zooplankton, should have higher encounter rates than zooplankton.

3.3. The Rothschild–Osborn Model

As shown in Section 2, the oceanic environment is highly turbulent (Gargett, 1989). At the spatial scales applicable to individual predator–prey encounters in the plankton (millimetres→centimetres) we can consider the distribution of turbulent energy to be isotropic and homogeneous (Denman and Gargett, 1995). Rothschild and Osborn (1988) suggested that small-scale turbulent motions could potentially increase predator–prey contact rates. In essence, they proposed that the relative velocity term of the original Gerritsen and Stricker (1977) model be revised to include a contribution to the relative motion of predator and prey made by microscale turbulence. They accomplished this by rewriting Equation 6 as:

$$A = \frac{(u^2 + 3v^2 + 4w^4)}{\sqrt{[3(v^2 + w^2)]}} \tag{7}$$

In this formulation u and v remain the swimming velocities (m s^{-1}) of the prey and predator, respectively, and w is the additional velocity (m s^{-1}) contributed by turbulence. More specifically, Rothschild and Osborn (1988) define w as "the root-mean-square turbulent velocity . . . that is uncorrelated with the turbulent velocity fluctuation at a point a distance r away". Thus, w decreases with decreasing r since, as two particles approach each other, the relative velocity difference between them necessarily goes to zero as the separation between the points goes to zero. Consequently, w is likely to be small over length scales appropriate to plankton encounter rates. Nonetheless, Rothschild and Osborn demonstrated mathematically that, given the formulation above in which w is raised to the fourth power, even modest values of w should result in appreciable increases in encounter rate. Moreover, as shown in Figure 2, the largest increases in encounter rates are expected to occur when the slowest-moving organisms are subjected to the highest levels of turbulence. In fact, for very small, slow-moving predators Rothschild and Osborn found that the turbulence contribution to the encounter rate was likely to be larger than the swimming component. Based on these findings, and the fact that contact rates experienced by a "population" of predators in the sea are unlikely to be the same as the contact rates experienced by predators of the same "population" in the laboratory (where turbulence levels are much reduced), Rothschild and Osborn (1988) recommended that the feeding requirements of plankton be reviewed. Subsequently, Evans (1989) suggested a revision of the Rothschild and Osborn (1988) formulation in order to prevent a breach of Gerritsen and Strickler's (1977)

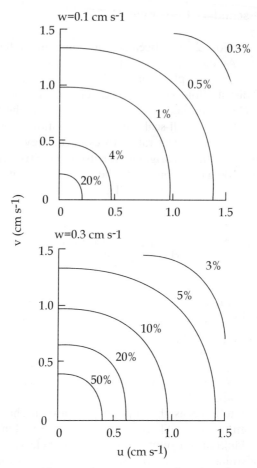

Figure 2 Contours of increased encounter as a function of u and v as predicted by Rothschild and Osborn (1988) (their Figures 2b and c). Contours for two different turbulent velocities, 0.1 and $0.3\,\text{cm}\,\text{s}^{-1}$, are shown. The increased encounter rates are expressed in terms of the non-turbulent rate and are thus independent of N, the prey density and R, the reactive distance.

original assumption of a uniform predator speed distribution. Evans's formulation is:

$$A = \frac{\surd[(u^2 + v^2 + 2w^2)]}{\surd[3(v^2 + w^2)]} \tag{8}$$

Evans reported that the two formulations differ by only 6% when $u = v$.

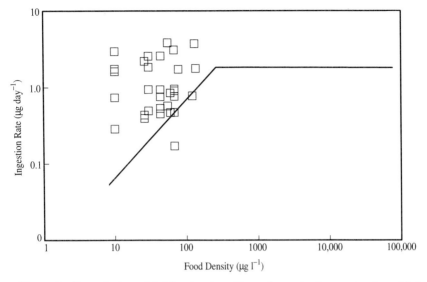

Figure 3 Compilation of field estimates of ingestion rates for marine larval fish as a function of prey density. The solid line represents the expected response of ingestion rate to prey density predicted from a compilation of laboratory studies. (Redrawn after MacKenzie *et al.*, 1990.)

3.4. Discrepancies between Laboratory and Field Results

Almost coincident with Rothschild and Osborn's proposal that microscale turbulence increased encounter rates between plankton, MacKenzie *et al.* (1990) reviewed the available data on the ingestion rates of larval fish and did indeed find a major discrepancy between laboratory and field studies. In general, they showed ingestion rates reported from laboratory experiments to be significantly lower than those reported in most field studies. Furthermore, field-derived ingestion rate data from larvae of eight marine fish species indicated that all were feeding at or near maximal rates (MacKenzie *et al.*, 1990), apparently independent of food concentration (Figure 3). Previously, Leggett (1986) had shown that the majority of laboratory estimates of the minimum prey densities required for growth and survival of larval fish also tended to be much higher than food concentrations commonly encountered in the field. This discrepancy had typically been resolved by invoking the probable reliance of larvae on locating patches of high food concentration in the wild (Cushing, 1972; Lasker, 1975). Subsequently, better-controlled laboratory experiments confirmed that larvae could indeed survive on the very low food concentrations commonly encountered in the field (Houde, 1978; Øiestad,

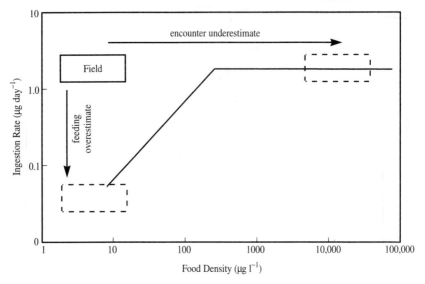

Figure 4 Possible explanations of the anomaly between laboratory and field derived estimates of ingestion rate in marine larval fish. (After MacKenzie *et al.*, 1990.)

1985), thus calling into question the notion that survival of larval fish is strongly tied to co-occurrence with prey patches (Leggett, 1986). Based on their comparison of field and laboratory data, MacKenzie *et al.* (1990) proposed two possible explanations for the higher larval ingestion rates observed in the field. The first was that field measurements might overestimate larval feeding rates (Figure 4). MacKenzie *et al.* (1990) suggested that this could occur if, for example, field sampling was biased towards the collection of the most effective foragers (the least effective presumably having either starved or been eaten), thereby artificially inflating the mean ingestion rate. Alternatively, they proposed that the higher ingestion rates observed in the field could result from underestimation of actual encounter rates experienced by larvae in the ocean. Likely candidates for factors which could increase encounter rates were micro-scale turbulence and the aforementioned small-scale prey patchiness (MacKenzie *et al.*, 1990).

4. MODELING TURBULENCE-DEPENDENT FEEDING RATES

To translate the impact of turbulence on encounter we must now incorporate the effects of turbulence on both the entire predation cycle and the distribution of prey. For example, if turbulence acts to increase encounter but simultaneously decreases rates of attack and pursuit success, the overall effect of turbulence on feeding may be negative. Consequently, the effects of turbulence on overall feeding rates cannot be predicted from a consideration of its effects on encounter alone. All of the components of Equation 3 must be included when modeling the impact of turbulence on overall larval feeding rates.

4.1. Components of the Feeding Cycle

As previously discussed, feeding can be broken down into a series of contingent events including encounter, attack and capture. Two features of this cycle are important. Failure at any one stage guarantees failure overall and, hence, no feeding. In contrast, success at any one stage does not guarantee successful feeding, as the overall outcome can be modified by events that follow. Furthermore, it should similarly be clear that demonstrating a link between a single event of the predation cycle and turbulence does not guarantee that a similar link will exist between turbulence and the overall feeding rate. This latter point lies at the heart of the analyses by Jenkinson and Wyatt (1992) and MacKenzie et al. (1994) of the impact of turbulence on overall feeding rates. Although the two papers utilize different analytical techniques, the conclusions reached (i.e. that turbulence-induced increases in encounter may not necessarily be translated into increases in ingestion) are similar.

Jenkinson and Wyatt (1992) define a dimensionless parametre, the Deborah number, to determine the residence times of prey within a predator's attack field. The Deborah number, De, may be defined as the ratio of the characteristic duration of the predation cycle to the deformation rate of the parcel of water in which the predator and prey are contained. When $De \ll 1$, the predator and prey are effectively contained within the same parcel of water for the entire predation cycle. When $De \gg 1$, shear deforms the parcel so rapidly that the prey is advected out of the attack field before the attack can be completed. Jenkinson and Wyatt were able to define a maximum value of De that permitted potential prey capture. Values of De beyond the maximum level required chemosensory or visual navigation for the larva to follow and eventually capture the prey. Importantly, Jenkinson and Wyatt (1992) concluded that the precise nature of the search behavior of the larva is an important determinant of success.

MacKenzie et al. (1994) took a different approach to analyzing the potential of post-encounter processes to influence overall feeding success. They assumed that turbulence acted only on the prey, and that larvae required a minimum time, t, in which to execute all of the behaviors necessary to pursue and capture prey. During this period of time, turbulence has the potential to move the prey anywhere in a sphere of radius ωt, termed the excursion sphere (where ω is the turbulent velocity). In circumstances where the final position of the prey within its excursion sphere lies outside the encounter volume of the larva, the larva "loses sight" of the prey and no feeding occurs. The probability of successful pursuit is therefore the overlap between the prey excursion sphere and the larva's encounter sphere. The degree of overlap between the two, and hence the probability of successful pursuit, is a non-linear function of ω for fixed values of t and R (the larval reactive distance). Finally, by combining the Rothschild and Osborn (1988) model of encounter with their model of successful pursuit, MacKenzie *et al.* (1994) were able to show that overall feeding rates were characterized by a domed function of turbulent velocity (Figure 5).

Recently, Jenkinson (1995) has compared these two approaches. The principal difference is that Jenkinson and Wyatt (1992) modeled turbulent velocity as a linear shear close to the fish larva, whereas MacKenzie *et al.* (1994) considered the turbulence field to be stochastic. This difference in approach reflects different initial viewpoints. Jenkinson and Wyatt's model starts at sub-Kolmogorov processes and scales upwards; MacKenzie *et al.* start at super-Kolmogorov scales and scale downwards. However, Jenkinson (1995) concluded that the two approaches agree well at the interface between their respective domains. Both models predict that (although increasing turbulence leads to increasing rates of encounter) not all encounters can be translated into captures. Overall, these results suggest that while initial increases in turbulence are beneficial, subsequent increases may lead to a *decrease* in feeding success. The implications of these findings for field conditions are that larvae should select (or be spawned in) areas characterized by intermediate levels of turbulence if they are to maximize ingestion rates. MacKenzie *et al.* (1994) also suggested that because the minimum pursuit time was a sensitive parametre in the model, selection for prey for which ts is shorter may occur under turbulent conditions.

4.2. Predator Search Behavior

Initial behavioral studies of larval fish feeding described the process as one in which larvae continually scan the environment for food as they move through it. This allowed Rosenthal and Hempel (1970) to model larval herring feeding by considering the number of prey contained within a cylindrical volume of water defined by the reactive distance of the larva

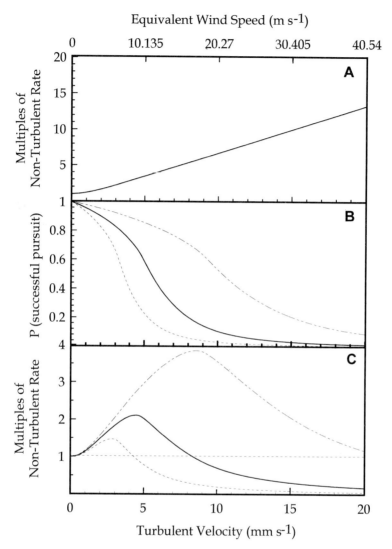

Figure 5 The influence of turbulence on: (A) encounter; (B) probability of successful pursuit; (C) relative ingestion rate for larval cod subject to turbulence. Parameter estimates given in MacKenzie *et al.* (1994). (From MacKenzie *et al.*, 1994.)

and its swimming speed (Figure 6). Subsequent modifications of the general framework recognized that predators attacked only prey in the upper half of the volume (Hunter, 1982). Indeed, for larvae of several species of fish, this appears to be an accurate description of their feeding

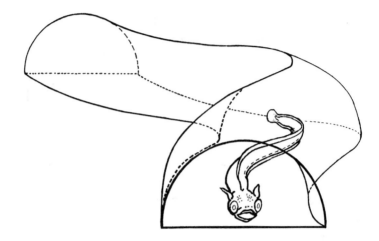

Figure 6   Model of a cruise-searching larval predator. Only those items which
stay in the tube are available. The fish is considered to be sweeping out a search
volume of radius *R* and length *vt*. (After Rosenthal and Hempel, 1970.)

behavior. Examples of such species include herring (Rosenthal and
Hempel 1970), northern anchovy *Engraulis mordax* (Hunter, 1972), red
drum *Sciaenops ocellatus* (Fuiman and Ottey, 1993) and clown fish
Amphiprion perideraion (Coughlin *et al.*, 1992). However, a series of
papers reporting the results of behavioral observations, which mapped the
location of prey in the larval encounter volume when they were attacked,
led O'Brien and co-workers to challenge the primacy of the cruise model
(Evans and O'Brien, 1988; O'Brien *et al.*, 1986, 1989, 1990; Browman and
O'Brien, 1992a, b). These studies suggested that prey were not always at
the periphery of the encounter volume when first attacked as predicted
by the cruising strategy. Rather, O'Brien and colleagues found that many
prey were well within the encounter volume when attacked. This led them
to propose a new model of foraging in fishes which they termed "saltatory
searching" (O'Brien *et al.*, 1990). In this model, larvae search only when
stationary in the water column. If prey is located during this pause, it will
be attacked. However, if no prey are located, the larva moves on, stops
and recommences scanning the environment. This predation strategy is
also known from other taxa where it is often termed "pause–travel
searching". Examples of species known to adopt this behavior as larvae
are Atlantic cod (MacKenzie and Kiørboe, 1995), white crappie (Browman
and O'Brien, 1992a), golden shiner *Notemigonus crysoleucas* (Browman
and O'Brien, 1992b) and striped bass *Morone saxatilis* (Chesney, personal
communication).

The two searching modes have important consequences for models of larval feeding rates. Most obviously, by assuming that larvae are continually searching the environment as they move, encounter rates derived from models of cruise strategies will overestimate encounter rates for pause–travel predators (Browman and O'Brien, 1992a; MacKenzie and Kiørboe, 1995). Furthermore, a consideration of optimal movement patterns in pause–travel predators suggests that a much more directed search is likely in these animals than in cruising predators.

The impact of turbulence on these alternative search strategies has recently been examined experimentally and theoretically (MacKenzie and Kiørboe, 1995; Werner *et al.*, 1995, 1996). In a series of detailed observational studies in the laboratory, MacKenzie and Kiørboe compared the responses to microscale turbulence of a known pause–travel predator (cod) and a known cruise predator (herring). Larvae of both species increased their rates of attack under turbulent conditions, although the increase was greater for cod. Moreover, for both species the enhancement due to turbulence was greatest at the lowest prey concentrations. However, the two species differed in the amount of time spent swimming. At higher turbulent levels cod increased the amount of time spent swimming, while herring larvae exhibited decreased activity. This difference resulted from the fact that, in cod, pause durations decreased under turbulent conditions. However, interpause travel distances remain constant and hence the proportion of the total time-budget spent in swimming increases. In herring, larvae swim shorter distances between encounters under turbulent conditions, and hence total time spent swimming decreases.

It is clear from the above considerations that the application of an encounter model based upon cruising strategies is inadequate to model encounter rates of a pause–travel predator. To overcome this problem, MacKenzie and Kiørboe (1995) proposed a modification of the Gerritsen and Strickler (1977) encounter model suitable for pause–travel predators: they assume that the larva searches a semi-spherical volume (of radius equal to the reactive distance) centered on its snout. However, they include an additional term for prey moved into the semi-sphere during the time that the larvae pauses. This second term is modeled by calculating the number of prey contained within a cylindrical volume of which: (1) the length is given by the pause duration and (2) the radius is a function of the prey movement as influenced by both the prey swimming speed and the turbulent velocity. However, this approach fails to capture fully the geometry of the larval prey interactions and can therefore only approximately represent the encounter process.

Although the inclusion of predator search behavior into the conceptual framework has undoubtedly improved our understanding of the role of turbulence-induced patterns in ingestion, our understanding is still inadequate. For example, although behavioral observations demonstrate that larval encounter volumes are often irregular, all models to date have assumed a regular sphere or semi-sphere. Detailed studies on the visual field of larval fish have suggested that this assumption is a gross simplification (Browman *et al.*, 1990): evidence shows that larval reactive volumes are likely to be more restrictive and are better characterized as pie-shaped wedges. Moreover, all of the approaches we have thus far discussed assume independence between encounters. Specifically, they assume that the behavior of the larva does not depend upon its success in previous encounters. Yet empirical evidence suggests that larvae often exhibit area-restricted searching (Vlymen, 1977; Hunter and Kimbrell, 1980). In this behavior rates of turning are increased following successful encounters thus serving to restrict the larvae to areas in which prey have previously been found. In contrast, when no prey are encountered rates of turning decline, leading the larva to search fresh areas of the water column. Incorporation of such behaviors is precluded from existing conceptual models. The assumptions used to date would be adequate if prey were randomly, or uniformly, distributed. It is unlikely that either of these assumptions is valid (Mackas *et al.*, 1985; Davis *et al.*, 1991). Currently, both the spatiotemporal structure of turbulent fields and the response of organisms to this structure are incompletely known. However, the interaction of these two processes likely impacts both the prey concentration experienced and the search behavior exhibited by larval fish.

4.3. Predictive Models of Feeding under Turbulent Conditions

In one of the first attempts to consider the effects of turbulence on feeding rates in the field, MacKenzie and Leggett (1991) modeled the impacts of wind-induced and tidally induced turbulence on ingestion rates of first-feeding cod larvae. Estimates of turbulent velocity, ω, were derived empirically (see Section 2) and then used in the Rothschild and Osborn (1988) model to predict contact rates between larval cod and their copepod prey. MacKenzie and Leggett concluded that failure to include the influence of small-scale turbulence could cause contact rates to be underestimated by up to 11-fold in tidal regions, and by more than two-fold in wind-mixed areas.

However, implicit in these conclusions are several critical assumptions. As already discussed, the fact that encounter rates increase does not necessarily imply that ingestion rates also increase. In fact, MacKenzie and Leggett (1991) noted that, although encounter rates would be highest at the sea surface for larvae in an environment experiencing wind-generated turbulence, feeding success might actually be *lower* due to reduced pursuit success (and consequently, reduced ingestion rate) experienced by larvae under highly turbulent conditions. Indeed, further consideration of this problem has recently led to the conclusion that, under realistic parametre estimates, peak ingestion rates should occur at depth (MacKenzie *et al.*, 1994).

Additionally, MacKenzie and Leggett's (1991) model failed to incorporate the response of prey to the turbulent field. In their original analyses, prey were assumed to be either uniformly mixed or concentrated at a hypothetical thermocline. In the latter case, even though the concentration varied with depth, this variation was not induced by the turbulence but rather it was externally imposed by the model framework.

Other researchers have sought to address these assumptions in their models. Using a Lagrangian approach to model turbulence-dependent ingestion, Yamazaki *et al.* (1991) tracked about 4000 model particles in a model volume in both turbulent and non-turbulent flow fields. Particles were separated evenly into eight groups by altering the parametres that defined each group's random walk. Each of the parametre values was chosen to simulate a different feature of plankton behavior. Calculated contact rates between each pairwise combination of groups were used to investigate the effects of both predator and prey behavior. They concluded that turbulence always increases the contact rate, regardless of predator or prey behavior. Furthermore, Yamazaki *et al.* (1991) suggested that predators that maintain constant velocities for longer periods of time show higher contact rates than those that turn frequently, and that greater reactive distances also increase contact rates. Finally, Yamazaki *et al.* (1991) compared predicted contact rates from the Gerritsen and Strickler (1977) model with the results of their Lagrangian simulation and coagulation theory models. These comparisons suggested that all three approaches produced qualitatively similar predictions of contact rate. Thus, the three approaches may be interchangeable.

Werner *et al.* (1995, 1996) included dynamic responses of larvae and prey to turbulence, and different larval searching strategies in a series of models designed to investigate growth and survival of larval cod and haddock (*Melanogrammus aeglefinus*) on George's Bank. In the George's Bank system, cod and haddock spawn on the northeast margin of the bank in late spring. Empirical evidence suggests that larvae are retained within a clockwise circulation (Lough and Bolz, 1989). Other evidence suggests that tidally generated turbulence dominates the system. Previous deterministic models had failed to adequately reproduce the observed growth pattern of both species on the bank (Werner *et al.*, 1993). Recent efforts addressed this failure by including individual variability among larvae, and

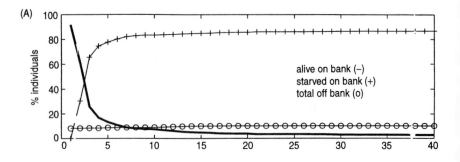

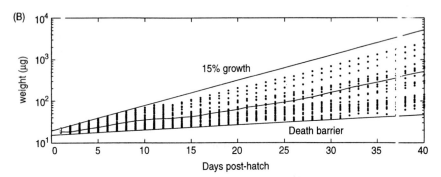

Figure 7 Predicted post-hatch history of cod larvae in simulations in which turbulence was included. (A) The percentage of larvae alive (solid line), starved on-bank (solid crossed line) and advected off the bank (solid circled line). (B) The daily size distribution of survivors. Individual points represent the weight of individual survivors. (From Werner *et al.*, 1996.)

the potential impact of small-scale turbulence on ingestion rates (Werner *et al.*, 1995, 1996). Werner and colleagues used an individual-based bioenergetics model to map captures onto growth and survival. In model runs without turbulence all larvae starved. However, in the presence of turbulence, the predicted contact rates were 2–5 times higher (than under non-turbulent conditions) and some larvae survived. Moreover, those that survived exhibited growth characteristics similar to those of field-caught larvae (Figure 7). Analysis of larval trajectories through space and time showed that the survivors were drawn from individuals that were spawned and remained below the surface layer (25 m) within the 60 m isobath throughout the simulation. By hindcasting the conditions experienced by the survivors, Werner *et al.* were able to show that survival resulted from turbulence-induced increases in encounter. In an attempt to include the known pause–travel behavior of cod larvae, Werner *et al.* (1995) subsequently modified the cruise–searching algorithm from their original paper. This "more realistic" simulation model yielded a similarly qualitative pattern of results, but underestimated larval growth rates. Thus, even if the model is conceptually correct, accurate parametre estimates remain important.

Similarly, Davis *et al.* (1991) modeled the relative importance of turbulence and microscale patchiness to growth and recruitment of planktonic communities, focusing particularly on haddock larvae. They concluded that spatial and temporal scales of patchiness in food concentrations enhance growth when larvae are capable of accumulating within the patch. In a subsequent series of model runs, Davis *et al.* investigated the interaction between patch integrity and wind-generated turbulence. They concluded that the initial effect of microscale turbulence (i.e. the initiation of wind-mixing following a calm period) is negative because it leads to the disruption of prey patches. Using parametre estimates to represent haddock larvae feeding on copepods, Davis *et al.* (1991) suggested that wind speeds of $0–10 \, \text{m s}^{-1}$ lead to disruption of patches and, ultimately, to declines in growth rates. However, as wind speeds exceeded $10 \, \text{m s}^{-1}$, declines in growth rates were reversed as larvae began to benefit from turbulence-induced increases in encounter and, hence, successful pursuits. In these simulations the reductions in growth resulting from patch dissolution were completely offset by gains in ingestion in turbulent conditions at wind speeds approximating $15 \, \text{m s}^{-1}$ (Figure 8). Finally, Davis *et al.* suggested that, in situations in which size-dependent mortality mechanisms determine recruitment, the variations in growth rates resulting from the interaction of patch integrity and turbulence-induced ingestion are of sufficient magnitude to affect year-class strength.

As demonstrated by these modeling examples, our understanding of the mechanisms underlying the role of turbulence in regulating feeding in larval fish has increased. However, our knowledge is still insufficient in several key areas. Most of the current models exclude the potential of spatial and temporal patterns in the distribution of turbulence, or in the responses of predators and prey to this patchy distribution. If larvae are routinely exposed to turbulence in their environments, there is no reason why they could not have evolved the ability to detect this patchiness, avoiding those areas that are detrimental to survival and seeking out those areas beneficial to survival. However, similar arguments could be made regarding the prey, leading to an evolutionary "arms race" between larvae and their prey. These models also assume that any increase in ingestion realized by exploiting turbulent environments comes without cost. Even if potential bioenergetics costs do not outweigh the gains from ingestion, they may act further to modify the turbulent regime selected by larvae

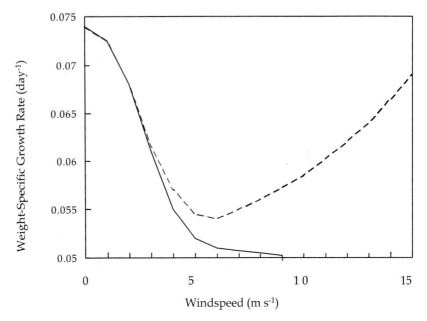

Figure 8 Weight-specific growth rate of larval cod when exploiting patchily distributed prey that are either experiencing turbulence-induced diffusion (dashed line) or remain stationary within the patch (solid line). (After Davis *et al.*, 1991.)

as optimal for survival. The potential impact of these assumptions will be discussed in Section 7.

5. LABORATORY STUDIES OF THE EFFECT OF TURBULENCE ON FEEDING ECOLOGY

5.1. Larval Fish

Few laboratory studies have directly assessed the influence of turbulence on feeding in larval fishes. Landry *et al.* (1995) examined the relationship between turbulence and ingestion rates in larval fathead minnows (*Pimephales promelas*). One experiment, involving grid-generated turbulent velocities of 0, 0.65, 1.31 and 1.96 cm s^{-1} (equivalent to surface layer turbulence generated by wind speeds of 0, 1.6, 3.2 and 5 m s^{-1}) and a constant prey concentration of 30 prey l^{-1} yielded a significant non-linear

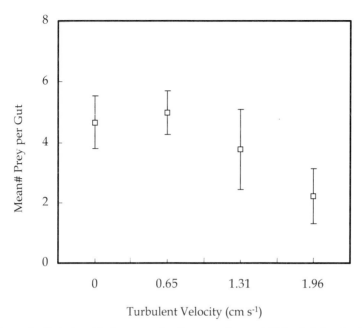

Figure 9 Relationship between feeding rate (mean ± SE) of fathead minnow larvae and grid-generated turbulence. (Redrawn from Landry *et al.*, 1995.)

response of feeding rate to turbulence following a 15 min exposure with a non-significant enhancement at the lowest turbulence level ($0.65 \, \text{cm s}^{-1}$) and a significant reduction in feeding at the highest level ($1.96 \, \text{cm s}^{-1}$) (Figure 9). A second experiment, again involving 15 min exposures, simultaneously evaluated the influence of prey densities (15, 50 and 500 prey l^{-1}) and level of turbulent velocity. Significant, but different, turbulence responses occurred at 15 and 500 prey l^{-1}. At the highest prey density the response was non-linear. Feeding rate was enhanced at the $0.65 \, \text{cm s}^{-1}$ turbulence levels but declined at higher turbulence. At the lowest prey density, feeding rate declined linearly with turbulence. While these findings clearly indicate an influence of turbulence on feeding rate, the responses are not fully consistent with the predictions of theory. For example, while suppressed feeding is predicted at high turbulence levels (MacKenzie *et al.*, 1994), the predicted enhancement of feeding at intermediate levels of turbulence was observed in only one trial (500 prey l^{-1}).

MacKenzie and Kiørboe (1995) examined the feeding rates of larval cod and herring under calm and turbulent conditions, and at prey concentrations ranging from 4 to 1500 nauplii l^{-1}. The turbulence level used was equivalent to that generated at 20 m depth by a surface wind of 6–7 m s^{-1}

($\varepsilon = 7.4 \times 10^{-8}\,\mathrm{Wkg}^{-1}$). This turbulence level is below that predicted to yield maximal increases in feeding rates in larval cod—approximately a wind of $15\,\mathrm{m\,s}^{-1}$ (MacKenzie *et al.*, 1994). Consistent with theory, feeding rate was consistently higher under turbulent conditions, and the magnitude of this effect was greatest at low prey densities. For example, at prey concentrations < 35 prey l^{-1} attack rates were 2.2- to 4.7-fold higher under turbulent conditions. This compares well with a 3.5-fold increase predicted by MacKenzie and Leggett (1991) for a similar combination of predator, prey, turbulence level and prey concentration.

Jeffrey and Dower (unpublished data) appear to be the first to have directly assessed the influence of microscale turbulence on larval growth. In a series of experiments involving larval zebrafish (*Brachydanio rerio*), they assessed growth over 15 days at turbulent velocities of 0, 1.6 and $2.5\,\mathrm{cm\,s}^{-1}$, equivalent to surface layer (0–20 m) turbulence generated by wind speeds of 0, 6 and $10\,\mathrm{m\,s}^{-1}$, respectively. Turbulence was generated by a variable speed oscillating grid. Larvae were reared on *Paramecium* at a concentration of 2000 prey l^{-1}. After 15 d, larvae in the intermediate turbulence treatment were significantly larger than larvae in the zero and high turbulence treatments. Growth rates of larvae in the latter two treatments did not differ. This hints at the possibility that the beneficial effect of enhanced encounter rates may be offset by an increased energetic cost. Clearly, this possibility warrants further attention.

Two additional studies provide insight into the influence of turbulence on larval growth and behavior. Reynolds and Thomson (1974) exposed post-metamorphic grunion (*Leuresthes sardina*) to a gradient of four turbulence levels specified as nil, slight, moderate and extreme, and observed their distribution over time. Grunion statistically favored the intermediate turbulence level, spending $> 50\%$ of their time in this turbulence condition. Chesney (1989) examined the interaction between light, turbidity and turbulence on ingestion in larval striped bass. Turbulence was generated by a stream of air bubbles released from the bottom of a cylindrical tank. Average turbulent velocities were $\sim 6.3\,\mathrm{cm\,s}^{-1}$, with a maximum of $10\,\mathrm{cm\,s}^{-1}$ (equivalent to a wind speed of almost $20\,\mathrm{m\,s}^{-1}$). Chesney concluded that reducing light or adding turbulence both decreased the growth of larval striped bass. Furthermore, although no significant differences were found, these data also suggest that larvae may pay an energetic cost to feed in turbulent environments.

5.2. Other Zooplankton

Several laboratory studies have explored the influence of microscale turbulence on copepod feeding, development rates, activity, behavior and metabolism. A consistent turbulence effect has been shown. Marassé *et al.* (1990), Saiz *et al.* (1992), Saiz and Alcaraz (1992a) and Saiz and Kiørboe (1995) have all reported increases in feeding rates in zooplankton exposed to microscale turbulence relative to non-turbulent conditions. Consistent with theory (MacKenzie *et al.*, 1994), these responses have been greatest, or have occurred, only in experimental conditions in which the prey density was below levels required to create satiation in non-turbulent conditions. Saiz and Kiørboe (1995) observed that low levels of turbulence increased clearance rates by a factor of 4 over the non-turbulent condition, but that higher turbulence rates depressed clearance rates. However, even at high levels of turbulence, clearance rates were still greater than those observed under zero turbulence. Saiz *et al.* (1992) also reported enhanced feeding rates by copepods at low turbulence and depression of feeding rates under high turbulence conditions. This dome-shaped response to turbulence is predicted by theory (MacKenzie *et al.*, 1994; Kiørboe and MacKenzie, 1995).

Shortened development times of zooplankton raised in turbulent versus non-turbulent conditions have also been noted by several authors (Oviatt, 1981; Alcaraz *et al.*, 1988; Saiz and Alcaraz, 1991). Saiz and Alcaraz (1991) found growth in the copepod *Acartia grani* to be quite variable, with naupliar growth rates being higher under turbulent conditions and copepodite growth rates being higher under non-turbulent conditions.

Metabolic and activity rates also increase under turbulent conditions (Saiz and Alcaraz 1992a, b; Alcaraz *et al.*, 1994; Saiz, 1994) leading to the potential for negative energy gain even under enhanced feeding. However, Hwang *et al.* (1994) have shown that the copepod *Centropages hamatus* modifies its escape and foraging behavior in the face of episodic turbulent conditions (such as typically occur in nature) to maximize feeding while minimizing the energy cost associated with escape responses induced by turbulence. It may be that variable behavioral responses help ensure that the overall energetic effect of turbulence is positive under natural conditions. Once again, however, this finding highlights our lack of knowledge regarding the behavioral repertoires of zooplankton under natural conditions.

6. FIELD STUDIES OF THE EFFECT OF TURBULENCE ON FEEDING

The potential for alterations in the feeding success of larval fishes due to wind-induced turbulence has been recognized for some time. Lasker (1975, 1978, 1981) hypothesized that the feeding success of first-feeding anchovy larvae was dependent on a stable (i.e. very low turbulence) upper mixed layer which persisted sufficiently long for the development of a subsurface chlorophyll maximum, within which the larvae could feed. Peterman and Bradford (1987) validated Lasker's (1975) "stable ocean hypothesis" by demonstrating a significant positive relationship between larval survival and the frequency of calm, low wind speed periods ("Lasker events") during the anchovy spawning season.

More generally, the longstanding view that "the mean density of larval food organisms in the ocean is generally too low to support reasonable survival of fish larvae through to metamorphosis" (Lasker, 1975) led to the view that the association of larval fishes with local patches of high prey concentration was critical to larval growth and survival (Vlymen, 1977). Owen (1989) clearly established that fine-scale and microscale patchiness of microplankton populations were greater at low wind speeds (although some patchiness did persist even at wind speeds in excess of $10 \, m \, s^{-1}$), and in the more stable layers of the seasonal pycnocline. Hence, small planktonic predators moving as little as 2 m vertically are likely to encounter non-colonial microplankton populations that range up to 140% higher (or lower) than their average density in the upper mixed layer when turbulence is low. These studies, and many others, generated and sustained the view that turbulence in the ocean was likely to have a *negative* effect on the feeding and survival of larval fishes.

In contrast, Cury and Roy (1989) and Roy *et al.* (1992) found a dome-shaped relationship between recruitment and Ekman-type (wind-generated) upwelling when analyzed over temporal scales of years and spatial scales of hundreds to thousands of kilometres. The systems studied were Peruvian anchoveta (*Engraulis ringens*), Pacific sardine (*Sardinops sagax*), West African sardine (*Sardina pilchardus*) and sardinellas (*Sardinella aurita* and *S. maderensis*), and Californian northern anchovy (*Engraulis mordax*). In such Ekman-type situations, recruitment peaked at wind speeds of approximately $5–6 \, m \, s^{-1}$. It is at these speeds that wind stress begins to exert a measurable mixing effect on the surface layer of near-shore waters, and at which desegregation of phytoplankton patchiness begins (Theriault and Platt, 1981; Demers *et al.*, 1986). Cury and Roy (1989) and Roy *et al.* (1992) concluded that the dome-shaped response of recruitment to upwelling systems was the result of the interacting effects of the stimulation of primary/secondary production by upwelling on the one hand, and the desegregation of plankton patchiness by winds greater than $5–6 \, m \, s^{-1}$ on the other (Figure 10). Ware and Thomson (1991) reported similar findings for the recruitment of Pacific sardine off southern California, but concluded that the optimum wind speed in their system was in the range of $7–8 \, m \, s^{-1}$.

The work of Rothschild and Osborn (1988), and the subsequent

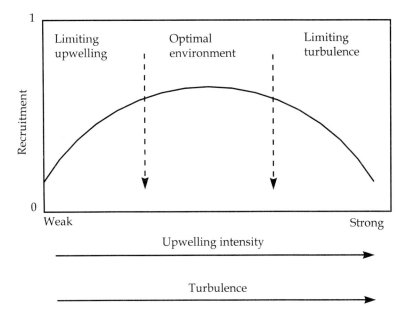

Figure 10 Relationship between turbulence and recruitment success hypothesized by Cury and Roy. (Redrawn from Cury and Roy, 1989.)

development of their ideas for larval fish (Sundby and Fossum, 1990; Davis *et al.*, 1991; MacKenzie and Leggett, 1991, 1993; Jenkinson and Wyatt, 1992; MacKenzie *et al.*, 1994; Muelbert *et al.*, 1994; Sundby *et al.*, 1994; Kiørboe and MacKenzie, 1995; MacKenzie and Kiørboe, 1995; Saiz and Kiørboe, 1995), has provided a different perspective from which to view these results. These analyses suggest that the prior assumption of an association of larval fishes with prey patches as prerequisite to successful feeding and survival may be flawed. Recent work by Munk (1995), in which cod larvae were shown to be capable of feeding at high rates even at prey concentrations < 10 nauplii l^{-1}, reinforces this idea.

Moreover, these analyses also suggest that the dome-shaped response of recruitment to wind-driven upwelling and mixing observed by Cury and Roy (1989) and Roy *et al.* (1992) may be a function of the effect of wind-driven turbulence on the interacting encounter and capture probabilities of feeding larvae (MacKenzie *et al.*, 1994; MacKenzie and Kiørboe, 1995). We note, however, that the latter interpretation in no way negates the importance of wind-induced upwelling (see Ware and Thomson, 1991) as it still regulates the *average* prey concentrations available to the larvae on an annual basis.

Sundby and Fossum (1990) were the first to directly evaluate the

Rothschild and Osborn "turbulence theory" in the field. Using historical feeding data for Arcto-Norwegian cod larvae plus simultaneously sampled nauplii (prey) concentrations and wind data, they found that contact rates between larvae and prey were elevated by a factor of ~ 2.8 as the average wind speed increased from 2 to $6\,\mathrm{m\,s}^{-1}$. The corresponding increase predicted under the Rothschild and Osborn (1988) model was 2.2-fold. Furthermore, Sundby and Fossum found that, at wind speeds of $> 4\,\mathrm{m\,s}^{-1}$, turbulent velocity contributed more to contact rate than did larval swimming speed. They concluded that interannual variation in the intensity of wind mixing "must be an important regulatory mechanism for the formation of year-class strength". Subsequent field studies (Sundby et al., 1994) have confirmed the positive influence of wind-induced turbulence on feeding rates in larval cod and indicate that the highest feeding rates occur at wind speeds $> 10\,\mathrm{m\,s}^{-1}$. MacKenzie et al. (1994) later incorporated these data into their encounter-rate model and revised this predicted optimum wind speed for cod larvae feeding rates upwards to $\sim 15\,\mathrm{m\,s}^{-1}$.

In an elegant combination of field and modeling work, Matsushita (1992) analyzed the interacting effects of prey particle separation and average particle velocity, induced by the combination of swimming speed and turbulent motion, on prey capture success by larval fishes. The results of his analysis of his meticulous field observations conform well with the MacKenzie et al. (1994) model and also predict a dome-shaped capture success function with an optimum turbulence level resulting from winds in the range of 15–$22\,\mathrm{m\,s}^{-1}$. Matsushita (1992) estimated the optimal turbulence level for larval clupeoid fishes under the conditions he observed in the field to be $0.5\,\mathrm{cm\,s}^{-1}$, which corresponds to an average wind speed of $18\,\mathrm{m\,s}^{-1}$.

Using literature data derived from laboratory and field studies, Muelbert et al. (1994) tested the hypothesis that turbulent, tidally well-mixed regions (previously identified by Sinclair and Iles (1985) as spawning and nursery areas for Atlantic herring) constitute a preferential feeding environment for herring larvae. They concluded that at existing natural food densities in coastal Nova Scotia, Canada, larvae in tidally well-mixed regions require prey concentrations that are approximately one order of magnitude lower than would be required in stratified regions if equivalent feeding rates were to be achieved. MacKenzie and Leggett (1991) predicted that prey encounter rates at prey densities equivalent to those reported by Muelbert et al. (1994) would be 5–12 times higher in tidally well-mixed zones relative to stratified zones. Muelbert et al. also concluded that the feeding environment in stratified waters off southwest Nova Scotia is detrimental for herring larvae, and that high mortality would be expected for larvae overwintering in such areas. These findings are consistent with their hypothesis and suggest that the retention hypothesized by Sinclair and Iles (1985) and Sinclair (1988) may actually be the product of increased survival in tidally well-mixed (turbulent) zones relative to the adjacent less turbulent stratified zones leading to the *appearance* of retention.

7. SYNTHESIS

From the preceding sections it is evident that substantial progress has been made since Rothschild and Osborn (1988) first proposed a role for microscale turbulence in plankton trophodynamics. Nonetheless, many key questions remain unresolved, particularly with respect to population-level responses (such as survival and recruitment) to turbulence. We maintain that only through the continued integration of laboratory and field results into predictive models will we advance the understanding of biophysical coupling at the scales relevant to larval fish and their prey. In the final section of the paper we: (1) summarize the state of our present knowledge by considering the evidence in support of a hierarchy of hypotheses relating microscale turbulence to larval fish ecology; (2) examine ways in which existing knowledge might be better used to enhance the understanding of these issues; (3) identify areas where further research and, possibly, new theoretical frameworks will be required to advance us past the point where we now stand.

7.1. The State of our Existing Knowledge

7.1.1. Turbulence Increases Encounter Rates between Predator and Prey

Beginning with Rothschild and Osborn (1988), numerous models and simulations have now shown that it is mathematically possible for microscale turbulence to increase encounter rates between a wide variety of planktonic predators and prey (Davis *et al.*, 1991; MacKenzie and Leggett, 1991; Yamazaki *et al.*, 1991; MacKenzie *et al.*, 1994; Muelbert *et al.*, 1994; Kiørboe and MacKenzie, 1995; Kiørboe and Saiz, 1995; MacKenzie and Kiørboe, 1995; Werner *et al.*, 1996). This overwhelming body of theoretical work has, no doubt, contributed to the recent ascendancy of the "turbulence theory" in fisheries ecology. However, despite this, very few studies have *empirically* demonstrated turbulence-mediated increases in encounter rates. Marassé *et al.* (1990) used video-cinematography to show that the copepod *Centropages hamatus* encounters more phytoplankton (per unit time) under turbulent conditions. For larval fish, MacKenzie and Kiørboe (1995) directly observed that cod and herring larvae feeding on *Acartia tonsa* nauplii assumed significantly more "attack positions" under turbulent conditions and that, by inference, encounter rates must also have been higher. To date, there is no direct evidence of increased encounter rates between larval fish (or other zooplanktonic predators) and their prey in a field setting.

7.1.2. *Turbulence Increases Ingestion Rates of Predators*

As in the previous case, there is mathematical evidence to support this hypothesis both for larval fish (MacKenzie *et al.*, 1994; Kiørboe and MacKenzie, 1995) and for copepods (Kiørboe and Saiz, 1995). However, apart from observations by Landy *et al.* (1995), empirical evidence of turbulence-related increases in larval fish ingestion rates is lacking. Likewise, only a few laboratory studies have documented turbulence-mediated increased ingestion rates in other planktonic predators (Saiz *et al.*, 1992; Peters and Gross, 1994; Saiz and Kiørboe, 1995; Shimeta *et al.*, 1995). To date, the primary field evidence in support of this hypothesis remain the studies by Sundby and Fossum (1990) and Sundby *et al.* (1994) documenting increased gut fullness in larval cod in response to increasing wind speed. An important caveat to these studies, however, is that increased ingestion rate may not be due solely to increased encounter rate; it is conceivable that either the escape abilities of prey are reduced under turbulent conditions (thereby making prey easier to capture), or that larvae switch to smaller, slower, easier-to-capture prey (as predicted by MacKenzie *et al.*, 1994) under turbulent conditions.

We must also consider the possibility that, in the field at least, the effect of turbulence on larval fish ingestion or gut fullness may be a second-order effect. If the effect of wind is to homogenize prey patches throughout the water, and if larval fish originally co-occurred with those prey patches, then homogenization by wind-induced turbulence could negatively affect larval ingestion rates as proposed by Davis *et al.* (1991) for wind speeds of ~ 6–$10\,\mathrm{m\,s}^{-1}$. However, as already mentioned, it remains far from certain whether larval fish regularly co-occur with high-density prey patches (Leggett, 1986; Taggart and Leggett, 1987a, b; Govoni *et al.*, 1989). Consequently, if larvae do not routinely co-occur with prey patches, the homogenization of patches by wind-induced turbulence could, conceivably, lead to both an increase in the average prey concentration perceived by the larvae and to increased larval encounter and ingestion rates.

7.1.3. *Turbulence Enhances Growth of Larval Fish*

Although largely unexplored (at least with respect to larval fish), the implications of this hypothesis could be far-reaching. Davis *et al.* (1991) show that, in theory, predator growth *can* be enhanced by increased turbulence via increased encounter rates, leading to increased ingestion rates. Empirically, turbulence has also been shown variously to increase, decrease, or have no discernible effect on the growth rates and growth efficiencies of copepods (Saiz and Alcaraz, 1991; Saiz *et al.*, 1992) and protozoans (Peters and Gross, 1994; Shimeta *et al.*, 1995). To date, the observations by Jeffrey and Dower (unpublished data) of enhanced growth of zebrafish under intermediate levels of turbulence represent the only empirical evidence that turbulence affects the growth of larval fish. To our knowledge, there is no field evidence which shows increased growth of larval fish resulting from the (direct or indirect) effects of turbulence.

On the contrary, Maillet and Checkley (1991) showed that otolith growth rates in 3–15-day-old Atlantic menhaden (*Brevoortia tyrannus*) were significantly *negatively* correlated with wind speed and suggested that storm-induced turbulence reduced prey availability to larvae as suggested by Lasker (1978, 1981). This picture is complicated, however, by Maillet and Checkley's further observation that otolith growth in 16–50-day-old larvae was significantly negatively correlated with wind speed *and* heat flux. Furthermore, whereas the correlation between otolith growth and heat flux was maximal at a 2-day lag (i.e. decreased otolith growth lagging 2 d behind increased heat flux), the correlation between otolith growth and wind speed was maximal at a 4-day lag, suggesting otolith growth (in 16–50-day-old larvae) responded more rapidly to decreasing temperature than to increasing wind speed. Alternatively, the greater lag in older larvae could indicate that larger (i.e. older) larvae are less sensitive to highly turbulent conditions than are smaller (i.e. younger) larvae. This could occur if, for instance, the faster swimming speeds and larger perceptive radii of older larvae act to diminish the detrimental effects of high turbulence (i.e. the slower swimming speeds and shorter perceptive radii of younger larvae could reduce pursuit success under high turbulence). Clearly, the potential for interactions among turbulence, temperature and larval size deserves more attention than it has heretofore received.

7.1.4. *Turbulence Increases Recruitment of Larval Fish*

Even minor increases (e.g. < 10%) in the growth and survival rates of larval fish can lead to major increases in recruitment (Rice *et al.*, 1993). By inference, the models of Davis *et al.* (1991) and Werner *et al.* (1995, 1996) predict that, assuming constant juvenile mortality rates, turbulence-mediated growth of larvae should lead to increased survival and, hence, recruitment. Note, however, that although increased larval growth and survival may be requisites for high recruitment in a given year, the outcome is by no means guaranteed (see, for example, Bailey and Spring, 1992 and Bailey *et al.*, 1995).

Empirically, there is no support for the hypothesis that increased turbulence enhances recruitment. However, in recent years a sizeable body of field evidence has suggested that interannual variability in recruitment is related to interannual variability in wind-induced turbulent mixing (Peterman and Bradford, 1987; Cury and Roy, 1989; Ware and Thomson, 1991; Roy *et al.*, 1992; Bailey and Macklin, 1994), and generally takes the form of a dome-shaped relationship. Moreover, most of these studies propose that the net effect of high wind speeds is to reduce the availability of prey to larvae; the result being low survival and recruitment. As mentioned previously, however, this contention rests heavily on the assumption that larvae rely on micropatches of prey in the field. Alternatively, a dome-shaped response could also be generated if the net effect of strong turbulent mixing were to reduce the pursuit success of larval fish (i.e. Figure 5: although encounter rates might be high, the probability of successful pursuit could decline) as proposed by MacKenzie *et al.* (1994).

Interestingly, these studies converge in their suggestion that wind speeds of ~ 5–8 m s^{-1} are optimal for recruitment of a range of species (Peruvian anchoveta, Pacific sardine, West African sardine, Pacific hake, walleye pollock) across a range of environments (Peruvian upwelling, western North American upwellings, western African coastal upwellings and the Gulf of Alaska). In contrast, the model of Davis *et al.* (1991) predicts that feeding rates should *decrease* over wind speeds of 5–10 m s^{-1}, purportedly due to the breakdown of prey patches (but see Owen, 1989). Davis *et al.* (1991) show that wind speeds $\leq 5 \text{ m s}^{-1}$ enhance formation of prey patches that can be exploited by predators and that turbulence from wind speeds $\geq 10 \text{ m s}^{-1}$ increases encounter rates between predator and prey. In fact, perhaps the main difference between these field studies and the encounter and feeding rate models reviewed in Section 4 is that while both predict dome-shaped relationships between turbulence (or wind speed) and larval fish feeding, the feeding models generally predict optimum wind speeds that are much higher (usually 15–20 m s^{-1}) than the values reported from the field (Davis *et al.*, 1991; Matsushita, 1992; MacKenzie *et al.*, 1994). Among the possible explanations for this discrepancy is the reality that most models: (1) assume homogeneous prey distributions; (2) do not allow for prey to be mixed out of the system altogether; (3) assume that larvae are never mixed deep enough for predatory abilities to become light-limited.

7.2. What Else can be Done Without "New" Knowledge?

7.2.1. *Modeling*

Attempts to model the effects of turbulence on encounter rates and feeding success have been predicated on a number of restrictive assumptions, both physical and biological. Physically, existing encounter rate models are largely steady-state approximations in that they assume a "frozen" turbulent field (i.e. turbulent dissipation rates are either constant throughout the entire model domain or assume fixed values at different depths). In other words, we have not yet examined the effects of spatio-temporally variable turbulence on encounter and growth of larvae. Of particular importance is the need to understand better how turbulence "spins up" ("winds down") in response to increasing (decreasing) wind speeds, particularly in coastal areas where fetch may be important (and where many commercially important fish species typically spawn). For example, Frank and Leggett (1981) and Taggart and Leggett (1987a, b) have demonstrated that, in coastal Newfoundland, the food environment encountered by newly hatched capelin larvae (*Mallotus villosus*) is strongly affected by the *direction* from which the wind is blowing. Both studies showed that the microzooplankton prey suitable for first-feeding capelin larvae are more abundant when onshore winds prevailed. Under such conditions, we might predict the encounter rates experienced by larval fish to be quite different for winds of the same magnitude but of opposite directions.

Another area where our physical knowledge needs to be improved is in the calculation of ω from ε. We have previously presented some of the competing ideas relating to the definition of r in Equation 2 (Section 2). The impact of these alternative definitions on estimated feeding rates is significant. Kiørboe and MacKenzie (1995) compared estimated encounter rates for cod feeding under turbulent conditions using the two most common definitions of r: (1) the average prey separation distance; or (2) the larval reactive distance. We have compared all four definitions suggested in the literature and find that, under the conditions reported by Sundby and Fossum (1990) in their original paper, the various definitions produce feeding rate estimates that differ by almost an order of magnitude. Clearly, identification of the correct approach is key to enabling interpretation of both laboratory and field data.

Chief among the biological assumptions are those of homogeneous prey distribution, random predator and prey movement, and the uniform search volume of predators. Although useful starting-points from which to consider interactions between turbulence and feeding, we also know that these assumptions do not accurately reflect the distribution of larval fish

and their prey in the field. For instance, it has long been known that
plankton patchiness occurs over sub-metre scales (Cassie, 1963; Haury *et
al.*, 1978; Mackas *et al.*, 1985; Owen, 1989) and that larval fish also display
non-random small-scale distributions (Fortier and Leggett, 1984). As
pointed out in the previous section, discrepancies between optimal wind
speeds observed in the field and those generated by encounter and feeding
models may result from our failure to incorporate realistic descriptions of
predator and prey distributions and movements into the models. Furthermore, until we understand both the spatiotemporal structure of oceanic
turbulence and the response of organisms to it we are unlikely to develop
adequate models.

A second area where knowledge is incomplete is in our understanding
of the effects of turbulence on larvae of different species. As an example
of how this deficiency can be addressed, we have modified the model of
MacKenzie *et al.*'s (1994) such that the terms in the model are all functions
of larval size using equations given in Miller *et al.* (1988) and Miller (1990).
In these simulations we considered a range of species of larvae that
hatched over a range of sizes from 2 to 12 mm standard length. All fish
ate the same "copepod nauplii" prey. Parametres for prey size and
swimming were those used in MacKenzie *et al.* (1994). To adapt the
MacKenzie *et al.* model we replaced estimates of reactive distance,
swimming speed and minimum pursuit time with size-dependent functions
given in Miller *et al.* (1988) and Miller (1990).

Two trends are apparent in the model output (Figure 11). First, as larval
size increases, so too does the range of turbulent velocities that result in
increased feeding rates. Thus, we might expect species with larger larvae
to exploit a wider range of turbulent environments. Second (and contrary
to previous observations), species with larger larvae derived slightly
greater benefits from exploiting turbulent environments. We caution,
however, that these simulations are based upon functions derived from
interspecific generalizations and, thus, may not compare accurately
individual pairs of species. Moreover, these simulations (based as they are
on interspecific comparisons) may not reflect how turbulence-induced
feeding rates may respond *within* a species as individuals grow.

As a further example, future models should also consider the role of
vertical migration (diel and otherwise). Such behavior is widespread
amongst larval fish (Neilson and Perry, 1990; Heath, 1992) and crustacean
zooplankton (Cushing, 1951; Ohman, 1990) and has obvious implications
for interactions with microscale turbulence: whether we are dealing with
wind-induced (top-down) or tidally induced (bottom-up) turbulence, we
know that turbulent dissipation rates are depth-variable (Gargett, 1989).
Consequently, the turbulence-affected encounter rates experienced by a
larval fish will vary partly as a function of its vertical position in the water

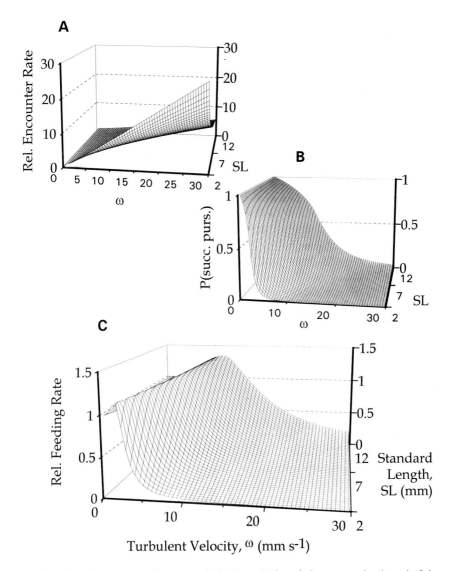

Figure 11. Size-dependency of turbulence-induced increases in larval fish feeding. We generated a size-dependent version of the MacKenzie *et al.* (1994) model using relationships for reactive distance, swimming velocity and minimum pursuit time (R, v and t, respectively) given by Miller *et al.* (1988) and Miller (1990). These are: $R = 1.1 \cdot l$; $v = 10^{1.07 \cdot \log(l) - 1.11}$; and $t = 1.33 \cdot e^{-0.0035 \cdot l}$, where l is the larval standard length (mm). Note that the *x*- and *y*-axes are the same for all panels, that all panels are standardized against non-turbulent conditions, and that model larvae all ate the same "copepod nauplii" prey. (A) Relative increase in encounter rate as a function of larval size. (B) Probability of successful pursuit as a function of larval size. (C) Relative feeding rate as a function of larval size.

column (Heath, 1992). Although the effect of depth-dependent encounter rates has been considered in steady-state formulations (MacKenzie and Leggett, 1991; MacKenzie *et al.*, 1994), what has yet to be done is to integrate the range of encounter conditions experienced by an individual larva as it changes position during vertical excursions.

What has been shown, however, is that larvae *do* appear to actively modify their position in the water column in response to changing turbulent conditions. In a detailed time-series study of herring larvae, Heath *et al.* (1988) found that the mean depth of the larval population was strongly correlated with the intensity of wind-induced turbulent mixing: under strong wind conditions, the center of density of the larval distribution shifted deeper in the water column. Furthermore, Heath *et al.* (1988) point out that, when compared to the vertical distribution of the less mobile copepod nauplii from the same site, the magnitude of the change observed in the vertical distribution of herring larvae was large enough to be unlikely to have resulted from passive mixing alone. This brings up the interesting possibility that larval fish may *avoid* high turbulence situations via active swimming. Certainly, the findings of Heath *et al.* (1988) and those of Ellertsen *et al.* (1984), who noted reduced upward migration of cod larvae during highly turbulent conditions, hint at such a possibility. Along a similar line, Mackas *et al.* (1993) have recently provided evidence that vertical habitat partitioning by certain calanoid copepods may be maintained via species-specific preferences for different turbulence regimes. Certainly, the behavioral responses of larval fish to varying levels of turbulence warrant further investigation.

Likewise, the behavior of prey species under turbulent conditions needs to be better quantified. MacKenzie *et al.* (1994) predicted that if pursuit success by larval fish is an inverse function of turbulence, larvae would be expected to prey more heavily on smaller, slower-moving prey under turbulent conditions (a possibility with implications not only for the present topic but also for questions of optimal foraging strategy). Certainly, empirical observations have shown that copepods are more active under turbulent conditions, displaying higher rates of both feeding and escape responses (Marassé *et al.*, 1990; Saiz and Alcaraz, 1992a, b; Hwang *et al.*, 1994). Whether this increased activity in response to turbulence makes copepods more (or less) difficult for larval fish to capture remains to be seen. Of course, key to testing any hypothesis regarding larval selectivity for specific prey types is the need to develop realistic scenarios in which predators can exploit more than one prey type. To our knowledge, the models by Werner *et al.* (1995, 1996) remain the only studies in which larval fish, in this case cod and haddock, are "offered" a realistic prey field.

This leads to the question of how best to model predator search behaviors. Recent work suggests that the perceptive volume of larval fish is more accurately described as an upward-directed pie-shaped wedge (Browman et al., 1990; Browman and O'Brien, 1992a) rather than as the spherical or hemi-spherical formulations employed in most encounter models. If this is confirmed as a general result, the use of the latter two formulations would tend to overestimate larval contact rates, perhaps by as much as 5–10-fold (Browman, personal communication). Furthermore, it has been shown that search behavior significantly affects the ability of larvae to remain near food in non-homogeneous prey fields (Vlymen, 1977), and that larvae change their foraging behavior in response to different prey densities and composition (Munk, 1992, 1995). Each of these subjects clearly deserves further attention.

7.2.2. Laboratory Work

Many of the points raised in the previous section, especially those dealing with the actual encounter process and the issue of predator response to multiple prey types, can only be fully explored in a laboratory setting. Indeed, in a recent paper revisiting the issue of how best to model encounter, pursuit and ingestion, Kiørboe and MacKenzie (1995) pointed out the necessity of detailed observations of larval feeding behavior, both to fine-tune parametre estimates and to validate model predictions.

With respect to higher levels of our hypothesis hierarchy (i.e. regarding larval growth and recruitment) there are also improvements that can be made to future empirical studies. The first issue is that, almost invariably (and largely for logistical reasons), empirical studies have been conducted in small-volume aquaria in which turbulence is generated by an oscillating grid. The problems with this approach are twofold. First, despite assumptions of uniform conditions, turbulent velocities no doubt vary throughout the tanks, a problem compounded by the fact that it is very difficult to measure turbulent dissipation rates in small aquaria. Recently, a number of novel approaches have been developed to overcome the latter, for example, particle-tracking via video-cinematography (Marassé et al., 1990; Kiørboe and Saiz, 1995) and the use of three-dimensional acoustic microscale current metres (MacKenzie and Kiørboe, 1995). Technological issues aside, however, the question remains as to how closely we should expect the turbulence regimes to which larval fish are typically exposed in the laboratory to approximate conditions encountered in the spatio-temporally variable turbulent fields that exist in nature. It may well be that repeating some of the earlier feeding experiments (under laboratory conditions) with either smoothly varying or intermittent turbulence, either of which is a more likely approximation of natural conditions, would prove to be a useful exercise. Further to this point, however, the fact remains that even with variable turbulence conditions, the small size of most enclosure studies still precludes motion at larger scales (e.g. upwelling, advection) which undoubtedly play a role in determining the feeding environments encountered by larvae in the ocean.

The second point has to do with the integrative effect of exposure to turbulence. Until recently, empirical studies have dealt with the original Rothschild and Osborn (1988) hypothesis concerning, specifically, the effect of turbulence on encounter rate. Consequently, these experiments have generally exposed larvae to turbulent conditions for very short time periods: Landry *et al.* (1995) starved larvae overnight and then ran feeding trials under turbulent conditions for only 15 min before examining larval gut contents. Similarly, MacKenzie and Kiørboe (1995) observed feeding behaviors of individual larvae for only a few minutes at a time. Although useful for studying the behavioral responses of larvae to turbulence, these "instantaneous" measures offer little insight into the issue of whether prolonged exposure to turbulence affects larval growth and survival.

From a fisheries ecology point of view, however, it is the extension of the Rothschild and Osborn (1988) hypothesis across longer time-scales that is of primary interest. If microscale turbulence is beneficial to larval feeding success, then it is over time-scales of days to weeks (rather than minutes and hours) that turbulence-mediated changes in feeding success should translate into differences in larval growth and survival. Towards this end, then, we propose that it is time to begin conducting such longer-term experiments, using realistic prey concentrations, that are designed to detect cumulative benefits that larvae may derive from prolonged exposure to turbulence. Furthermore, should such experiments fail to detect any long-term benefits to larvae (either in terms of increased growth or survival), we suggest that a re-examination of some of the assumptions underlying the "turbulence theory" may be required.

7.2.3. *Field Studies*

Many of the shortcomings of field studies that attempt to link turbulence to feeding success in larval fish relate to the question of the appropriate spatio-temporal scale over which to integrate and sample. Physically, we are still hampered by the difficulty of effectively measuring turbulence in the field. Although new techniques for estimating dissipation rates from CTD data look promising (Galbraith and Kelley, 1996) measuring turbulence profiles is never likely to become as routine as measuring temperature or salinity profiles. Biologically, this is an important issue, especially as related to the previously discussed need to understand "spin up" and "wind down" times, since it hinders our ability to establish the time-scales over which turbulence and feeding interact. Simply stated, our inability to readily measure the evolution of the turbulent velocity fields experienced by larvae makes it difficult to decide over which time-scales to integrate our biological sampling.

For instance, how soon after a change in wind direction or wind speed should we expect to see a concomitant increase (or decrease) in encounter rates? Furthermore, how long should we expect such enhanced (or depressed) feeding conditions to last? These are valid questions because, to a large extent, the temporal scale over which turbulence and feeding are coupled determines the appropriate biological variables that should be monitored. If changes in wind speed produce changes in turbulent velocity with characteristic time-scales of only a few hours, then we might expect either gut fullness or digestive enzyme activity to be good indicators of very recent increased feeding success. This issue is complicated, however, by recent work on cod (van der Meeren and Næss, 1993; Munk, 1995) showing that larval feeding behavior changes in response to available food concentration. One possible consequence of these results is that observed gut fullness may have little to do with prey concentration *per se*. On the contrary, if the characteristics of the turbulence field change more slowly (e.g. days), then perhaps a more integrative measure such as the ratio of larval RNA : DNA, reflective of feeding conditions experienced over longer periods, is more appropriate (Ferron and Leggett, 1994).

We suggest that future field efforts could be improved through the adoption of multispecies approaches. To our knowledge, field studies to date have focused on the effects of turbulence on single target species (usually cod or herring). A multispecies approach could have the added advantage of decreasing the likelihood that linkages between turbulence and feeding are obscured by species-specific behavioral responses. As an example, suppose we want to measure larval gut fullness in response to increased wind-induced turbulence. As noted already, laboratory observations of the behavior of larvae under turbulent conditions are rare (but see Munk and Kiørboe, 1985; MacKenzie and Kiørboe, 1995; Munk, 1995). Consequently, monitoring several species simultaneously may reduce the possibility that our results are biased by some previously unknown response to turbulence in a single species. Furthermore, if the linkage between turbulence and feeding is a general feature of planktonic food webs (Kiørboe, 1993), then we should expect to see parallel responses among different planktonic predators (e.g. larval fish, copepods, jellyfish, chaetognaths) even though predator strategy (e.g. pause–travel versus cruise) might moderate the magnitude of the response amongst different taxa.

7.3. What Have We yet to Do?

The most important unanswered questions regarding the role of turbulence in larval fish ecology are those related to our latter two hypotheses: does turbulence affect larval growth and, ultimately, recruitment? We have suggested the need for empirical studies to determine the cumulative influence of exposure to turbulence on larval fish, and a re-examination of some of the underlying assumptions of the "turbulence theory". Let us now consider this latter point.

The first assumption requiring consideration is that microscale turbulence, through its effect on encounter rates, is necessarily a net benefit to larval fish. One prospect still awaiting study is the potential for an energetic cost to larvae associated with foraging in highly turbulent environments. MacKenzie et al. (1994) and Kiørboe and MacKenzie (1995) have suggested dome-shaped relationships between turbulence and feeding success, but the possibility that microscale turbulence might exact an energetic cost from larvae has received little attention. Increased metabolic rates and decreased growth efficiencies have been observed in copepods raised under high turbulence conditions (Marassé et al., 1990; Saiz and Alcaraz, 1992a; Alcaraz et al., 1994), suggesting an energetics cost that might potentially outweigh the benefits of increased encounter rates. These results are probably due, in part, to two things. First, copepods detect prey and (importantly) predators by mechanoreception; hence, increased movement and metabolic rates are likely to be a *direct* response to the stimulus of turbulent motion. Larval fish are primarily visual predators (but see Batty and Hoyt, 1995) and are, therefore, less likely to respond to turbulence directly (i.e. by increasing metabolic rates, etc.). Nevertheless, there is evidence that in a confined laboratory setting, under turbulent conditions, larval fish spend more time swimming (MacKenzie and Kiørboe, 1995), possibly in response to increased encounter rates with prey. If larval pursuit success declines under turbulent conditions (MacKenzie et al., 1994), the energetic cost of foraging may exceed the benefit derived from the increased encounter rates.

A second assumption that must be re-examined is the question of food availability. Throughout this review we have stressed the point that what we refer to as the "turbulence theory" rests heavily on the assumption that food abundance *per se* directly influences larval feeding success, growth and, ultimately, recruitment. To establish whether the turbulence–feeding relationship is important beyond the level of increasing encounter rates we must first establish whether larvae regularly experience food-limiting conditions in nature. A review of the minimum food concentrations required for larval survival (Leggett, 1986) indicates that larvae may only rarely encounter truly food-limited conditions. This idea has recently been reinforced by Munk's (1995) findings that larval cod can grow and survive at naupliar concentrations $< 10 \, l^{-1}$. In addition, it has been shown that larval fish (specifically yolk-sac and early exogenous feeding stages) can compensate for lack of preferred food (i.e. copepod nauplii) by feeding on prey from other trophic levels, including protozoans and phytoplankton (Houde and Schekter, 1980; van der Meeren and Næss, 1993), providing further evidence that food-limitation may not be of primary importance to larval fish survival.

In fact, it could be argued that if food availability *per se* (i.e. in the absence of other encounter/feeding enhancing processes) is *not* limiting, then there remains little incentive to conduct further research on the role of microscale turbulence in larval fish ecology, as such work is unlikely to bring us any closer to answering the larger question of what controls interannual variability in recruitment in marine fish populations. Obviously, this is an extreme view, and we offer it here merely to point out the primacy of establishing a link between food availability and larval survival, not only with respect to "turbulence theory" but for the field of fisheries ecology in general. In fact (despite the aforementioned evidence to the contrary), evidence pointing to a strong link between food concentration and cohort strength does continue to emerge (Fortier *et al.*, 1995, 1996; Gotceitas *et al.*, 1996).

Third, let us assume that hypotheses 1, 2 and 3 of our hierarchy are correct: that turbulence increases encounter rates, ingestion rates and growth rates of larval fish. Let us also assume that there is no energetic cost to larvae of foraging in turbulent environments. Even under such conditions, however, it remains unclear whether increased growth (and hence, size-at-age) will necessarily result in increased survival, the prerequisite for improved recruitment. The importance of growth rate and size-at-age as regulators of survival during the larval stage are themselves the subject of active debate under the headings of the "bigger is better hypothesis" (Litvak and Leggett, 1992) and the "stage duration hypothesis" (Houde, 1987). This research, too, is leading to fundamental re-thinking of long-held assumptions which go well beyond the scope of this review. The results of this work will be highly relevant, however, given the possibility that microscale turbulence could differentially influence larval growth rates, both within populations and between years.

7.4. Conclusions

In this review, we have explored the effects of microscale turbulence on the feeding ecology of marine fish larvae. We have shown that great strides have been made in our understanding of the potential effects of turbulence in the 8 years since Rothschild and Osborn first proposed a role for turbulence in planktonic food webs. In the intervening period the effects of microscale turbulence on encounter and ingestion have been clearly documented. Current research is now beginning to establish the relationship between turbulence and other aspects of the feeding ecology of fish larvae, such as patterns of prey selectivity and growth. These as yet unanswered questions must be addressed before we can, and indeed should, attempt to draw inferences about the further link between turbulence, survival and recruitment. We believe, however, that with the new technologies that are rapidly becoming available and with the understanding currently available, resolution of these problems can be achieved in the foreseeable future.

ACKNOWLEDGEMENTS

We wish to thank Professor Alan Southward, Brian MacKenzie and Ed Hill for reviewing this manuscript. Brad de Young provided useful comments on the physics section. JFD wishes to thank Howard Browman for interesting discussions on larval search behaviors. This work was supported by grants from the Natural Sciences and Engineering Research Council of Canada to W. C. Leggett. This is contribution number 2727 from the Center for Environmental and Estuarine Studies of the University of Maryland.

REFERENCES

Alcaraz, M., Saiz, E., Marassé, C. and Vaque, D. (1988). Effects of turbulence on the development of phytoplankton biomass and copepod populations in marine microcosms. *Marine Ecology Progress Series* **49**, 117–125.
Alcaraz, M., Saiz, E. and Calbert, A. (1994). Small-scale turbulence and zooplankton metabolism: effects of turbulence on heartbeat rates of planktonic crustaceans. *Limnology and Oceanography* **39**, 1465–1470.
Bailey, K. M. and Houde, E. D. (1989). Predation on eggs and larvae of marine fishes and the recruitment problem. *Advances in Marine Biology* **25**, 1–83.
Bailey, K. M. and Macklin, S. A. (1994). Analysis of patterns in larval walleye pollock *Theragra chalcogramma* survival and wind mixing events in Shelikof Straits, Gulf of Alaska. *Marine Ecology Progress Series* **113**, 1–12.
Bailey, K. M. and Spring, S. M. (1992). Comparison of larval, age-0 juvenile and age-2 recruit abundance indices of walleye pollock, *Theragra chalcogramma*, in the western Gulf of Alaska. *ICES Journal of Marine Science* **49**, 297–304.

Bailey, K. M., Canino, M. F., Napp, J. M., Spring, S. M. and Brown, A. L. (1995). Contrasting years of prey levels, feeding conditions and mortality of larval walleye pollack *Theragra chalcogramma* in the western Gulf of Alaska. *Marine Ecology Progress Series* **119**, 11–23.

Batty, R. S. and Hoyt, R. D. (1995). The role of sense organs in the feeding behaviour of juvenile sole and plaice. *Journal of Fish Biology* **47**, 931–939.

Browman, H. I. and O'Brien, W. J. (1992a). The ontogeny of search behaviour in the white crappie, *Pomoxis annularis*. *Environmental Biology of Fishes* **34**, 181–195.

Browman, H. I. and O'Brien, W. J. (1992b). Foraging and prey search behaviour of golden shiner (*Notemigonus crysoleucas*) larvae. *Canadian Journal of Fisheries and Aquatic Sciences* **49**, 813–819.

Browman, H. I., Gordon, W. C., Evans, B. I. and O'Brien, W. J. (1990). Correlation between histological and behavioural measures of acuity in a zooplanktivorous fish, the white crappie (*Pomoxis annularis*). *Brain Behaviour and Evolution* **35**, 85–97.

Cassie, R. W. (1963). Microdistribution of plankton. *Oceanography and Marine Biology Annual Review* **1**, 223–252.

Chambers, R. C. and Leggett, W. C. (1987). Size and age at metamorphosis in marine fishes: an analysis of laboratory-reared winter flounder (*Pseudopleuronectes americanus*) with a review of variation in other species. *Canadian Journal of Fisheries and Aquatic Sciences* **44**, 1936–1947.

Chesney, E. J. Jr (1989). Estimating the food requirement of striped bass *Morone saxatilis*: effects of light, turbidity and turbulence. *Marine Ecology Progress Series* **53**, 191–200.

Coughlin, D. J., Strickler, J. R. and Sanderson, B. (1992). Swimming and search behaviour in clownfish, *Amphiprion perideraion*, larvae. *Animal Behaviour* **44**, 427–440.

Cury, P. and Roy, C. (1989). Optimal environmental window and pelagic fish recruitment success in upwelling areas. *Canadian Journal of Fisheries and Aquatic Sciences* **46**, 670–680.

Cushing, D. H. (1951). The vertical migration of planktonic crustacea. *Biological Reviews* **26**, 159–192.

Cushing, D. H. (1972). The production cycle and the numbers of marine fish. *Symposia of the Zoological Society of London* **29**, 213–232.

Cushing, D. H. (1975). "Marine Ecology and Fisheries." Cambridge University Press, Cambridge, UK.

Cushing, D. H. (1990). Plankton production and year-class strength in fish populations: an update of the match/mismatch hypothesis. *Advances in Marine Biology* **26**, 249–293.

Davis, C. S., Flierl, G. R., Wiebe, P. H. and Franks, P. J. S. (1991). Micropatchiness, turbulence and recruitment in plankton. *Journal of Marine Research* **49**, 109–115.

Demers, S., Legendre, L. and Therriault, J.-C. (1986). Phytoplankton responses to vertical tidal mixing. *In* "Tidal Mixing and Plankton Population Dynamics" (M. J. Bowman, C. M. Yentsch and W. T. Peterson, eds), pp. 1–140. Lecture Notes on Coastal and Estuarine Studies. Springer, Berlin, Germany.

Denman, K. L. and Gargett, A. E. (1995). Biological–physical interactions in the upper ocean: the role of vertical and small scale transport processes. *Annual Review of Fluid Mechanics* **27**, 225–255.

Ellertsen, B., Fossum, P., Solemdal, P., Sundby, S. and Tilseth, S. (1984). A case study on the distribution of cod larvae and availability of prey organisms in relation to physical processes in Lofoten. *In* "The Propagation of Cod (*Gadus morhua* L.)" (E. Dahl, S. Danielssen, E. Moksness and P. Solemdal, eds), pp. 453–478. Flødevigen Rapportserie, vol. 1. Skein, Norway.

Evans, B. I. and O'Brien, W. J. (1988). A reevaluation of the search cycle of planktivorous arctic grayling, *Thymallus arcticus*. *Canadian Journal of Fisheries and Aquatic Sciences* **45**, 187–192.

Evans, G. T. (1989). The encounter speed of moving predator and prey. *Journal of Plankton Research* **11**, 415–417.

Ferron, A. and Leggett, W. C. (1994). An appraisal of condition measures for marine fish larvae. *Advances in Marine Biology* **30**, 217–304.

Fortier, L. and Leggett, W. C. (1984). Small-scale variability in the abundance of fish larvae and their prey. *Canadian Journal of Fisheries and Aquatic Sciences* **41**, 502–512.

Fortier, L., Ponton, D. and Gilbert, M. (1995). The match/mismatch hypothesis and the feeding success of fish larvae in ice-covered southeastern Hudson Bay. *Marine Ecology Progress Series* **120**, 11–27.

Fortier, L., Gilbert, M., Ponton, D., Ingram, R. G., Robineau, B. and Legendre, L. (1996). Impact of freshwater on a subarctic coastal ecosystem under seasonal ice (southeastern Hudson Bay, Canada). III. Feeding success of marine fish larvae. *Journal of Marine Systems* **7**, 251–265.

Frank, K. T. and Leggett, W. C. (1981). Wind regulation of emergence times and early larval survival in capelin (*Mallotus villosus*). *Canadian Journal of Fisheries and Aquatic Sciences* **38**, 215–223.

Fuiman, L. A. and Ottey, D. R. (1993). Temperature effects on spontaneous behavior of larval and juvenile red drum *Sciaenops ocellatus*, and implications for foraging. *Fishery Bulletin US* **91**, 23–35.

Galbraith, P. S. and Kelley, D. E. (1996). Identifying overturns in CTD profiles. *Journal of Atmospheric and Oceanic Technology* **13**, 688–702.

Gargett, A. E. (1989). Ocean turbulence. *Annual Review of Fluid Mechanics* **21**, 419–451.

Gerritsen, J. and Strickler, J. R. (1977). Encounter probabilities and community structure in zooplankton: a mathematical model. *Journal of the Fisheries Research Board of Canada* **34**, 73–82.

Gotceitas, V., Puvanendran, P., Leader, L. L. and Brown, J. A. (1996). An experimental investigation of the "match/mismatch" hypothesis using larval Atlantic cod. *Marine Ecology Progress Series* **130**, 29–37.

Govoni, J. J., Hoss, D. E. and Colby, D. R. (1989). The spatial distribution of larval fishes about the Mississippi River plume. *Limnology and Oceanography* **34**, 178–187.

Greene, C. H. (1985). Planktivore functional groups and patterns of prey selectivity in pelagic communities. *Journal of Plankton Research* **7**, 35–40.

Haury, L. R., McGowan, J. A. and Wiebe, P. H. (1978). Patterns and processes in the time-space scales of plankton distributions. *In* "Spatial Patterns in Plankton Communities" (J. H. Steele, ed.), pp. 277–327. Plenum Press, New York, USA.

Heath, M. R. (1992). Field investigations of the early life stages of marine fish. *Advances in Marine Biology* **28**, 1–174.

Heath, M. R., Henderson, E. W. and Baird, D. L. (1988). Vertical distribution of herring larvae in relation to physical mixing and illumination. *Marine Ecology Progress Series* **47**, 211–228.

Hill, P. S., Nowell, A. R. M. and Jumars, P. A. (1992). Encounter rate by turbulent shear of particles similar in diametre to the Kolmogorov scale. *Journal of Marine Research* **50**, 643–668.

Hjort, J. (1914). Fluctuations in the great fisheries of northern Europe viewed in the light of biological research. *Rapports et Procès-verbaux des Réunions. Conseil Permanent International pour l'Exploration de la Mer* **20**, 1–228.

Holling, C. S. (1959). The components of predation as revealed by a study of small mammal predation of the European pine sawfly. *Canadian Entomologist* **91**, 293–320.

Houde, E. D. (1978). Critical food concentrations for larvae of three species of subtropical marine fishes. *Bulletin of Marine Science* **28**, 395–411.

Houde, E. D. (1987). Fish early life dynamics and recruitment variability. *American Fisheries Society Symposium* **2**, 17–29.

Houde, E. D. and Schekter, R. C. (1980). Feeding by marine fish larvae: developmental and functional responses. *Environmental Biology of Fishes* **5**, 315–334.

Hunter, J. R. (1972). Swimming and feeding behaviour of larval anchovy *Engraulis mordax*. *Fishery Bulletin* **70**, 821–838.

Hunter, J. R. (1982). Feeding ecology and predation of marine fish larvae. *In* "Marine Fish Larvae: Morphology, Ecology and Relation to Fisheries" (R. Lasker, ed.), pp. 33–77. Washington Sea Grant Program, University of Washington Press, Seattle, USA.

Hunter, J. R. and Kimbrell, C. A. (1980). Early life history of Pacific mackerel *Scombrus japonicus*. *Fishery Bulletin US* **78**, 89–101.

Hwang, J. S., Costello, J. H. and Strickler, J. R. (1994). Copepod grazing in turbulent flow; elevated foraging behaviour and habituation of escape responses. *Journal of Plankton Research* **16**, 421–431.

Jenkinson, I. R. (1995). A review of two recent predation-rate models: the dome-shaped relationship between feeding rate and shear rate appears universal. *ICES Journal of Marine Science* **52**, 605–610.

Jenkinson, I. R. and Wyatt, T. (1992). Selection and control of Deborah numbers in plankton ecology. *Journal of Plankton Research* **14**, 1697–1721.

Jonsson, P. R. and Tiselius, P. (1990). Feeding behaviour, prey detection and capture efficiency of the copepod *Acartia tonsa* feeding on planktonic ciliates. *Marine Ecology Progress Series* **60**, 35–44.

Kiørboe, T. (1993). Turbulence, phytoplankton cell size, and the structure of pelagic food webs. *Advances in Marine Biology* **29**, 1–72.

Kiørboe, T. and MacKenzie, B. (1995). Turbulence-enhanced prey encounter rates in larval fish: effects of spatial scale, larval behaviour and size. *Journal of Plankton Research* **17**, 2319–2331.

Kiørboe, T. and Saiz, E. (1995). Planktivorous feeding in calm and turbulent environments with emphasis on copepods. *Marine Ecology Progress Series* **122**, 135–145.

Kundu, P. K. (1990). "Fluid Mechanics." Academic Press, San Diego, California, USA.

Landahl, M. T. and Mollo-Christensen, E. (1986). "Turbulence and Random Processes in Fluid Mechanics." Cambridge University Press, Cambridge, UK.

Landry, M. R., Lehner-Fournier, J. N. and Fagerness, V. C. (1985). Predatory feeding of the marine cyclopoid copepod (*Orycarus anglicus*). *Marine Biology* **85**, 163–169.

Landry, F., Miller, T. J. and Leggett, W. C. (1995). The effects of small-scale turbulence on the ingestion rate of fathead minnow (*Pimphales promelas*) larvae. *Canadian Journal of Fisheries and Aquatic Science* **52**, 1714–1719.

Larson, R. J. (1988). Feeding and functional morphology of the lobate ctenophore *Mnemiopsis maccradyi*. *Estuarine and Coastal Shelf Science* **27**, 495–502.

Lasker, R. (1975). Field criteria for survival of anchovy larvae: the relation between inshore chlorophyll maximum layers and successful first feeding. *Fishery Bulletin US* **73**, 453–462.

Lasker, R. (1978). The relationship between oceanographic conditions and larval anchovy food in the California Current: identification of factors contributing to recruitment failure. *Rapports et Procès-verbaux des Réunions. Conseil Permanent International pour l'Exploration de la Mer* **173**, 212–230.

Lasker, R. (1981). Factors contributing to variable recruitment of the northern anchovy (*Engraulis mordax*) in the California Current: contrasting years, 1975 through 1978. *Rapports et Procès-verbaux des Réunions. Conseil Permanent International pour l'Exploration de la Mer* **178**, 375–388.

Lazier, J. R. N. and Mann, K. H. (1989). Turbulence and diffusive layers around small organisms. *Deep-Sea Research* **36**, 1721–1733.

Leggett, W. C. (1986). The dependence of fish larval survival on food and predator densities. *In* "The Role of Freshwater Outflow in Coastal Ecosystems" (E. Skreslet, ed.), pp. 117–137. NATO ASI Series G7. Springer, Berlin, Germany.

Leggett, W. C. and DeBlois, E. (1994). Recruitment in marine fishes: is it regulated by starvation and predation in the egg and larval stages? *Netherlands Journal of Sea Research* **32**, 119–134.

Litvak, M. K. and Leggett, W. C. (1992). Age and size-selective predation on larval fishes: the bigger-is-better hypothesis revisited. *Marine Ecology Progress Series* **81**, 13–24.

Lough, R. G. and Bolz, G. R. (1989). The movement of cod and haddock larvae onto the shoals of Georges Bank. *Journal of Fish Biology* (Suppl. A) **35**, 71–79.

Mackas, D. L., Denman, K. L. and Abbott, M. R. (1985). Plankton patchiness: biology in the physical vernacular. *Bulletin of Marine Science* **37**, 652–674.

Mackas, D. L., Sefton, H., Miller, C. B. and Raich, A. (1993). Vertical habitat-partitioning by large calanoid copepods in the oceanic subarctic Pacific during spring. *Progress in Oceanography* **32**, 259–294.

MacKenzie, B. R. and Kiørboe, T. (1995). Encounter rates and swimming behaviour of pause travel and cruise larval fish predators in calm and turbulent environments. *Limnology and Oceanography* **40**, 1278–1289.

MacKenzie, B. R. and Leggett, W. C. (1991). Quantifying the contribution of small-scale turbulence to the encounter rates between larval fish and their zooplankton prey: effects of wind and tide. *Marine Ecology Progress Series* **73**, 149–160.

MacKenzie, B. R. and Leggett, W. C. (1993). Wind-based models for estimating the dissipation rates of turbulent energy in aquatic environments: empirical comparisons. *Marine Ecology Progress Series* **94**, 207–216.

MacKenzie, B. R., Leggett, W. C. and Peters, R. H. (1990). Estimating larval fish ingestion rates: can laboratory derived values be reliably extrapolated to the wild? *Marine Ecology Progress Series* **67**, 209–225.

MacKenzie, B. R., Miller, T. J., Cyr, S. and Leggett, W. C. (1994). Evidence for a dome-shaped relationship between turbulence and larval fish ingestion rates. *Limnology and Oceanography* **39**, 1790–1799.

Maillet, G. L. and Checkley, D. M. (1991). Storm-related variation in the growth rate of otoliths of larval Atlantic menhaden *Brevoortia tyrannus*: a time series analysis of biological and physical variables and implications for larva growth and mortality. *Marine Ecology Progress Series* **79**, 1–16.

Mann, K. H. and Lazier, J. R. N. (1991). "Dynamics of Marine Ecosystems: Biological–Physical Interactions in the Oceans." Blackwell Scientific, Boston, MA, USA.

Marassé, C., Costello, J. H., Granata, T. and Strickler, J. R. (1990). Grazing in a turbulent environment: energy dissipation, encounter rates and the efficacy of feeding currents in *Centropages hamatus*. *Proceedings of the National Academy of Science* **87**, 1653–1657.

Matsushita, K. (1992). How do fish larvae of limited mobility encounter nauplii in the sea? *Bulletin of the Plankton Society of Japan*, Special Volume, 251–270.

Miller, T. J. (1990). Body size and the ontogeny of feeding in fishes. PhD Dissertation. North Carolina State University, USA, 276 pp.

Miller, T. J., Crowder, L. B., Rice, J. A. and Marshall, E. A. (1988). Larval size and recruitment mechanisms in fishes: toward a conceptual framework. *Canadian Journal of Fisheries and Aquatic Sciences* **45**, 1657–1670.

Muelbert, J. H., Lewis, M. R. and Kelley, D. E. (1994). The importance of small-scale turbulence in the feeding of herring larvae. *Journal of Plankton Research* **16**, 927–944.

Munk, P. (1992). Foraging behaviour and prey size spectra of larval herring *Clupea harengus*. *Marine Ecology Progress Series* **80**, 149–158.

Munk, P. (1995). Foraging behaviour of larval cod (*Gadus morhua*) influenced by prey density and hunger. *Marine Biology* **122**, 205–212.

Munk, P. and Kiørboe, T. (1985). Feeding behaviour and swimming activity of larval herring (*Clupea harengus*) in relation to density of copepod nauplii. *Marine Ecology Progress Series* **24**, 15–21.

Neilson, J. D. and Perry, R. I. (1990). Diel vertical migrations of marine fishes: an obligate or facultative process? *Advances in Marine Biology* **26**, 115–168.

Oakey, N. S. and Elliott, J. A. (1980). An instrument to measure oceanic turbulence and microstructure. *Bedford Institute of Oceanography, Report Series* BI-R-77-3, 52 pp.

Oakey, N. S. and Elliott, J. A. (1982). Dissipation within the surface mixed layer. *Journal of Physical Oceanography* **12**, 171–185.

O'Brien, W. J., Evans, B. I. and Howick, G. L. (1986). A new view of the predation cycle of a planktivorous fish, white crappie (*Pomoxis annularis*). *Canadian Journal of Fisheries and Aquatic Sciences* **43**, 1894–1899.

O'Brien, W. J., Evans, B. I. and Browman, H. I. (1989). Flexible search tactics and efficient foraging in saltatory searching animals. *Oecologia* **80**, 100–110.

O'Brien, W. J., Browman, H. I. and Evans, B. I. (1990). Search strategies of foraging animals. *American Scientist* **78**, 151–160.

Ohman, M. D. (1990). The demographic benefits of diel vertical migration by zooplankton. *Ecological Monographs* **60**, 257–281.

Øiestad, V. (1985). Predation on fish larvae as a regulatory force, illustrated in mesocosm studies with large groups of larvae. *Northwest Atlantic Fisheries Organization Scientific Council Studies* **8**, 25–32.

Oviatt, C. A. (1981). Effects of different mixing schedules on phytoplankton, zooplankton and nutrients in marine microcosms. *Marine Ecology Progress Series* **4**, 57–67.

Owen, R. W. (1989). Microscale and finescale variations of small plankton in coastal and pelagic environments. *Journal of Marine Research* **47**, 197–240.

Peterman, R. M. and Bradford, M. J. (1987). Wind speed and mortality rate of a marine fish, the northern anchovy, *Engraulis mordax. Science, NY* **235**, 354–356.

Peters, F. and Gross, T. (1994). Increased grazing rates of microplankton in response to small-scale turbulence. *Marine Ecology Progress Series* **115**, 299–307.

Pond, S. and Pickard, G. L. (1983). "Introductory Dynamical Oceanography", 2nd edn. Pergamon Press, Oxford, UK.

Price, J. F., Terray, E. A. and Weller, R. A. (1987). Upper ocean dynamics. *Review of Geophysics* **25**, 193–203.

Reynolds, W. W. and Thomson, D. A. (1974). Responses of young Gulf grunion, *Leuresthes sardina*, to gradients of temperature, light, turbulence and oxygen. *Copeia* **1974**, 747–758.

Rice, J. A., Miller, T. J., Rose, K. A., Crowder, L. B., Marshall, E. A. and DeAngelis, D. L. (1993). Growth rate variation and larval survival: implications of an individual-based size-dependent model. *Canadian Journal of Fisheries and Aquatic Sciences* **50**, 133–142.

Rosenthal, H. and Hempel, G. (1970). Experimental studies in feeding and food requirements of herring larvae (*Clupea harengus*). *In* "Marine Food Chains" (J. H. Steele, ed.), pp. 344–364. Oliver and Boyd, Edinburgh, UK.

Rothschild, B. J. and Osborn, T. R. (1988). Small-scale turbulence and plankton contact rates. *Journal of Plankton Research* **10**, 465–474.

Roy, C., Cury, P. and Kifani, S. (1992). Pelagic fish recruitment success and reproductive strategy in upwelling areas: environmental compromises. *In* "Benguela Trophic Functioning" (A. I. L. Payne, K. H. Brink, K. H. Mann and R. Hilborn, eds). *South African Journal of Marine Science* **12**, 135–146.

Saiz, E. (1994). Observations of the free-swimming behaviour of *Acartia tonsa*: Effects of food concentration and turbulent water motion. *Limnology and Oceanography* **39**, 1566–1578.

Saiz, E. and Alcaraz, M. (1991). Effects of small-scale turbulence on development time and growth of *Acartia grani* (Copepoda: Calanoida). *Journal of Plankton Research* **13**, 873–883.

Saiz, E. and Alcaraz, M. (1992a). Free-swimming behaviour of *Arcartia clausii* (Copepoda: Calanoida) under turbulent water movement. *Marine Ecology Progress Series* **80**, 229–236.

Saiz, E. and Alacaraz, M. (1992b). Enhanced excretion rates induced by small-scale turbulence in *Acartia* (Copepoda: Calanoida). *Journal of Plankton Research* **14**, 681–689.

Saiz, E. and Kiørboe, T. (1995). Predatory and suspension feeding of the copepod *Acartia tonsa* in turbulent environment. *Marine Ecology Progress Series* **122**, 147–158.

Saiz, E., Alcaraz, M. and Paffenhöfer, G. (1992). Effects of small-scale turbulence on feeding rate and growth efficiency of three *Acartia* species (Copepoda: Calanoida). *Journal of Plankton Research* **14**, 1085–1097.

Shimeta, J., Jumars, P. A. and Lessard, E. J. (1995). Influences of turbulence on suspension feeding by planktonic protozoa: experiments in laminar shear fields. *Limnology and Oceanography* **40**, 845–859.

Sinclair, M. (1988). "Marine populations." Washington Sea Grant Program, Seattle, USA.

Sinclair, M. and Iles, T. D. (1985). Atlantic herring (*Clupea harengus*) distributions in the Gulf of Maine–Scotian Shelf area in relation to oceanographic features. *Canadian Journal of Fisheries and Aquatic Sciences* **42**, 880–887.

Smith, T. D. (1994). "Scaling Fisheries: The Science of Measuring the Effects of Fishing, 1855–1955." Cambridge University Press, Cambridge, UK.

Sundby, S. and Fossum, P. (1990). Feeding conditions of Arcto-Norwegian cod larvae compared with the Rothschild–Osborn theory on small-scale turbulence and plankton contact rates. *Journal of Plankton Research* **12**, 1153–1162.

Sundby, S., Ellertsen, B. and Fossum, P. (1994). Encounter rates between first-feeding cod larvae and their prey during moderate to strong turbulent mixing. *International Council for the Exploration of the Sea, Marine Science Symposium* **198**, 393–405.

Taggart, C. T. and Leggett, W. C. (1987a). Short-term mortality in post-emergent larval capelin *Mallotus villosus*. I. Analysis of multiple *in situ* estimates. *Marine Ecology Progress Series* **41**, 205–217.

Taggart, C. T. and Leggett, W. C. (1987b). Short-term mortality in post-emergent larval capelin *Mallotus villosus*. II. Importance of food and predator density, and density-dependence. *Marine Ecology Progress Series* **41**, 219–229.

Therriault, J.-C. and Platt, T. (1981). Environmental control of phytoplankton patchiness. *Canadian Journal of Fisheries and Aquatic Science* **38**, 638–641.

Tiselius, P. and Jonsson, P. R. (1990). Foraging behaviour of six calanoid copepods: observations and hydrodynamic analysis. *Marine Ecology Progress Series* **66**, 23–33.

Townsend, A. A. (1976). "The Structure of Turbulent Shear Flow", 2nd edn. Cambridge University Press, Cambridge, UK.

van der Meeren, T. and Næss, T. (1993). How does cod (*Gadus morhua*) cope with variability in feeding conditions during early larval stages? *Marine Biology* **116**, 637–647.

Vlymen, W. J. (1977). A mathematical model of the relationship between larval anchovy (*Engraulis mordax*) growth, prey microdistribution, and larval behaviour. *Environmental Biology of Fishes* **2**, 211–233.

Ware, D. M. and Thomson, R. E. (1991). Link between long-term variability in upwelling and fish production in the northeast Pacific Ocean. *Canadian Journal of Fisheries and Aquatic Sciences* **48**, 2296–2306.

Werner, F. E., Page, F. H., Lynch, D. R., Loder, J. W., Lough, R. G., Perry, R. I., Greenberg, D. A. and Sinclair, M. M. (1993). Influences of mean advection and simple behaviour on the distribution of cod and haddock early life stages on Georges Bank. *Fisheries Oceanography* **2**, 43–64.

Werner, F. E., Perry, R. I., MacKenzie, B. R., Lough, R. G. and Naime, C. E. (1995). Larval trophodynamics, turbulence and drift on Georges Bank: a sensitivity analysis of cod and haddock. International Council for the Exploration of the Sea, Annual Science Conference, 21–26 September 1995. Aalborg, Denmark, ICES CM 1995/Q:26.

Werner, F. E., Perry, R. I., Lough, R. G. and Naime, C. E. (1996). Trophodynamics and advective influences on Georges Bank larval cod and haddock. *Deep-Sea Research* **43**.

Yamazaki, H. and Osborn, T. R. (1988). Review of oceanic turbulence: implications for biodynamics. *In* "Toward a theory on Biological–Physical

220

Interactions in the World Ocean" (B. J. Rothschild, ed.), pp. 215–234. Kluwer, Dordrecht, Netherlands.

Yamazaki, H., Osborn, T. R. and Squires, K. D. (1991). Direct numerical simulation of planktonic contact in turbulent flow. *Journal of Plankton Research* **13**, 629–643.

Adaptations of Reef Corals to Physical Environmental Stress

Barbara E. Brown

Department of Marine Sciences and Coastal Management, University of Newcastle upon Tyne, Newcastle upon Tyne NE1 7RU, UK

I speculated whether a species very liable to *repeated* and great changes in conditions might not assume a fluctuating condition ready to be adapted to
Darwin (1881)

ADVANCES IN MARINE BIOLOGY VOL. 31
ISBN 0-12-026131-6

1. INTRODUCTION

Geological history has shown that coral reefs have a considerable capacity to withstand environmental stresses, having survived and evolved during periods of major climatic upheaval over the past 260 million years (Grigg, 1994). The ability (or otherwise) of corals to adapt to environmental stress is fundamental to understanding evolutionary change over geological time and to prediction of coral reef survival under increasingly stressful conditions of altered climate and pollution.

In recent years there has been increasing recognition (Done, 1992; Woodley, 1992; Veron, 1995) that coral reefs are *not* the stable communities living in benign environments, freed from seasonal fluctuations, as suggested by Newell (1971); instead, they are ecosystems subject to frequent disturbances on time-scales of minutes to years (Figures 1 and 2). However, there has been relatively little attempt to collate adaptational processes shown by reef corals as they respond to an environment that changes over these time-scales.

The aims of this review are to examine the nature of selected environmental fluctuations on reefs through time, and the ability of modern reef corals to adapt to these changing conditions. While it is recognized that life-history traits (e.g. growth, reproductive effort, generation time) have critical roles to play in adaptive responses, the main focus of this review is physiological adaptation to physical stress as reflected at the cellular and organism levels of organization.

Almost all reviews on environmental stress invariably commence with an evaluation of the term "stress". Increasingly, a common view is taken that stress describes an environmental stimulus which, by exceeding a threshold value, disturbs animal or plant function by a detectable departure from a steady state (Bayne, 1985; Bradshaw and Hardwick, 1989; Grime, 1989; Hoffman and Parsons, 1991). Bayne (1975) defined stress as a "measurable alteration of a physiological (or behavioural, biochemical or cytological) steady state which is induced by environmental change, and which renders the individual (or the population or community) more vulnerable to further environmental change".

When organisms respond to environmental stress they do so by expressing physiological and other attributes, which may be adaptive. Such adaptations may compensate in part for loss in performance but may also

Figure 1 Coral reefs disturbed by extreme physical environment factors. (A) An intertidal coral reef at Phuket, Thailand, in June 1991 showing extensive bleaching of the branching coral *Acropora aspera* in response to elevated sea water temperatures. Six months later these corals had recovered their algal symbionts and were fully pigmented. (B) Shallow water reef corals at Pulau Pari, Indonesia, in May 1983 showing species-specific bleaching in response to elevated sea water temperature resulting from the ENSO event of that year.

Figure 2   Storm damage to branching *Acropora cervicornis* at 8 m depth on the fore-reef at Discovery Bay, Jamaica, 3 years after Hurricane Allen in 1980. (Photograph courtesy of Dr T. Scoffin.)

be less than fully effective and so represent a reduction in fitness relative to optimal conditions (Bayne, 1985). Adaptation to environmental stress consists of a genetic and a non-genetic, or phenotypic, component. Non-genetic or phenotypic adaptations usually involve changes in the performance of an individual during its lifetime. Bradshaw and Hardwick (1989) predicted that genetic adaptation to stress would result where selection was operating consistently, while phenotypic adaptation would be more likely when selection operated in a temporary or fluctuating manner. Such adaptive phenotypic responses are able to occur at a speed which matches the speed of environmental change and would be produced within a single genotype through phenotypic variability or plasticity.

   In corals, very specific morphological and physiological phenotypic responses can be found in relation to different stresses, with responses operating over a wide range of biological time-scales. Photoadaptation is an example of a phenotypic adaptation which involves a remarkable spectrum of responses ranging from changes at the cellular level to behavioural and growth form variations. However, in corals, as in plants (Bradshaw and Hardwick, 1989), most interest has been in morphological rather than physiological phenotypic plasticity. Veron and Pichon (1976) have described corals as highly plastic organisms, referring particularly to the considerable range of variations that they exhibit in their skeletal characters. The authors did not distinguish, however, between genotypically and phenotypically determined variations in their definition of "coral ecomorphs". This term was adopted for coral colonies that showed wide intraspecific skeletal variation in response to specific ecological conditions.

Corals, then, show a wide array of morphologies within species; they also exhibit a diverse range of feeding strategies (Porter, 1976; Lasker, 1976, 1981); reproductive patterns (Sammarco, 1982; Richmond, 1985; Harrison and Wallace, 1990) and other life-history traits (Jackson and Hughes, 1985; Jackson, 1991; Hughes *et al.*, 1992). Much of this variation may be attributed to genetic differences between species. The level of phenotypic adaptation remains largely unexplored, apart from the evaluation of morphological responses of coral colonies to the environment, where some species have been shown to exhibit wide phenotypic plasticity (Foster, 1979, 1980; Willis, 1985). In others, however, there is limited phenotypic plasticity and a strong association between genotype and growth form (Willis, 1985; Willis and Ayre, 1985; Ayre and Willis, 1988; Weil, 1993). It has also been argued that for a number of coral species reports of phenotypic plasticity have been confused by taxonomic errors (Knowlton and Jackson, 1994).

Difference in the extent of phenotypic variation between species and populations are usually explained in terms of natural selection adjusting plasticity levels in response to the degree of environmental variability experienced (Hoffman and Parsons, 1991). For corals it is likely that the major influences on adaptive processes, both genetic and phenotypic, are the products, firstly, of fluctuating abiotic (and biotic) factors, which have varied both in geological and ecological time and, secondly, the heterogeneity of the coral reef environment.

2. FLUCTUATING ABIOTIC INFLUENCES: VARIATIONS OVER TIME

2.1. Environmental Fluctuations in Geological Time

Tropical areas are much less environmentally stable over geological time than originally thought (Taylor, 1978; Colinvaux, 1987; Paulay, 1990) with reefs (broadly defined here as including all biologically produced, three-dimensional bodies of sediment raised above the surrounding sea floor—see Copper, 1994) being destroyed by the major extinction events in the last 500 million years (Sheehan, 1985). Patterns in Figure 3 show that the extinction profiles of reef faunas and other faunas are similar and suggest some common external factors. In other words, extinctions were being physically rather than biologically driven (Raup and Boyajian, 1988).

However, it is important in considering the selectivity or uniformity of extinction of reef organisms to define precisely what is meant by reef or reefal. Rosen and Turnsek's (1989) analysis of extinctions, in contrast to that of Raup and Boyajian (1988), suggests a more selective extinction pattern for reef corals based on a division between zooxanthellate (those corals harbouring symbiotic algae or zooxanthellae) and non-zooxanthellate species. They propose that zooxanthellate corals were much more susceptible to extinction than non-zooxanthellates, especially in the late Cretaceous (approximately 70 million years ago). They consider that anoxic conditions and climate cooling were probably unimportant in the demise of zooxanthellate corals at that time and that extinction patterns would more likely result from climatic warming, reduced illumination due to dust in ocean and atmosphere, and tectonic/eustatic regression.

Reconstructions of long-term variations in climate suggest smooth and gradual changes in parameters such as ocean temperature and atmospheric carbon dioxide concentrations (Wilkinson and Buddemeier, 1994), but this may be a false impression based on low chronological precision and incomplete stratigraphic records (Buddemeier and Hopley, 1988). For this reason reconstruction of variability of abiotic factors during geological time is restricted in Figure 3 to the last 150 000 years, for which more complete records are available.

Figure 4a shows a reconstruction of global sea level over the last 150 000 years based on analyses of uplifted Pleistocene reef terraces in New Guinea (Chappell, 1983). This profile should be interpreted as a global average curve, the amplitude of which is expected to vary from one part of the world to another as a result of changing water-ice distributions on the shape of the earth itself (Clark et al., 1978). Over this time-scale, sea level has oscillated through a range of 100 metres, with rates of change often exceeding 10 metres per 100 years. The sea level curve through geological history may be interpreted as a low-frequency cycle of major glacial and interglacial periods with a periodicity of more than 100 000 years and a smaller higher frequency oscillation every 20 000 years (Chappell, 1981). Potts (1984a) has proposed a model in which populations of corals in shallow water habitats were continually and severely disrupted, primarily by the smaller fluctuations during the latter part of the Quaternary period of glacial–interglacial cycles (the last 2 million years). He argues that this was the outcome of the mean amplitude of the fluctuations, which were greater (approximately 25 metres) than the bathymetric zone (< 20 metres) supporting most coral growth, and the frequency of these disturbances, which provided too little time for relatively long-lived corals

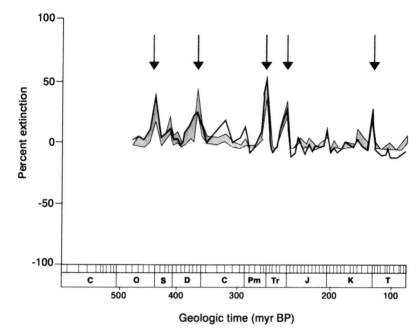

Figure 3 Extinction profile for 417 reef genera (shown by hatch band) compared with all other genera (shown by solid line). Arrows show the timing of major extinction events during geological time: C, Cambrian; O, Ordovician; S, Silurian; D, Devonian; C, Carboniferous; Pm, Permian; Tr, Triassic; J, Jurassic; K, Cretaceous; T, Tertiary; myr BP, millions of years before the present. (After Raup and Boyajian, 1988.)

to respond, in an evolutionary sense, to the rapidly changing environment.

Simultaneously through this period, palaeoclimatic records show considerable variability in sea surface temperature (SST), which has been described as the result, at least in part, of orbital forced variations in trade wind and monsoon strength (Prell and Kutzbach, 1987; McIntyre *et al.*, 1989). Figure 4b–e shows the fluctuations in monsoon strength over the last 150 000 years as evidenced by the monsoon pollen index in Gulf of Aden marine sediments (van Campo *et al.*, 1982), monsoon upwelling (Prell, 1984), salinity changes in the Bay of Bengal (Cullen, 1981) and SST values obtained from the South Indian Ocean (Hutson and Prell, 1980). These and other palaeoclimatic records indicate four strong monsoon-related events at 10, 82, 104 and 126×10^3 years BP according to Prell and Kutzbach (1987).

While there may be controversy about the absolute SST recorded from proxies such as foraminifera (used above) and corals (CLIMAP, 1981;

Adams *et al.*, 1990; Anderson and Webb, 1994; Crowley, 1994; Guilderson *et al.*, 1994) it is clear that there were not only marked fluctuations in SST but also in seasonality (Figure 4f) in some areas of the tropical ocean during the last 150 00 years (McIntyre *et al.*, 1989).

Fluctuations in SST in the past would result not only from major influences such as climate changes but also from interannual phenomena like El Niño southern oscillation (ENSO) — a complex atmospheric and oceanographic feature which sporadically alters large-scale atmospheric pressure systems, wind systems, rainfall patterns, ocean currents and sea level variations in the tropical and subtropical oceans (Bjerknes, 1969; Wyrtki, 1975, 1979, 1985; Cane, 1983, 1986; Rasmusson, 1985). ENSO events may have profound effects on coral reefs through elevation of sea water temperatures, depression of mean sea levels, high irradiance during calm conditions and cyclonic disturbances (Brown, 1987). In the eastern Pacific, Glynn and Colgan (1992) estimated that during the last 6000 years, between 12 and 30 ENSO events of severe magnitude may have interrupted eastern Pacific reef building. Similar short-term disruption of reefs may have also resulted from hurricane activity in geological time. Ball *et al.* (1967) estimated that 160 000–320 000 hurricanes have hit the Florida Keys during the course of the last 2 million years.

During the last 2 million years certain families of scleractinian corals have survived the rigours of climate change and unpredictable fluctuations in abiotic conditions but many others have become extinct. Survivors with a long geological history include the branching corals from the families Pocilloporidae and Acroporidae, the massive corals of the family Poritidae, the Faviidae, Siderastreidae, Agariciidae and solitary corals of the family Fungiidae—almost all of which have had a history spanning several million years. Veron and Kelley (1988) have calculated that the rates of evolution and extinction of Indo-Pacific species over the last 2 million years have been similar, with a value of 4.4% of species per million years. They further argue that although Plio–Pleistocene glacio-eustatic changes have had catastrophic effects on coral reefs, they have not had

---

Figure 4 Reconstruction of long-term variation in climate over the last 150 000 years. (a) Relative sea level (m) (after Chappell, 1983). (b) Monsoon strength as deduced from the monsoon pollen index (after van Campo *et al.*, 1982). (c) Arabian Sea upwelling index as deduced from planktonic assemblages (after Prell, 1984). (d) Salinity changes in the Bay of Bengal as deduced from foraminiferan records (after Cullen, 1981). (e) Indian Ocean sea surface temperature (SST) as deduced from oxygen isotope stratigraphy (after Hutson and Prell, 1980). (f) Seasonality measured as the annual range of sea surface temperature (SST) variability as deduced from oxygen isotope stratigraphy from a core (RC24-16) taken in the eastern equatorial Atlantic (after McIntyre *et al.*, 1989).

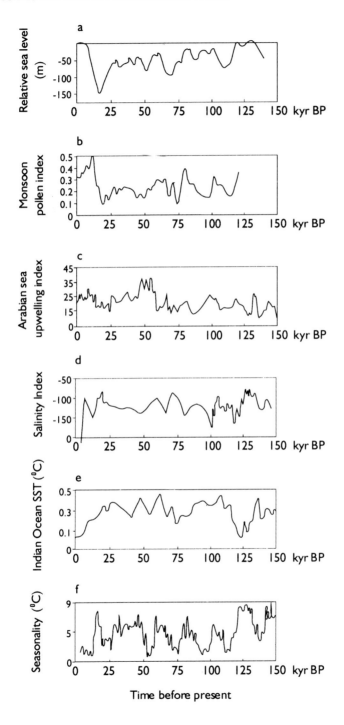

a major effect on the diversity of reef coral species (with the possible exception of the Acroporidae), at least over the past 2–3 million years.

In contrast to the Indo-Pacific, the Caribbean fauna has changed markedly over this period, with the extinction of some genera and also the addition of new groups such as the agariciids (Veron and Kelley, 1988). During the Plio–Pleistocene turnover 2–3 million years ago, the genera *Pocillopora*, *Stylophora* and *Goniopora* became extinct in the Caribbean, the last to become extinct being *Pocillopora*, identified on the youngest reef terraces of Barbados (Geister, 1978). Extinction of these genera has been attributed to global cooling and eustatic sea level change (Frost, 1977) but recent studies of the fossil record (Budd *et al.*, 1993) suggest that neither cooling nor global changes in sea level were likely causal factors. The fact that faunal stability (Mesolella, 1967; Mesolella *et al.*, 1970; Jackson, 1992) prevailed throughout the high-frequency temperature and sea level fluctuations of the Pleistocene also argues against sea level and temperature being primary causes of coral extinction. Instead, Budd *et al.* (1993) suggest that more regional abiotic factors, such as altered oceanic circulation patterns resulting from the closure of the Isthmus of Panama 3.5 million years ago, may have been ultimately responsible for the demise of selected coral genera. Despite the Plio–Pleistocene turnover, species richness changed little and reef communities were common across the Caribbean during the Pleistocene.

It has been argued by Potts and Garthwaite (1991) that the survival of most coral taxa in both the Caribbean and Indo-Pacific over the last 2–3 million years must be attributed, at least in part, to their extensive intraspecific genetic variation, permitting rapid local adaptation to environmental conditions encountered at a particular site in any one generation. But just how variable and predictable were environmental changes on coral reefs: on a time-scale of decades or centuries? Some idea of this variability can perhaps be gleaned from closer examination of fluctuations of abiotic factors on the modern reef.

2.2. Environmental Fluctuations on the Modern Reef

The earlier literature abounds with statements on the relative stability of physical variables in the tropical latitudes, many of these relating particularly to temperature (Wallace, 1878; Vernberg, 1962; Newell, 1971) and also lack of seasonality (Newell, 1971; Rosen, 1981). Subsequently, Rosen (1984) qualified the idea of stability of tropical climates, highlighting the influence of monsoonal regimes in some tropical areas and the distinctly seasonal climatic patterns of regions such as the Great Barrier Reef and northern Red Sea. A careful examination of a selection of historical environmental data and information resulting from continuous monitoring of environmental variables on coral reefs clearly indicates that these habitats are subject to significant fluctuations in physical conditions on an internannual, annual, seasonal and daily basis (Mayer, 1914; Orr, 1933a, b; Wells, 1952; Glynn and Stewart, 1973; Potts and Swart, 1984; Porter, 1985; Wolanski and Pickard, 1985; Cubit et al., 1989; Cook et al., 1990; Coles and Fadlallah, 1991; Hayashibara et al., 1993; Brown et al., 1994a; Lough, 1994).

2.2.1. Interannual/Annual Fluctuations

The major interannual/annual abiotic variations that significantly affect coral reefs include long-term solar radiation variability due to orbital changes of the earth, short-term changes in total solar radiant flux, ENSO events, fluctuation in monsoonal strength and hurricane activity.

2.2.1.1. *Variations in solar radiation*   There are two sources of solar radiation variability (see Shine et al., 1990, for review). The first acts on time-scales of 10 000–100 000 years and is caused by changes in the Sun–Earth orbital parameters. Variations in climate on these time-scales are thought to be initiated in the Earth's orbital parameters, which in turn influence the latitudinal and seasonal variation of solar energy received by the Earth (the Milankovitch effect) and which have been implicated in initiating the major glacial–interglacial cycles in geological time. The second source of variability results from changes in the energy output of the sun in response to changes in solar surface activity (sunspots and faculae), which appears to follow in phase with the 11 year sunspot cycle (Kerr, 1991). This second source of variability, which manifests itself in very small changes in the solar irradiance at the earth's surface, is unlikely to have any marked effect upon the physical environment of coral reefs (Thomson, 1995). Any change in solar radiation would have a dual effect on coral reefs: first, indirectly through climate change (sea temperature, cloudiness and seasonality) and, second, directly on the quantum energy available for photosynthesis.

2.2.1.2. *ENSO*   The El Niño southern oscillation (ENSO) phenomenon is recognized as an irregular oscillation of the coupled ocean–atmosphere system in the tropical Pacific, occurring approximately every 3–5 years. During the peak of the El Niño, SSTs in the eastern tropical Pacific can be several degrees above the climatological mean. The climatic influences of the El Niño are not restricted to the Pacific alone; for example, it has long been recognized that the Asian monsoon and ENSO are linked (Quinn, 1992). During the El Niño phase of the oscillation, the Asian south-west monsoon tends to be weak (Gregory,

1989a). The correspondence of precipitation anomalies between the eastern and central Pacific and Indonesia and northern Australia have also been clearly established (Fennessey and Shukla, 1988). In addition, mean sea level may be significantly depressed during ENSO years at sites outside the Pacific with links between the southern oscillation index (SOI) and mean sea level shown for sites in the Indian Ocean and South China Sea (Bray *et al.*, 1996). Recent work (Tourre and White, 1996) has also indicated El Niño-related sea water warming in the Indian Ocean with possible links to sea water warming in the Atlantic 18 months later.

The most influential environmental fluctuations for coral reefs which result from ENSO activity are increased sea temperatures, possibly higher solar irradiance as a result of altered weather patterns, and lowered sea levels. Over the period 1728–1983, Quinn *et al.* (1987) noted 47 ENSO events which they described as strong or very strong, using SOI values. The 1982–83 event was an example of a very strong effect which resulted in considerable coral mortality at sites across the Pacific from reefs in Panama (50–90% mortality) (Glynn, 1984), French Polynesia (Salvat, in Glynn, 1984), Tokelau Islands in the central Pacific (Perez, in Glynn, 1984), Southern Japan (Yamaguchi, in Glynn, 1984) and Indonesia (80–90% mortality on sheltered, shallow reef flats) (Brown and Suharsono, 1990).

Recovery of reefs may take at least several decades, and for those affected by the 1982–83 El Niño recovery has been slow. Delayed or long-term effects resulting from the 1982–83 ENSO in the Pacific have continued at least up until 1991. These have included altered corallivore foraging patterns, a shift in dominance from live coral to algal turf, macroalgae and other non-reef building taxa and increases in sea urchin abundance on dead reef frameworks which have led to severe bioerosion (Glynn, 1988, 1993; Glynn and D'Croz, 1990). On several eastern Pacific reefs bioerosion now exceeds coral framework production (Eakin, 1992).

In French Polynesia sea water warming associated with ENSO events in 1986–87, 1991 and most recently in 1993 has interrupted the recovery of reefs affected by the 1982–83 El Niño by causing further coral mortality and sublethal effects that ultimately might affect growth and reproduction (Gleason, 1993b). Delayed recovery of coral reefs, following the 1982–83 ENSO, has also been observed on the shallow reef flats of Indonesia. Here, lowered sea levels in 1987 and 1990 and redistribution of shingle resulting from coral mortalities in 1982–83 have caused significant declines in coral cover on reefs that had begun to show signs of recovery over the period 1985–87 (Brown, 1996).

Coral reefs thus appear to be repeatedly subject to physical variations in their environment which ensue as a result of ENSO activity. Wolter and Hastenrath (1989) have shown that over the past 40 years the negative southern oscillation phase and its manifestations throughout the global tropics have become more common. An increase in the frequency and/or strength of ENSO events could have profound implications for coral reefs, as recent patterns of coral mortality indicate.

2.2.1.3. *Monsoon variability* The monsoon, especially the Asian monsoon, displays significant seasonal variation and interannual variability with the onset and retreat of the summer monsoons in Asia and Australia being associated with abrupt changes in the atmospheric general circulation (Yeh *et al.*, 1959; McBride, 1987). An earlier or later monsoon onset, or a longer or shorter duration, usually causes flood or drought (Gates *et al.*, 1990). More significantly for coral reefs in monsoonal regions, alterations in the monsoon timing and intensity cause marked changes in coastal upwelling, water circulation patterns, solar irradiance and turbidity due to increased/decreased freshwater run-off. Part of the interannual variability of the summer monsoon appears to be associated with anomalies in SST, both local (Kershaw, 1988) and remote (Meehl, 1994). For example there is an apparent correlation between the strength of the Indian monsoon and SST in the eastern tropical Pacific, in that a poor monsoon is generally associated with a warm SST in the east Pacific (Gregory, 1989b). Variability in ENSO signals have been recorded in the geochemistry of living corals from Galapagos (Shen *et al.*, 1992) and from living and fossil corals from Papua New Guinea (Tudhope, 1993; Tudhope *et al.*, 1994) while variability in the south-west monsoon has been noted from the geochemical record of living corals from Oman (Tudhope, 1993).

2.2.1.4. *Hurricanes* Another major intermittent disturbance in coral reef areas is hurricane activity (Scoffin, 1993). Frequent hurricanes occur in belts 7–25° north and south of the equator. Outside these areas hurricanes are rare and of lesser intensity. Reefs growing in hurricane belts suffer periodic damage from hurricane storms and surge. Most damage is caused by massive waves in a depth zone of 0–20 m but coral boulders falling down steep slopes cause damage at much greater depths, for example, > 100 m (Harmelin-Vivien, 1985).

It should be noted that not all hurricanes are of equal force and although the annual average of hurricanes hitting an area may be as much as 5–8 per year for the Great Barrier Reef (Done, 1992) or Caribbean (Woodley, 1992), respectively, any particular reef may only be struck by a severe hurricane 4–8 times a century (Harmelin-Vivien, 1985). There are of course exceptions to this—in French Polynesia six hurricanes ravaged the reefs between December 1982 and April 1983. In this example hurricane intensities were similar to those reported in 1903–05 which wrought catastrophic damage to the Polynesian Islands (Harmelin-Vivien and Laboute, 1986). Hurricane-induced waves exceeding 10 m height broke coral colonies at depths of 18–20 m and many were thrown up as boulder ramparts on the outer reef flats. Other broken corals rolled down the steep slope, shattering deeper, more fragile and foliaceous corals at depths below 40 m, resulting in increasingly destructive effects with depth.

The time for reefs to recover to pre-hurricane conditions varies between 5 and 40 years, depending on the degree of initial damage to the reef and providing that there is no intervening disturbance (Scoffin, 1993).

2.2.2. Seasonal Fluctuations

Seasonality affects all reefs as evidenced by the seasonal banding patterns in massive coral skeletons worldwide (Buddemeier and Kinzie, 1976). Seasonality is a major factor in monsoonal regions where 70% of all coral reefs are found (Potts, 1983; Alongi, 1987; McClanahan, 1988).

The geographical distribution of seasonal range in physical variables has frequently been misunderstood for, as Moore (1972) noted: "it has been stated that there is no seasonal change of temperature in tropical waters". Such statements arise because the database used refers only to oceanic rather than coastal waters. Also, the limited data for coastal areas have not, until recent years, allowed access to mean monthly maximum and minimum temperatures which may be more relevant (as extremes) than mean monthly or annual temperatures in interpreting physiological tolerances of organisms (Hoffman and Parsons, 1991). This latter point is likely to be particularly important for tropical marine organisms. Mayer noted as early as 1914 that tropical biota live closer to their upper lethal temperature limits than organisms at higher latitudes—a feature widely acknowledged to be important in the marked susceptibility of reef corals to elevations in sea water temperature above the seasonal maximum (Jokiel and Coles, 1990).

The effects of latitude on the salinity range in coastal waters, on spring/neap tidal ranges, on total solar radiation, on precipitation and on ozone levels are shown in Figure 5. For solar radiation (Figure 5a) the greatest ranges are observed in higher temperate latitudes when compared with the tropics—nevertheless, tropical areas show considerable seasonal variability. For salinity (Figure 5b) ranges in the tropics are similar to those in the temperate latitudes (Moore, 1972). Coral reefs are strongly affected by tidal influences. On intertidal reefs seasonal variations in the range and timing of the tides have important consequences for periods of subaerial exposure, whilst on reefs which are not subject to emersion, solar heating over shallow water reef areas may be intense at low tide (Glynn and Stewart, 1973; Potts and Swart, 1984). A measure of the tidal variability

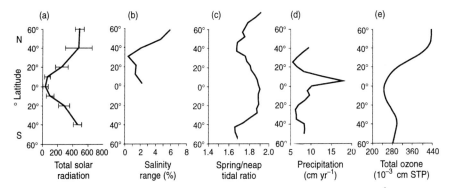

Figure 5   The effects of latitude on selected physical parameters. (a) Total solar radiation (figures calculated from Budyko, 1974). Means shown ± standard deviation. (b) Salinity range for northern latitudes of the coast of the United States (after Moore, 1972). (c) Spring/neap tidal ranges from 5000 localities covering all seas (after Moore, 1972). (d) Average values of precipitation for all oceans (after Sverdrup *et al.*, 1942). (e) Average total ozone averaged over the period 1957–67 for the month of January (after Cutchis, 1982).

encountered not only on a daily basis but also on a monthly basis is the ratio of spring/neap tidal ranges. Moore (1972) has shown that the ratio is higher in tropical than at temperate latitudes (Figure 5c) and that associated stresses may therefore be considerable in tropical waters.

Precipitation is also maximal in the tropical latitudes (Figure 5d) (Sverdrup *et al.*, 1942). With high evaporation levels in these regions, the net effect of evaporation and precipitation would be to cause fluctuations in salinity in those tropical waters that receive considerable run-off from land. Finally, there is a marked difference in ozone concentrations with latitude (Figure 5e), which in turn results in significantly higher levels of ultraviolet radiation in the tropical belt compared with temperate latitudes (Cutchis, 1982).

Kinsey and Domm (1974) and Kinsey (1985) noted the marked nature of seasonality in community metabolism of coral reefs and suggested that the phenomenon did not appear to be strongly correlated with latitude despite a greater seasonality of temperature with increasing latitude. It is worth looking at the evidence for greater seasonality of temperature within tropical latitudes. Interpretation of the data is obviously constrained by a number of factors that include differences in instrumentation, the comparison of continuous and discontinuous sampling techniques, the particular hydrography of selected sites and the varying depth ranges from which measurements have been taken. Nevertheless, a plot of seasonal maximum and minimum temperatures (and, where available, seasonal monthly temperature ranges) reveals very little latitudinal pattern even

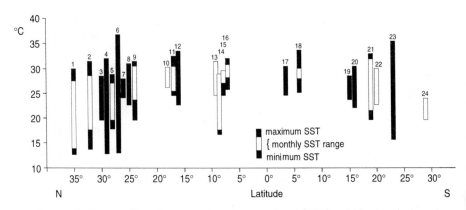

Figure 6 Seasonality of sea surface temperature (SST) within tropical and subtropical latitudes. Data derived from the following sources: 1, Tribble and Randall (1986); 2, Cook *et al.* (1990); 3, Fishelson (1973); 4, Downing (1985); 5, Jokiel and Coles (1990); 6, Coles and Fadlallah (1991); 7, Hayashibata *et al.* (1993); 8, Fitt *et al.* (1993); 9, Ogden *et al.* (1994); 10, Edmunds and Davies (1989); 11, Glynn (1973); 12, Yap *et al.* (1994); 13, Cubit *et al.* (1989); 14, Wellington and Glynn (1983); 15, Wellington and Glynn (1983); 16, Dunne (1993a); 17, Anderson and Sapulette (1981); 18, Dunne (1993b); 19, Wolanski and Pickard (1985); 20, Wolanski and Pickard (1985); 21, Kenny (1974); 22, Babcock *et al.* (1994); 23, Potts and Swart (1984); 24, Babcock *et al.* (1994). Maximum and minimum SST shown together with monthly SST range where available.

when sites known to be particularly shallow are excluded (Figure 6). There appears to be very little evidence of increased seasonality with increasing latitude in the waters overlying coral reefs. What is reflected is the considerable variation at both higher and lower latitudes in SST ranges—a factor which is more likely to be a function of the choice of site. It is noteworthy that seasonal SST ranges are quite marked in the low latitudes and annual temperature ranges of 4 to 5 centigrade degrees are not uncommon in shallow waters (3–5 m) surrounding reefs close to the equator in Indonesia and Thailand.

While temperature is often the most cited environmental variable in discussion of seasonality, seasonal differences in solar radiation, sedimentation regimes, mean sea level, wave energy and nutrient concentrations may also be marked for many reefs (Figure 7). Seasonal fluctuations in salinity values, current regimes and tidal patterns are also commonplace on coral reefs (Andrews and Pickard, 1990).

2.2.3. *Daily Fluctuations*

Daily variations in abiotic factors in reef waters are poorly documented yet they hold an important key to the understanding of the physiological tolerances of reef organisms and their scope for adaptation. In general, most physical data pertain to sea water temperatures (see Andrews and Pickard, 1990, for review of diurnal changes in physical parameters). Some of the earliest references suggest that corals in pools may survive brief exposure, at low tide, to temperatures of 56°C (Gardiner, 1903). Such temperatures are well above those reported by Mayer (1918) as the upper temperature limit (36–38°C) for corals. At other sites temperatures of 40°C attained on reef flats at midday may cause severe mortality to reef organisms (Glynn, 1968).

Tidal influences (both diurnal and semi-diurnal) are clearly important factors governing daily fluctuations in SST, salinity, water movement and circulation and irradiance, particularly in shallow waters. Since most measurements in published work refer solely to one parameter, it is important to recognize that daily maxima for certain factors, for example, water temperature, irradiance and salinity, may coincide thus exposing corals to a range of conditions which may have synergistic physiological effects. Furthermore, since most measurements are derived from discontinuous sampling most authors acknowledge that the ranges in temperature expressed are not necessarily the maximum ranges encountered (Potts and Swart, 1984).

The greatest daily temperature ranges recorded in reef waters are those over shallow reef flat areas where ranges of 6 to 14 centigrade degrees are commonplace (Orr, 1933a; Cloud, 1952; Wells, 1952; Endean et al., 1956; Glynn, 1973; Potts and Swart, 1984). Much smaller daily ranges, between 0.1 and 0.8 centigrade degrees, have been recorded in the deeper waters of lagoons (Munk et al., 1949; Wiens, 1962; Pugh and Rayner, 1981; Andrews et al., 1984; Pickard, 1986).

As important as the daily ranges are the maximum and minimum temperatures achieved in a 24-hour period, particularly when these might approach upper and lower lethal limits for corals. Orr (1933b) reported a range of temperatures from 25.3 to 34.9°C for a tidal pool on the Great Barrier Reef over the period 25–26 November 1928; the lowest temperature ever recorded in these pools was 18°C but no daily range for that day was given. Maximum and minimum temperatures for the moat at Heron Island were given as 16 and 35°C but these do not represent the fluctuations that took place in a single day (Potts and Swart, 1984). The lowest temperature ever encountered in this 3 year study was 13.5°C on the inner reef flat in 1974 and 13.5°C on the outer reef flat in 1976.

An increasing number of continuous temperature recordings are now being made on coral reefs. These data are mainly being collected in shallow waters (0.5–1.6 m below chart datum) where diurnal variations of 1.5 centigrade degrees (irrespective of season) have been recorded on the upper slopes of fringing reefs at Madang, Papua New Guinea (Tudhope, unpublished) and Phuket (Dunne, 1994a). Greater diurnal temperature variations of 2.5–4.5 degree centigrade have been noted in shallow lagoonal waters in the Maldives and Indonesia where oceanic water is subject to solar heating as it passes over shallow sandflats during tidal exchange (Dunne, 1994b, c). Continuous temperature records show that shallow water corals can be subject to rapid and unpredictable fluctuations in temperature consisting of marked falls and subsequent increases of 3–4 degree centigrade over a 12 h period (Figure 8) which occur during overcast, stormy weather.

3. FLUCTUATING ABIOTIC INFLUENCES: VARIATIONS OVER SPACE AS REFLECTED IN REEF HETEROGENEITY

Reef heterogeneity will give rise to fluctuations in abiotic factors received by corals living in different habitats and microhabitats. Such differences may give rise to considerable variability in responses of individual coral colonies, of the same species, to physical factors. The highly heterogeneous nature of the reef environment, in terms of topography and physicochemical variations, has been highlighted by many reef workers (Stoddart, 1969; Chevalier, 1971; Veron and Pichon, 1976; Done, 1983). Processes operating at all spatial scales from water mass structure (macro) to water circulation around obstacles (micro) act in maintaining the heterogeneity of the reef habitat.

Within a horizontal traverse of 100 m, or less on some reefs, and certainly on a depth profile of 10 m there will be marked variations in the topography, wave action, current speed, irradiance and sedimentation pattern (Done, 1983). At an even finer scale different parts of a single coral colony will experience different conditions of illumination, hydrodynamics and sedimentation (Chevalier, 1971).

Figure 7 Seasonality in tropical coastal waters as reflected by selected physical parameters. (a) Average sea surface temperature at Phuket, Thailand, for 1951–1980 (after Dunne, 1994a). (b) Average monthly sun hours at Phuket, Thailand, for 1981–92 (after Dunne, 1994a). (c) Monthly mean sea level at Phuket, Thailand, for 1940–91 (after Dunne, 1994a). (d) Nutrients (dissolved inorganic nitrogen (DIN) and phosphate (P)) at Davies Reef, Great Barrier Reef, for 1990 (after Ayukai, 1993). (e) Average weekly sedimentation at Phuket, Thailand, for 1984 (after Brown *et al.*, 1986). (f) Mean wave height and frequency of wave heights greater than 4 m in the Bermuda region (after Morris *et al.*, 1977). (g) Monthly mean rainfall at Phuket, Thailand, for 1981–93 (after Dunne, 1994a). All means ± standard deviation apart from (d) where means ± standard error are shown.

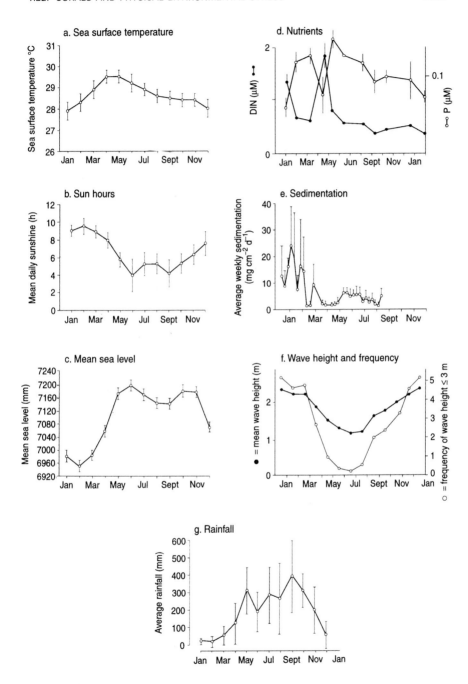

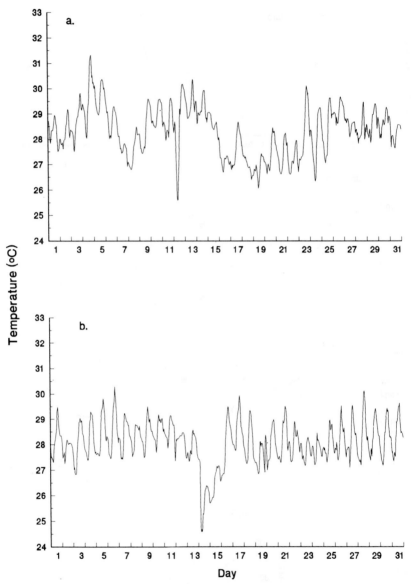

Figure 8 Hourly record of sea surface temperature (SST) from thermistors located 0.5 m below lowest low water in 2 m maximum water depth on shallow coral reefs in: (a) Pulau Pari, Indonesia, for January 1995 (after Dunne, 1994b); and (b) Furanafushi Island, Maldives, for July 1993 (after Dunne, 1994c). Both sites are exposed to continuous oceanic flushing apart from periodic extreme low tides.

Gradients in physicochemical conditions across shallow reef areas and down reef slopes are reflected in zonation patterns of corals where it has been noted that the growth forms of corals show a clearer zonation than the species (reviewed in Done, 1983). Gradients in water movements (waves and currents) and irradiance (with depth and turbidity) are acknowledged to be the major determinants of zonation (Sebens and Done, 1993), although these influences may be modified by competitive/predatory interactions depending on the biota.

3.1. Water Flow

Water flow acts as a major selective force to determine the composition of coral assemblages favouring particular growth forms on different parts of the coral reef (Brown and Dunne, 1980; Sebens and Done, 1993). In the latter study on Davies Reef on the Great Barrier Reef, corals within lagoonal habitats experienced flows which were about one-third to one-half of those of corals on the windward or leeward sides of reef flats. Corals below 8 m on the windward and leeward sides of the reef experienced flows comparable with those on the shallowest areas of lagoonal habitats.

Variations in the wave pattern and current strength in coral reef environments are also clearly shown by Roberts and Suhayda (1983) on the east coast of Nicaragua where average reef flows were in the range of $10\text{--}20 \text{ cm}^{-1}$, although individual wave surges produced values of up to 180 cm^{-1}. The wave filtering effect of the reef during high and low tides was very clear in this work. In both cases a significant proportion of the incoming wave energy was extracted at the reef crest, providing a much modified wave state in the back reef compared with the fore reef. Currents measured at the reef crest varied at least twofold depending on the tidal state from 7.1 cm s^{-1} at high tide to 17.3 cm s^{-1} at low tide. Similar studies at Grand Cayman (Roberts et al., 1977) indicated the greater relative importance of current forces over wave forces at different points on the reef; current forces on the reef at the shelf edge were approximately equivalent to the wave forces on the shallow shelf at a depth of approximately 3 m. The boundary between wave-dominated and current-dominated zones corresponded to distinct changes in the coral community, the wave-dominated zone being characterized by thickly branched, bladed and encrusting growth forms, whereas delicately branched, massive and plate-like growth forms prevailed in the current-dominated zone. Some coral genera such as *Agaricia* were common to both zones, although the current zone was dominated by *Montastrea* and *Acropora cervicornis* while the wave zone was dominated by *Acropora palmata*.

Detailed knowledge of water movements around reefs is still restricted to a small number of locations—mainly the Great Barrier Reef and some sites in the Caribbean (Andrews and Pickard, 1990). Even for these areas where models can quite reliably describe circulation in shallow well-mixed waters, Wolanski and Hamner (1988) have shown that in a calm weather complex, small-scale, three-dimensional reef-induced water circulation patterns prevent normal diffusion processes. Water circulation patterns on two reef flats in Guam (Marsh *et al.*, 1981) also revealed considerable patchiness with the direction of wave-driven water movement varying across the reef margin and reef flat on a rising tide, while some areas of the reef flat showed very sluggish water movements with extremely restricted circulation.

At the microhabitat level close to the surface of any coral in an area of unidirectional flow, there will be a turbulent boundary layer, the nature of which will depend on the surface roughness, shape and flow speed above that surface (Wainwright and Koehl, 1976; Sebens and Done, 1993). The latter authors also describe a larger gradient of flow speed which can be several metres in extent—from unobstructed mainstream flow well above the bottom, down to the cavities and channels between coral heads and buttresses. In addition, shallow subtidal reefs are frequently dominated by wave-induced oscillatory flow which prevents the formation of a steady-state boundary layer. Thus a massive coral several metres high on the reef front could, depending on its position and proximity to neighbouring colonies, experience vastly different hydrodynamic regimes on its apical and lateral surfaces at any one time. Further, temporal variations in flow may be considerable: water flow speeds measured at a height of 10 cm over a *Montastrea annularis* head at a depth of 15 m over a 25 min interval ranged from < 5 to > 25 cm s^{-1} (Patterson *et al.*, 1991) over this short period. Such differences in water flow may be significant in terms of carbon production either by different parts of a colony or between different colonies. Strong water flows may trigger retraction of polyps and shading of zooxanthellae with consequent effects on photo-tsynthesis (Fabricius, 1995); in the soft coral *Dendronephthya hemprichi* in the Red Sea polyps contracted at currents greater than 25 cm s^{-1}.

3.2. Irradiance

The penetration of solar radiation into tropical seas has been the subject of a number of studies (Jerlov, 1968, 1976; Dustan, 1982; Kirk, 1983, 1992; McFarland, 1991). The effects of increasing depth result in all the infrared and visible red (> 780 nm) being absorbed by water molecules within the first 0.5 m (Jerlov, 1976) reducing the solar energy available to 50% of the above-surface value. Attenuation of solar radiation with depth results in the rapid disappearance of red and yellow wavelengths with blue and green wavelengths penetrating to depths as great as 200 m in clearest ocean waters.

It is commonly stated that dissolved substances and particulate matter are in low concentrations in tropical waters (Falkowski *et al.*, 1990; Falkowski *et al.*, 1990; McFarland, 1991), yet many coral reefs thrive in waters with a high dissolved organic and suspended particulate load. Here, the optical properties of sea water are further modified by increased concentration of phytoplankton and suspended particulate material (Jeffrey, 1980). Dissolved yellow substances from land run-off, detritus and marine humic materials modify the submarine irradiance field by strongly absorbing in the violet and blue-green region of the spectrum. This shifts the underwater irradiance from blue-green (480 nm) to green and yellow (550 nm) wavelengths. In the turbid coastal waters of fringing reefs of South-East Asia, blue and green wavelengths are reduced to <1% of surface values in depths less than 1 m (Dunne and Brown, unpublished data). A similar reduction in UVB (280–320 nm) is seen in these waters and in those of temperate coastal lagoons (Piazena and Häder, 1994). For coral reefs, then, the quantity and spectral quality of submarine irradiance vary with water depth and water clarity, the latter factor varying both seasonally and spatially over the reef.

At the microhabitat level, microtopography (particularly the angle of surfaces) is important in evaluating variation in irradiance (Ohlhorst and Liddell, 1988). Brakel (1979), measuring multidirectional irradiance on the fringing reef at Discovery Bay, Jamaica, showed that details of substratum, slope and exposure had a marked influence on the light available to benthic organisms. Thus, for example, changing the substratum slope from horizontal to 60°, or moving under a canopy of *Acropora cervicornis*, caused a decrease in irradiance of 50%. Sheppard (1981) similarly showed even greater decreases in downwelling irradiance under canopies of *Acropora hyacinthus* and *Acropora cythera* on the Great Barrier Reef.

Sun azimuth and angle are all-important in determining the total irradiance received by different surfaces on a coral colony. Recent work on aerially exposed corals shows that not only irradiance levels but also spectral quality of the irradiance differ on differently inclined faces of a single colony (Brown *et al.*, 1994b). Under clear sky conditions, spectroradiometer scans show that a horizontal surface received almost identical levels of short-wavelength ultraviolet radiation (280–320 nm) in the afternoon as the inclined irradiated surface facing the sun for solar altitudes of 65° or less. At longer wavelengths, irradiance is greater on the inclined surface rising to a maximum difference of 22%

photosynthetically active radiation (PAR). Underwater radiance measurements on differently angled surfaces also show marked variations over short distances and, wherever topography presents a rugged surface, shadows occur (Jaubert and Vasseur, 1974) with consequent alterations in radiance levels (Roos, 1967). The movement of the shadows is subject to both daily and seasonal influences that are in turn governed by sun altitude and azimuth.

3.3. Temperature

Spatial variation in sea water temperatures at any one time across reefs has been recorded by several authors (Glynn, 1973; Potts and Swart, 1984; Hayashibara *et al.*, 1993) although there are few detailed studies providing continuous data which might be of value in interpreting physiological tolerances of corals to temperature fluctuations. At Heron Island (Potts and Swart, 1984) the inner reef flat showed the most variable temperature over a 2-year period (range = 12.7°C), temperature variation was intermediate on the outer reef flat (range = 8.5°C) and lagoon (range = 6.8°C), and most stable on the slope where temperature ranges rarely exceeded 3 degree centigrade. Based on SST and water movements the authors concluded that at least two water masses affect the Heron Island reef: a shallow, surface water mass that is greatly influenced by prevailing weather conditions and a more oceanic water mass largely restricted to the outer slopes. Large ranges in water temperature observed on the reef crest were similar to those at other shallow sites, indicating that the crest was exposed to reef flat or lagoonal water moving off the reef. Conversely, the very small ranges of SST measured only 10 m away in 15 m of water suggest that the slope site was permanently exposed to oceanic water. Potts and Swart (1984) concluded that while corals on the crest and the foot of the slope are only 50 m apart, they appear to be surrounded by quite different water masses with very different water regimes. Such effects are likely to be maximal in the summer when waters in the reef lagoon are generally warmer than the shelf waters (Wolanski and Hamner, 1988).

In Aldabra atoll, where shallow waters are subject to intense solar heating because the timing of extreme low water results in maximum insolation (Farrow and Brander, 1971), ocean water enters the shallow reef each tide, acting to buffer temperature and salinity changes within the lagoon (Pugh and Rayner, 1981). The degree of mixing of lagoon and oceanic water in atolls is variable depending upon the lagoon volume, the open or closed nature of the atoll and seasonal effects (Andrews and Pickard, 1990). Mixing between lagoon and oceanic water is only partial during tidal inflow and as a result corals in lagoon and outer reef slopes will experience different water masses with varying temperature regimes.

4. ADAPTATIONS OF REEF CORALS TO TEMPERATURE CHANGE

For corals with symbiotic algae, effects of temperature change are complicated by the need to consider both the plant and the animal components of the symbiosis. Because of the limited literature available in some areas, it will be necessary to consider not only the thermal biology of corals but also other symbiotic coelenterates to provide a balanced viewpoint. Before addressing the scope for thermal adaptation it is important to outline a number of basic tenets of ecophysiology and thermal biology.

There is a considerable body of evidence to show that organisms display an impressive ability to adjust their physiology to overcome direct effects of temperature variations, by either phenotypic and/or genotypic adaptations (Cossins and Bowler, 1987). Phenotypic adaptations to natural temperature change are commonly known as acclimatizations and involve changes in the performance of an organism during its lifetime. Such adaptations include adjustments to daily and seasonal temperature regimes and are usually compensatory. Acclimatization involves changes in the phenotype, the limits of which are set by the genotype. These physiological adjustments, induced by environmental stimuli, are usually reversible but, in contrast to genotypic adaptations, they are not passed on to succeeding generations, although the capacity to make adjustments are inherited.

Acclimatization should be distinguished from acclimation which is reserved for laboratory adaptations (Prosser, 1973). The distinction is important; for example, there are major differences between an organism acclimated to high temperatures in the laboratory and one acclimatized in the field. The direction of seasonal cues, other than temperature, also creates the possibility of anticipatory adjustments which may "preadapt" the organisms to rapidly deteriorating conditions (Cossins and Bowler, 1987). One illustration is the possible involvement of photoperiod as a seasonal cue for temperature acclimatization (Roberts, 1964; Burns, 1975). For symbiotic corals living at higher latitudes, such a finding has interesting implications, particularly because zooxanthellae (shown to be sensitive to changing irradiance conditions) might have a role in mediating such responses. Laboratory acclimation very rarely, if ever, mimics either the sequence or the intensity of such environmental stimuli.

4.1. Temperature Acclimation/Acclimatization Responses in Corals and Symbiotic Anemones

Although it has been recently claimed that corals have limited abilities to acclimate to increased temperature (Jokiel and Coles, 1990; Smith and Buddemeier, 1992) the data supporting such conclusions are extremely limited. Information is restricted almost entirely to the pioneering work of Jokiel and Coles (1974, 1977) and Coles and Jokiel (1977, 1978) on the temperature biology of 3–5 species of reef corals from Hawaii and Enewetak, where acclimation responses were not the main objective of the studies. Critical examination of these experiments (Jokiel and Coles, 1977; Coles and Jokiel, 1978), show that the corals tested did show some acclimatory ability.

For example, some colonies of *Pocillopora*, *Fungia* and *Montipora* survived after acclimation to ambient summer temperatures of 26–27°C for 30 h, followed by exposure to 31°C for a further 13 d, 33–34.6°C for 10 d and return to 31–32°C (Jokiel and Coles, 1977). The results indicated not only marked interspecific differences in thermal tolerance of corals but also differences in the susceptibility of different colonies of the same species. In other marine invertebrates links have been made between different temperature tolerances of individuals in a species and their rates of protein turnover and energy metabolism (Hawkins *et al.*, 1987). The variability in the performance of individuals, their ability to respond to environmental stress and links with genotypic differences is a relatively new area of physiological endeavour (Koehn and Bayne, 1989; Hawkins, 1991; Clarke, 1993) and clearly one with particular implications for the study of temperature tolerances in corals. By ignoring individual variability we may be underestimating not only the scope of reef corals to adjust to temperature changes but also the mechanisms by which they do so.

In a subsequent acclimation experiment, Coles and Jokiel (1978) used the most thermally sensitive species studied in the earlier work, *Montipora verrucosa*. Corals were first maintained for 56 d at a range of temperatures (20, 24, 26 and 28°C) before transfer to 28°C, a gradual raising of this temperature within a day to 30°C followed by elevation over 2.5 h to 32–32.5°C and cooling to 24°C over 48 h. Remarkably, 74% of the *Montipora* colonies acclimated to 28°C actually survived this rigorous treatment. Not only did corals pre-exposed to high temperatures show a high survival but corals previously held at 26°C also displayed 61% survival, while those held at 24 and 20°C showed survival rates of 30 and 40%, respectively. Clearly, survival was achieved over this complex temperature regime by individual colonies pre-exposed to both high and low temperatures. These results indicate not only acclimatory ability but also the remarkable tolerance of some individual colonies to a fluctuating temperature regime.

Further evidence hitherto used to support the idea that corals may not be able to adapt rapidly to increased temperatures is the observation that repeated annual summer bleaching of the same coral colonies has been observed in the thermal plume of a power station in Hawaii (Coles, 1975; Jokiel and Coles, 1990). While such results may be suggestive of an inability to adapt to increased temperature, other interpretations are possible given the likelihood that the detailed thermal history of the corals prior to bleaching may have differed each year; that the thermal characteristics of the power station plume may also have varied on an annual basis; and that the summer paling of the corals may have been a product of increased temperature and summer irradiance, the latter factor also potentially varying from year to year. Conclusions on the significance of these findings must remain equivocal.

An earlier study on the thermal tolerance of corals from Hawaii (Clausen and Roth, 1975) is suggestive of thermal acclimatization, with corals selected for study at different times of the year (and therefore with different thermal histories) showing differing responses to increased temperature. More recent work by Al-Sofyani and Davies (1992) in Jeddah, Saudi Arabia, has shown that the coral *Echinopora gemmacea* demonstrates seasonal acclimatization. The response was a compensatory one in which there was no difference in respiratory rate between seasons, mean seasonal SST varying between 24°C (winter) and 30°C (summer). In contrast, *Stylophora pistillata* from the same site appeared to show no compensatory changes in physiological responses to summer conditions. *Stylophora* showed a higher photosynthetic rate and a higher metabolic expenditure in the high-light, high-temperature conditions of the summer when compared with the winter. Thus the physiological responses of the two coral species to seasonal changes in temperature and light are clearly quite different. The authors suggest that these differences might be attributed to differing nutritional strategies adopted by the two corals— *Stylophora* being described as more autotrophic and dependent on the activity of the zooxanthellae while *Echinopora* may be more dependent on particulate food.

4.2. Cellular Mechanisms of Compensation to Temperature Stress

4.2.1. *Enzymatic Adaptations*

At the cellular level the process of temperature acclimation is the result of complex reorganization of cell metabolism. Such changes depend upon the regulation of the activity of key enzymes. Hochachka and Somero (1984) have described three principal mechanisms by which regulation can be achieved: (1) by a change in the cellular concentration of the enzyme; (2) a change in the types of enzyme isoforms present; and (3) a modification of the kinetics of the pre-existing enzymes.

4.2.1.1. *Enzyme concentrations* The unequivocal demonstration of changes in enzyme concentration as a result of thermal acclimation has only been achieved in a few studies in fish (Cossins and Bowler, 1987). No specific studies on thermal acclimation in corals have combined measures of acclimation and quantitative estimates of enzyme activity, although Lesser *et al.* (1990) demonstrated increased activities of the enzymes superoxide dismutase, catalase and ascorbate peroxidase within the zooxanthellae of the zoanthid *Palythoa caribaeorum* in response to elevated temperature (29–33°C) compared with ambient (25–27°C). These increased enzyme activities were interpreted as a mechanism for detoxifying active forms of oxygen produced as a result of increased photosynthesis. In the host the changes in enzyme activities were less clear-cut but increased temperature also resulted in significantly higher catalase activity.

4.2.1.2. *Enzyme isoforms and alloforms* Where organisms have become adapted to a particular temperature over long periods of evolutionary time the structural and functional characteristics of their proteins are often optimized for that temperature. Many organisms, however, are able to function with variable tissue temperatures and adaptation mechanisms may involve the use of proteins that exist in a variety of different forms (isoforms) (Johnston, 1983).

It is important to distinguish between phenotypic variations arising from the expression of genes coding for different functional isoforms of individual proteins and those due to genetic variations in which conservative mutations may occur in one or more of these different genes. Protein variants arising from genetic variation within populations or races of a species are called "alloforms".

Very little work has been carried out on the relationship between temperature adaptation and the presence of enzyme isoforms and alloforms in coelenterates. Early work by Clausen and Roth (1975) proposed the existence of two enzyme isoforms in the coral *Pocillopora damicornis* from Hawaii and Enewetak as an explanation for the presence of two temperature optima in corals from varying thermal backgrounds. However, no investigation of enzymes was actually carried out during the course of this work.

Work on enzyme alloforms related to temperature in coelenterates is restricted to the sea anemone *Metridium senile* from the Atlantic coast of the USA (Hoffman, 1981, 1983, 1986). Here, the anemone occupies the steepest geographical thermal gradient in the world. Populations of the anemone show a graded sequence in differences in allele frequencies at several enzyme loci across this thermal gradient. One example is the phosphoglucose isomerase (PGI) locus where the products of alternative alleles are two PGI alloforms designated as electrophoretically "fast" and "slow". It appears that the fast allozyme is selectively favoured in warm water.

4.2.1.3. *Enzyme kinetics.* Temperature changes frequently have substantial effects upon the equilibrium constants of biochemical reactions, particularly those involving the reversible formation of non-covalent (or weak) chemical bonds. It would be expected that natural selection will have produced altered enzymes in response to changes in environmental temperature so that they may maintain normal function as much as possible. The conservation of kinetic properties of enzymes isolated from different biological systems has been demonstrated in a number of studies comparing the temperature responses of homologous proteins isolated from organisms adapted to different environments (Hochachka and Somero, 1984; Somero, 1986). Comparison of closely related species in habitats differing in temperature offers an experimental approach for investigating differences in the kinetic properties of homologous enzymes, although it is always difficult to rule out the possibility of other reasons for observed differences between species. In the absence of relevant research on corals, an example can be found in the work of Graves and Somero (1982) where they compared kinetic parameters for the enzyme lactate dehydrogenase in three barracuda species living in temperate, subtropical and tropical environments. They found that the K_m values (Michaelis–Menten constant) for the enzyme did not vary when compared at the mid-range temperatures experienced by the species in their natural habitats. The kinetic parameters did change at the highest temperature evaluated (25°C), which suggests that different thermal environments may select for enzymes that maintain constant, central metabolite concentrations irrespective of habitat temperature (Hoffman and Parsons, 1991).

Another parameter of enzyme kinetics is the activation energy (E_a) which is a measure of the enzyme's ability to overcome thermal thresholds in a reaction. Increasing the reaction temperature for a given enzyme decreases the activation energy required. Various studies discussed by Hochachka and Somero (1984) show that species evolutionarily adapted to low temperatures have enzymes with lower E_a values than do warm-adapted species and this appears to be borne out by work on coelenterates (Walsh and Somero, 1981; Walsh, 1985). Considering glutamate dehydrogenases in sea anemones, the lowest values of E_a were found in Antarctic species, intermediate values in cool-temperate species and highest values in the symbiont-containing anemone *Anthopleura elangtissima* which experienced a temperature range of 15–35°C.

4.2.2. *Heat-stress Proteins*

The expression of heat-shock proteins (hsps) is one aspect of cellular adaptation to temperature that is now being studied in coelenterates. It is usually assumed that the heat-shock response protects organisms from the disruptive effects of elevated temperature, including chaperoning unfolded proteins (Hemmingsen *et al.*, 1988), preventing aggregation problems during folding (Pelham, 1986), facilitating intercompartmental transport (Chirico *et al.*, 1988) and uncoating clathrin-coated vesicles in cells (Chappell *et al.*, 1986). It should be made clear that many hsps are stress proteins rather than specific responses to elevated or reduced temperatures, since they are induced by a whole variety of stressors (Hoffman and Parsons, 1991).

Hsp 70 has been found in the corals *Gonipora djiboutiensis* and *Goniopora pandoraensis*, both in laboratory heat-shock experiments and also in intertidal *G. pandoraensis* studied directly from the field (Sharp *et al.*, 1994). This same hsp, however, was not detected in other coelenterates. Bosch *et al.* (1988) reported that the major hsp in heat-shocked *Hydra vulgaris* was a 60-kDa hsp while two low-molecular mass hsp70 homologues (28 and 29 kDa) were identified in intertidal and subtidal populations of the temperate symbiotic sea anemone *Anemonia viridis* before and after heat shock (Sharp *et al.*, 1994). In the Caribbean coral *Montastrea faveolata*, seven different hsps with appropriate molecular weights of 95, 90, 78, 74, 33, 28 and 27 kDa were detected after heat shock, while in the tropical sea anemone *Aiptasia pallida* fewer hsps with different molecular weights (82, 72, 68 and 48 kDa) were synthesized during temperature stress (Black *et al.*, 1995).

Similar differences in the expression of hsps were observed between intertidal and subtidal anemones and corals. In the more thermally stressed intertidal anemones (*A. viridis*) and coral species (*G. pandoraensis*), high levels of hsp were observed compared with either none or very limited concentrations in subtidal anemones and corals (*G. djiboutiensis*), respectively (Sharp *et al.*, 1994; Sharp, 1995). Transplantation of the subtidal coral species to the intertidal area for 16 and 32 d resulted in a marked increase in the amounts of hsp produced, in response to the new environment (Figure 9), suggesting phenotypic variation in protein expression.

The hsp responses described above have been shown to be attributable to the anemone and coral hosts, and not directly to the symbiotic algae or zooxanthellae. No hsp 70 production has been observed in zooxan-

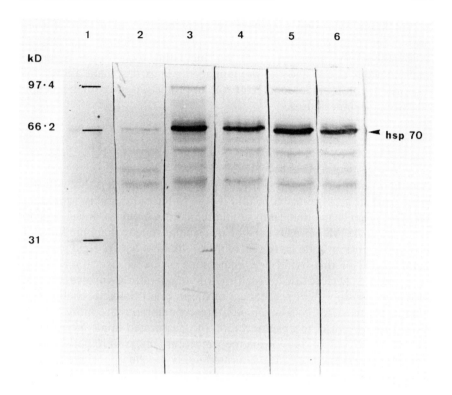

Figure 9 Immunoblots of subtidal *Goniopora djiboutiensis* prior to and post-transplantation to a pool on an intertidal reef flat for 16 and 22 days, at Phuket, Thailand (Sharp, 1995). Track 1 shows Mr standards. Track 2 shows control extract of *G. djiboutiensis* prior to transplantation. Tracks 3 and 4 show extracts from *G. djiboutiensis* following transplantation to the reef flat for 16 d. Tracks 5 and 6 show extracts from *G. djiboutiensis* following transplantation to the reef flat for 32 d. Note the increase in the expression of hsp 70-related proteins following transplantation to the intertidal reef flat for 16 and 32 d (tracks 3, 4 and 5, 6, respectively).

thellae extracted from the coelenterate hosts, either from the field or after heat-shock treatment (Bythell *et al.*, 1995). However, a highly abundant 33 kDa protein has been detected in zooxanthellae of the temperate sea anemone *A. viridis* and in the tropical corals *G. djiboutiensis*, *G. pandoraensis* and *G. stokesi*. Production of the 33 kDa protein appears to be stimulated by various environmental factors that include thermal shock, season, habitat and holding conditions, although its precise role is far from clear (Bythell *et al.*, 1995).

The study of the nature and significance of hsps in corals is still in its infancy. In general, heat-shock protein research to date has been centred in the laboratory; only relatively recently has the induction of these proteins been shown in the field (Spotila *et al.*, 1989; Sanders *et al.*, 1991). It is quite likely, however, given the fluctuating temperature regime of shallow water corals and the preliminary findings described above, that hsp production plays an important role in protection of cellular functions at extreme temperatures.

4.2.3. *Homeoviscous Adaptation*

An important feature of temperature-protein adaptations is that a protein must exist in a "semi-stable" state if it is to have the ability to undergo the changes in shape that catalysis and regulation require (Hochachka and Somero, 1984). Membranes must be in a similar state to maintain function; this represents a state between extreme fluidity and a rigid gel. The maintenance of this liquid crystalline state is known as "homeoviscous adaptation". Acclimation to high or low temperatures involves changes in the composition and percentage of unsaturated lipids to maintain the homeoviscous state. At low temperatures, membranes are more likely to enter the gel phase; the incorporation of unsaturated lipids increases the double bond content and lowers the transition temperature between fluid and gel phases. Incorporating saturated lipids has the opposite effect, and helps to maintain the homeoviscous state at high temperatures.

As expected, the lipids of temperate and cold-water anemones are more unsaturated than those in warm-water forms (Bergmann *et al.*, 1956; Blanquet *et al.*, 1979; Harland *et al.*, 1991). Kellog and Patton (1983), in a study of lipid droplets in the tropical symbiotic anemone *Condylactis gigantea*, commented on the remarkably high degree of saturation of lipid classes extracted from both host and zooxanthellae. Similar results have also been found in the coral *Pocillopora damicornis* (Patton *et al.*, 1977) although unsaturated lipids have also been detected. It has been suggested that the occurrence of unsaturated lipids in corals may reflect an external dietary lipid source whereas the saturated lipids are produced by the zooxanthellae (Patton *et al.*, 1977; Meyers, 1979). Such a conclusion reflects an inherent problem in all the above studies, namely that the analyses conducted are usually bulk rather than membrane fatty acids.

As well as playing a role in homeoviscous adaptation, saturated fats are very resistant to oxidation compared with unsaturated lipids. As such, they may be beneficial to shallow water tropical organisms living in high solar irradiance where photo-oxidative effects are a hazard (see Shick, 1991).

Although there is evidence of lipid differences and membrane properties associated with intraspecific variation in resistance to extreme temperatures in plants (Murata and Yamaya, 1984), nothing is known of the roles that these factors may play in corals and their symbiotic algae. Corals such as those living in the Arabian Gulf that experience an annual temperature range of 24.8 degree centigrade must surely have evolved a remarkable suite of cellular mechanisms to achieve such a wide temperature tolerance.

Somero (1978) considered that a complete understanding of temperature adaptation can be achieved only if one can encompass the enzymes, the solutions that bathe them and membrane lipids, together with interactions between these components. To this, Hoffman and Parsons (1991) added: "enzymes need to be studied in situations that are extreme if responses to environmental stress are to be understood at the protein level". It would seem, then, that for a better understanding of the thermal biology of corals, cellular mechanisms of adaptation offer a fertile area of research endeavour.

4.3. Inter-regional Comparisons of Thermal Tolerance in Reef Corals

Genotypic or evolutionary adaptations to temperature can be demonstrated by comparing the physiology of species that inhabit different thermal environments or latitudinally separated populations of the same species.

In all such cases it is critical that phenotypic adaptations or acclimatizations are excluded from comparisons. This is possible if the previous thermal experience of the experimental organism is carefully controlled. Another complicating phenomenon is "canalization", in which the physiological status or properties of an organism are influenced by the thermal exposure of early developmental stages, irrespective of its recent thermal experience. The importance of this process is largely unexplored, except in a few cases, but it does point to the ultimate necessity to breed and rear organisms under controlled conditions if clear genotypic adaptation to temperature is to be demonstrated (Cossins and Bowler, 1987).

Few studies have taken such factors into account (Vernberg and Costlow, 1966) and certainly no work of this sort has been carried out on corals. The implication for canalization in these organisms is interesting since planulae of some species appear to be more tolerant of elevated temperatures than adult corals (Edmondson, 1946; Coles, 1985).

Where latitudinal comparisons of coral thermal tolerance have been carried out, field observations and controlled experiments have been reported as showing that tropical corals from Enewetak had an upper lethal and sublethal limit that was about 2 degree centigrade higher than that of subtropical Hawaiian corals (Coles *et al.*, 1976). While this may be the case for the two common species under consideration, i.e. *Fungia scutaria* and

Pocillopora verrucosa (where *P. verrucosa* = *P. elegans* (Enewetak) and *P. meandrina* (Hawaii)) (Veron and Pichon, 1976), the other three species tested from each site were not conspecifics. The variations observed in these cases may therefore be species-specific rather than latitudinally based.

Other comparisons between different species from the same genus in different coral provinces (Marcus and Thorhaug, 1981) suggest that the Caribbean species *Porites porites* has a bleaching threshold about 1 degree centigrade higher than its congener *Porites compressa* from Hawaii. These findings must be viewed with caution for not only are different species involved in the comparison but also it is not clear at what time of year corals from Hawaii were collected for experimentation. As we have already seen, time of collection of specimens and their previous thermal history may be important in the final estimation of thermal tolerance.

The additional problem of intraspecific variability in bleaching thresholds also presents a difficulty in the analysis of such data sets. Adjacent colonies of otherwise similar massive *Porites* colonies in Thailand displayed substantial variation in their susceptibilities to bleaching during a period of elevated SST (Tudhope *et al.*, 1993), while intraspecific variation in bleaching thresholds, in response to high temperatures in Hawaii, has been attributed to genotypic differences between clones of *Porites compressa* (Jokiel and Coles, 1990).

While we might suspect geographic variations in lethal temperature tolerances between coral species, much more detailed work remains to be carried out to determine whether or not these differences are genotypically based. Clearly, corals are subject to much wider fluctuations in temperature than has previously been recognized and the evidence that corals (as a group) live close to their lethal limits is suggestive rather than strong.

4.4. The Nature and Significance of Coral Bleaching in Response to Increased Temperature

Coral bleaching, described primarily in response to increased sea water temperatures, has received widespread attention during the last decade (Glynn, 1993). The response is not unique to reef-building corals; other organisms that contain algal symbionts react similarly on exposure to stressful environmental conditions. Nor is high or low temperature the only stress which will evoke the bleaching response. Bleaching appears to be a generalized, and often reversible, response to a wide variety of environmental stress factors (Brown and Howard, 1985). While increased SSTs during ENSO years have been cited as the most likely factor responsible for worldwide bleaching during the past decade, the potential synergistic effects of increased irradiance and temperature cannot be ignored (Coles and

Jokiel, 1978), with many workers reporting bleaching during periods of low wind velocity, clear skies and low turbidity (Glynn, 1993).

The precise mechanism of bleaching has been variously described (Steen and Muscatine, 1987; Muscatine et al., 1991; Gates et al., 1992; Brown et al., 1995). Laboratory-based studies on cellular responses to cold and heat stress in a symbiotic anemone and coral suggest that bleaching involves the release of intact host endoderm cells containing viable zooxanthellae (Gates et al., 1992). However, subsequent work has indicated that a range of cellular mechanisms are involved in bleaching under natural conditions, the most predominant being the degradation of zooxanthellae in situ (Brown et al., 1995). An interesting feature in the early stages of bleaching appears to be the limitation of observable damage in the host endoderm to only those cells harbouring zooxanthellae (Le Tissier and Brown, 1996).

The trigger for the bleaching response in corals has not yet been unequivocally demonstrated. Iglesias-Prieto et al. (1992), working with cultured zooxanthellae, suggest that reduced photosynthesis at temperatures above 30°C, and resulting reduction in transfer of metabolites between algae and host, may be responsible. They propose that consequent to the above changes, disruption in chemical signals between host and symbiotic algae (Markell et al., 1992) may result in bleaching. An alternative mediator of the bleaching response might be active forms of oxygen resulting from high photosynthetic rates already described in symbiotic cnidarians exposed to high temperatures (Lesser et al., 1990). Both temperature and/or active species of oxygen could induce changes in the thylakoid membranes of the zooxanthellae chloroplasts disrupting photochemistry (Asada and Takahashi, 1987; Kyle, 1987; Ludlow, 1987) and producing high concentrations of active forms of oxygen that overwhelm the available enzymic defences. Recent work by Le Tissier et al. (unpublished) has suggested that the trigger for the bleaching response may lie in an inability of host cells, harbouring zooxanthellae, to osmoregulate in response to increased transfer of photosynthate which results from initial stimulation of photosynthesis during temperature/irradiance stress.

Buddemeier and Fautin (1993) have put forward a proposal that bleaching is an adaptive mechanism which allows the coral to be repopulated with a different "type" of symbiotic alga, possibly conferring greater stress resistance. They suggest that at least part of the intraspecific variability in bleaching patterns in the field may be due to different "types" of algae in conspecific hosts. While bleaching may offer an opportunity for repopulation of coral hosts by different algae, it should be recognized that severely bleached corals in the field retain a significant number of zooxanthellae which are capable of repopulating coral tissues (Szmant and

Gassman, 1990; Brown *et al.*, 1994b, 1995). Even corals which appear completely white may retain between 25 and 50% (i.e. $3\text{--}5 \times 10^6$ zooxanthellae cm^{-2}) of their original complement of zooxanthellae during a natural bleaching event (Brown *et al.*, 1995). The evidence for infection of corals by algae from an external source is limited to experimental studies, primarily on anemones and jellyfish (Kinzie and Chee, 1979; Schoenberg and Trench, 1980; Fitt, 1985). An improved understanding of algal taxonomy (using DNA techniques), the environmental tolerances of different symbiotic units, mechanisms of transmission of zooxanthellae and host infection are required before any firm conclusions on the adaptive significance of bleaching can be reached. The recent recognition of three taxa of zooxanthellae each found in two Caribbean coral species, *Montastrea annularis* and *Montastrea faveolata* (Rowan and Knowlton, 1995) living in shallow water (< 10 m depth), may be of fundamental significance in the interpretation of the bleaching response, should the different taxa be shown to exhibit different environmental tolerances.

An alternative way of considering the sublethal bleaching response may be as a damage-limitation process which has developed to carry corals through stressful periods. In examples of natural bleaching events in the field, many fully and partially bleached corals recover; for example, in the 1991 bleaching episode in Tahiti 75% of *Pocillopora* colonies recovered while 53% of *Acropora* showed recovery 5 months after the start of bleaching (Gleason, 1993b). In Thailand during a bleaching event in 1991, 28% of colonies noted as bleached in 1991 had recovered 1 year later with a further 72% showing partial mortality only (Brown *et al.*, 1993). Thus coral colonies affected by bleaching or partial bleaching do survive prolonged periods (up to 6 months) with a reduced complement of zooxanthellae (Glynn *et al.*, 1985b; Szmant and Gassman, 1990; Satapoomin, 1993; Brown *et al.*, 1995). During that time the coral may show reduced skeletal growth (Goreau and Macfarlane, 1990; Allison *et al.*, 1996), lowered lipid production (Glynn *et al.*, 1985a) and possible reallocation of energy reserves normally devoted to reproduction (Szmant and Gassman, 1990).

From the limited evidence available it seems that the thermal tolerance of the symbiotic algae may be less than that of the animal host. An apparent lack of compensatory acclimatization in photosynthetic rate of the algae at higher temperatures (Clark and Jensen, 1982; Shick, 1991) and loss of zooxanthellae prior to heat-shock protein production in the coral host (Sharp, 1995) would support such a contention.

If intracellular oxygen radicals are a major mediator in the bleaching process (Lesser *et al.*, 1990), then damage to surrounding host tissue could be expected to be very rapid (Mopper and Zhou, 1990). Any bleaching mechanism that acted to confine damage in the endoderm would obviously have benefits. Recent evidence from ultrastructural studies suggests that special host cells harbour degenerating zooxanthellae (Le Tissier and Brown, 1996) and in so doing may restrict damage to specific endodermal cells. In assessing the effects of increased temperature on the zoanthid *Palythoa caribaeorum*, Lesser *et al.* (1990) showed significant increases in the activity of enzymes associated with detoxification of active oxygen radicals in the symbiotic algae. The enzyme activity in the host was only "suggestive of similar changes in the zooxanthellae". This result might indicate that potential early damage is restricted to the zooxanthellae and the host cells that they occupy (Shick and Dykens, 1984).

Another indicator of possibly restricted effects of damage in the endoderm comes from the use of lysosomal latency as an indicator of subcellular stress in sea anemones. When symbiotic and aposymbiotic anemones were subject to temperature elevation, there were marked differences in the localization and quantification of lysosomal enzyme activity in the two groups (Suharsono *et al.*, 1993). In symbiotic anemones, after 48 h, most enzyme activity was associated with the outermost membrane of the zooxanthellae and the cytoplasm of the degenerate algal cell while in aposymbiotic anemones significantly lower levels of enzyme activity were detected in cells throughout the endoderm. Such results also suggest that immediate damage may be confined to zooxanthellae and their specialized host cells.

Clearly, there are important energetics costs for the host during bleaching in protecting cells in intimate contact with zooxanthellae, namely, defensive enzyme production and the costs of cell turnover for those cells involved in the "removal" of zooxanthellae. Nevertheless, the ability of many corals to retain a reduced resident population of zooxanthellae throughout a bleaching event, possibly replacing their nutritional contribution by catabolism of proteins (Szmant and Gassman, 1990), suggests that physiological plasticity and the development of a bleaching mechanism may be important survival attributes under stressful conditions.

5. ADAPTATION OF REEF CORALS TO FLUCTUATIONS IN SOLAR RADIATION

Except in controlled experiments in the laboratory, reef corals are rarely exposed to solar radiation of a constant flux or spectral quality. Diurnal and seasonal changes in photosynthetically active radiation (PAR) measured from 400 to 700 nm, range from darkness to 2300 μmol m^{-2} s^{-1} above water for intertidal corals at local solar noon in Phuket, Thailand

(Dunne, personal communication) to 1500 μmol m^{-2} s^{-1} for corals at 1 m depth at noon on a clear day on the Great Barrier Reef (Chalker et al., 1983). In addition, waves on the water surface create "flashes" of radiation that may be very intense and of short duration; Griffiths and Kinzie in Hawaii (in Falkowski et al., 1990) measured instantaneous PAR values in excess of 4000 μmol m^{-2} s^{-1} in shallow water due to flashes when PAR above the surface was 2100–2500 μmol m^{-2} s^{-1}.

Less predictable than diurnal and seasonal changes are the variations caused by cloudiness, storms which may reduce surface irradiance and increase turbidity and phytoplankton blooms (Falkowski et al., 1990). On three consecutive cloudy days at Phuket, Thailand, the maximum instantaneous value of above-surface downward PAR never exceeded 1385 μmol m^{-2} s^{-1} (Dunne, personal communication). On such days it has been calculated that reduced levels of photosynthesis would result in a negative energy balance for the coral Pocillopora damicornis (Davies, 1991) and the coral would have to draw upon lipid reserves. It has been calculated that such reserves would last for 28 consecutive cloudy days (Davies, 1991, 1992).

Generally speaking, the spectral quality of solar radiation and its effects upon coral physiology have received less attention than radiant flux. This may be because workers looking at adaptation of unicellular algae to radiation (Richardson et al., 1983) consider that most algal classes use a limited spectrum of solar radiation available in marine waters to meet their absolute energy requirements. They conclude that radiant flux plays a more important role than spectral quality in the control of microalgal distributions. In coral reef science greater attention to radiant flux may result from the fact that most experimental work is carried out using PAR, since zooxanthellae have relatively flat photosynthesis action spectra over the range 400–700 nm (Scott and Jitts, 1977). Nevertheless, alteration of spectral quality received by corals at different depths and in different habitats may have important consequences in terms of regulation of certain cellular processes. For example, low intensity blue-green radiation exerts a profound effect upon pigment content, chloroplast structure and photosynthetic capacity of free-living dinoflagellates (Jeffrey and Vesk, 1977; Vesk and Jeffrey, 1977). Similar marked effects of altered spectral quality have been shown for coral zooxanthellae in vitro and for the coral–algal symbiotic unit (Kinzie et al., 1984; Kinzie and Hunter, 1987).

Corals, and in particular their symbiotic algae, show a remarkable range of adaptations to varying flux and quality of solar radiation. Much of this capacity to adapt to such fluctuations resides in the flexibility of the photosynthetic apparatus of the zooxanthellae to adjust to different radiant environments (Jeffrey, 1980), although the coral may also modify the irradiance environment received by the zooxanthellae by behavioural and growth form alteration.

Physiological aspects of photoadaptation have been covered in recent reviews (Barnes and Chalker, 1990; Falkowski et al., 1990; Shick, 1991). It is not the intention of this paper to duplicate this effort, but rather to highlight the physiological plasticity of corals shown both on a daily/seasonal basis and through their ability to colonize extreme environments. Responses of corals will be evaluated as adaptations to changing levels of solar radiation (including photoadaptation, photoinhibition and daily and seasonal fluctuations), and responses to ultraviolet radiation (UVR) in particular.

5.1. Photoadaptation

Physiological changes in corals mirror the highly dynamic radiation environment around the colony, such changes generally acting to optimize the efficiency with which solar radiation is harvested and used in photosynthetic reactions (Falkowski et al., 1990). In its broadest sense, photoadaptation describes changes in photosynthetic performance with changes in radiation during growth. It encompasses behavioural, morphological, biochemical and physiological responses of the coral.

5.1.1. Behavioural Responses

Expansion and contraction of tissues provide a rapid and flexible means of regulating radiant flux reaching zooxanthellae as solar radiation changes from minute to minute and from day to day (Pearse, 1974). Corals and anemones show remarkable powers of expansion and retraction of their tissues in response to changes in radiant flux (Abe, 1939; Kawaguti, 1954; Pearse, 1974; Sebens and de Riemer, 1977; Lasker, 1979, 1981; Porter, 1980; Dykens and Shick, 1984; Shick and Dykens, 1984; Brown et al., 1994b).

Depending on the state of retraction of the tentacles, where most of the symbiotic algae are found, the endodermal cells take on various morphological shapes (Glider et al., 1980). In the retracted state the algal cells become self-shading while in the expanded state the algal cells receive flux. Self-shading thus affords some protection from high levels of solar radiation (Lasker, 1979).

5.1.2. Morphological Adaptation

Morphological adaptation may occur at two scales: at the organism and at the cellular levels. For corals, adaptation may involve alteration in colony growth form and/or changes in the internal structure and composition of the zooxanthellae chloroplasts.

The plating of *Acropora* species in shallow water (Falkowski *et al.*, 1990) and the changes from massive to plating forms observed on transplantation of other species from shallow to deep waters (Willis, 1985) have been described as ways in which a coral may optimize light capture while maintaining a high surface area for gas exchange (Jokiel and Morrisey, 1986). These authors considered an overtopping morphology in *Pocillopora damicornis* to be a photoadaptive characteristic that increases the efficiency of energy utilization by overcoming the energetic limitations of individual algal cells being saturated at low radiant flux.

To achieve plating morphologies many corals orient in the direction of maximum flux. Young corallites of *Galaxea fascicularis* have been shown, by experiment, to reorient to maximum flux by bending, the non-irradiated part of the skeleton calcifying and extending more rapidly than irradiated portions (Hidaka and Shirasaka, 1992).

At the cellular level, differences in the morphology of the zooxanthellae have been observed in organisms exposed to high and low flux. In the tentacles of the tropical sea anemone *Aiptasia pallida* shade-adapted zooxanthellae were larger and more densely packed than those in anemones growing in high flux (Muller-Parker, 1987). Zooxanthellae in shade anemones also contained more thylakoids (flattened lamellae-like structures in the chloroplast within which chlorophyll and carotenoids are bound to different pigment-proteins that make up the photosynthetic apparatus) than those at high irradiance. Berner *et al.* (1987) observed differences in zooxanthellae ultrastructure within shaded and irradiated portions of the same colony in the soft coral *Lithophyton arboreum* with shaded zooxanthellae showing an increased stacking of thylakoids compared to irradiated zooxanthellae. Dubinszky *et al.* (1984) observed similar results for zooxanthellae in shade and sun-adapted corals (*Stylophora pistillata*) from the Red Sea and suggested that greater concentrations of chlorophyll pigment in shade-adapted zooxanthellae required more extensive matrix material. The increase in stacking of thylakoids may be an adaptation not only to decreased flux but also to changes in spectral quality with blue-light chloroplasts having a lower packing density of thylakoids than red-light chloroplasts (Lichtenthaler and Meier, 1984).

5.1.3. *Biochemical and Physiological Adaptations*

Barnes and Chalker (1990) have reviewed the biochemical and physiological adaptations of corals to varying radiant flux. They noted that photoadaptation to decreasing flux generally results in a marked change in concentrations of photosynthetic pigments contained within the zooxan-thellae and highlight three common trends. The first involves an increase in chlorophyll *a* and the accessory pigments chlorophyll *c* and peridinin with decreasing flux. The second trend is the frequently higher concentra-tions of β-carotene and yellow xanthophylls dinoxanthin and diadinoxan-thin in shallow water corals (Jeffrey and Haxo, 1968; Titlyanov *et al.*, 1980). The third is the relatively high concentrations of compounds which absorb over the range of 286–340 nm in corals from shallow reefs.

Trends such as the first one described above result in changes in the photosynthetic apparatus of the zooxanthellae, specifically in the photosynthetic units within the algal cells, which act to increase the efficient use of available radiant energy. The photosynthetic units (PSU) consist of light-harvesting chlorophylls and other pigments which absorb the solar quantum energy, transfer it to a reaction centre in the PSU, which contains chlorophyll *a*–protein complexes, where it is transformed into chemical energy. Photoadaptation of corals may involve both an increase in size and/or number of PSUs (see Barnes and Chalker, 1990 and Falkowski *et al.*, 1990 for reviews). A more recent paper by Iglesias-Prieto and Trench (1994) measured photoacclimation in cultured zooxanthellae from three symbiotic cnidarian hosts (a jellyfish, a zoanthid and a coral), each of which harbour a different species of zooxanthella. Results indicated that each algal species had quite different photosynthetic characteristics despite being cultured under the same conditions. One common characteristic was that all three species acclimated to low quantum flux by increasing both the number and the size of their PSUs; however, the detailed nature of these changes differed between species.

It is important to note that although the majority of studies on corals suggest that photoadaptation to low radiant flux involves a general increase in size rather than number of PSUs, the above study involved cultured zooxanthellae and not the intact symbiotic association. Iglesias-Prieto and Trench (1994) argue that their conclusions may be influenced by using cultured zooxanthellae which are not nutrient-limited and where increas-ing PSU numbers may be an option not seen in intact associations because of energy constraints. Nevertheless, an important conclusion from this work is that the different photoacclimatory capabilities of the algae, which correlate well with respective ecological distributions, may be under genetic constraints. An important conclusion from this work is that the different photoacclimatory capabilities of the algae, which correlate well with respective ecological distributions, may be under genetic constraints (Rowan and Knowlton, 1995). Iglesias-Prieto and Trench suggest that the different photosynthetic capacities are probably defined by the differing degrees of predictability encountered in different habitats, with abilities limited by the concentration of carbon-fixing enzymes or electron transfer chains.

Perhaps the most remarkable adaptation shown by a coral to reduced radiant flux and changed spectral quality is that of *Leptoseris fragilis*, which lives at depths of 95–145 m (Schlichter and Fricke, 1990; Kaiser *et al.*, 1993) in the Red Sea. The efficient use of the available solar energy is effected not only by pigment changes in the symbiotic algae, but also fluorescent pigments sited within the coral host tissues. Short-wavelength radiation which would otherwise not be absorbed by algal pigments is absorbed by the coral pigments and fluoresced into longer wavelengths for harvesting by the zooxanthellae. Such adaptations enable *L. fragilis* to colonize a habitat which is barren of all other symbiotic coral species. Furthermore, this mechanism is potentially present to some degree in all corals at all depths. Recent work by Mazel (1995) suggests that fluorescence observed in Caribbean corals does not contribute to photosynthesis.

5.2. Photoinhibition

Photoinhibition has been described as a reversible reduction of photo-synthetic capacity generally induced by exposure to high solar radiation. While the phenomenon has been commonly observed at high radiant flux in cultured algae, including zooxanthellae isolated from tropical sea anemones (Muller-Parker, 1984), its occurrence in the intact host–algae association has yet to be shown unequivocally. Exposure of symbiotic anemones to very high flux did not induce photoinhibition (Fitt *et al.*, 1982; Muller-Parker, 1984). However, Gattuso and Jaubert (1984) attributed unexpectedly low rates of photosynthesis in corals at 1 m water depth to photoinhibitive effects, while Chalker (1983) found reduced calcification in the Caribbean coral *Acropora cervicornis* when transferred from 17 m depth (irradiance 390 μmol m^{-2} s^{-1}) to an irradiance of 1950 μmol m^{-2} s^{-1}, an effect which he presumed to be the outcome of photoinhibition.

Although there has been no definitive demonstration of photoinhibition in corals and symbiotic anemones, recent work on bleaching responses of these organisms poses the question whether or not it might occur under extreme flux conditions. Damage to zooxanthellae and consequent bleaching have been noted as a result of exposure to high solar radiation in the field (Brown *et al.*, 1994b); on transplantation of hosts from deep to shallow water depths (Gleason and Wellington, 1993) and (for anemones and corals in the laboratory) on exposure to elevated PAR (Hoegh-Guldberg and Smith, 1989; Lesser, 1989; Lesser and Shick, 1989; Lesser *et al.*, 1990). In some of these examples ultrastructural damage to the algal cell was noted (Le Tissier, unpublished); in others reduced chlorophyll concentrations per zooxanthella cell were recorded (Lesser *et al.*, 1990), both factors having implications for the effective functioning of the photosynthetic apparatus. None of the above studies actually investigated photosynthesis/irradiance responses in stressed corals, thus leaving open the question of whether or not photoinhibition takes place.

While earlier studies in plant physiology considered photoinhibition as damaging, it is important to recognize that the inhibition of photosynthesis is reversible and that the phenomenon involves protective mechanisms that serve to dissipate excess energy (Krause, 1988). There is general agreement that photoinhibition results primarily from an inactivation of the electron transport system in the reaction centres of the chloroplast. In high radiant flux conditions one of these reaction centres (known as PSI or photosystem I) is photoinhibited by the presence of oxygen. During photosynthesis, chlorophyll is excited to the triplet state which can then interact with oxygen to give rise to potentially toxic singlet oxygen. As discussed earlier, defensive enzymes such as superoxide dismutase (SOD), catalase and ascorbate peroxidase in zooxanthellae and SOD and catalase in the host, together with carotenoid pigments in the algae, act in concert to deactivate these toxic compounds. It is, however, possible for these protective reactions to be "swamped" with resulting photoinhibition.

Photoinhibition of another reaction centre (PSII or photosystem II) involves damage to the functional integrity of the centre and subsequent loss of a key 32 kDa protein known as the DI protein. This protein has been described as a suicidal polypeptide which has a very high turnover in the chloroplasts of algae during photoinhibition (Kyle *et al.*, 1984; Ohad *et al.*, 1988; Osmond, 1994).

A suite of regulatory mechanisms thus exists in photosynthetic organisms to protect them against the damaging effects of high radiant flux. In summary, these include: (1) the primary photosynthetic reactions which "unload" the excited pigments; (2) the "protective" mechanisms (defensive enzymes and sacrificial carotenoid pigments) which help avoid harmful effects of photochemical side-reactions; (3) repair mechanisms (DI protein production) which compensate for any damage that occurs. Should these processes fail then photo-oxidation (bleaching of chlorophyll pigments, a process not to be confused with the bleaching response of the coral) and consequent damage to the chloroplasts will ensue.

While some work has been carried out on defensive enzyme production in tropical zoanthids (Lesser *et al.*, 1990), the regulatory mechanism described above has not been investigated in reef corals living in shallow, well-illuminated waters. The potential for photoinhibition is clearly high in this environment; the actual mechanisms involved in protection of the photosynthetic machinery of zooxanthellae and their energetics costs (Raven, 1989, 1994) are unknown for reef corals. The question of possible photochemical mechanisms that might exist in corals to prevent photoinhibition was first raised by Chalker *et al.* (1983) who noted that corals growing at 1 m depth were 95% saturated at an irradiance which was only 20% of that normally encountered by corals at this depth at local solar noon. Clearly, few advances have been made in this area in the past 10 years.

5.3. Daily and Seasonal Fluctuations in Solar Radiation

Exogenous, daily rhythms in photosynthesis have been noted in cultured dinoflagellate algae (Prezelin *et al.*, 1977; Prezelin and Sweeney, 1978), in zooxanthellae in culture and those freshly isolated from the coral hosts (Chalker, 1977; Chalker and Taylor, 1978) and zooxanthellae in intact symbioses (Porter, 1980; Muller-Parker, 1984; Chalker *et al.*, 1985).

Figure 10 shows the diurnal rate of photosynthesis in branching *Acropora* spp. at different depths on the Great Barrier Reef as predicted by Barnes and Chalker (1990). It is clear that in shallow waters down to 15 m depth, maximum rates of photosynthesis rise soon after sunrise, remain relatively constant through the day and decline rapidly an hour before sunset. Only at depths of > 40 m (4% surface irradiance) did photosynthetic saturation not occur through a significant part of the day (Barnes and Chalker, 1990).

Corals are well adjusted to the predictable changes in solar radiation that occur on a daily basis, but surprisingly little is known about their seasonal photoadaptive abilities. Studies in the Red Sea showed seasonal photoadaptation in one coral species (*Echinopora gemmacea*) but not in the other (*Stylophora pistillata*) (Al-Sofyani and Davies, 1992). Seasonal photoadaptive responses have been observed by Chalker *et al.* (1984) for *Acropora granulosa* on the Great Barrier Reef, Australia. The transition from relatively high irradiance in summer to relatively low irradiance in winter produced analogous changes in parameters such as maximum photosynthetic rates and the initial slopes of the light saturation curve. Both factors increased in response to decreasing irradiance when expressed as a function of protein content of the coral (Chalker *et al.*, 1984; Barnes and Chalker, 1990). When the maximum gross rate of photosynthesis was expressed in terms of chlorophyll content, values decreased during winter as a result of increasing chlorophyll *a* content with decreasing irradiance (Barnes and Chalker, 1990). Similar seasonal photoadaptive changes have been observed during the wet and dry monsoon seasons for shallow reef flat corals in Phuket, Thailand (Bythell, unpublished).

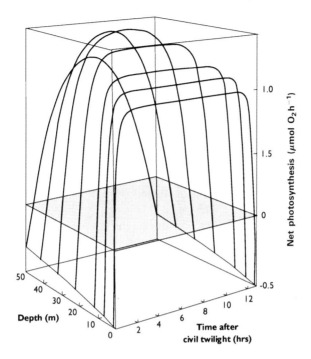

Figure 10 Calculated diurnal rates of net photosynthesis for *Acropora* spp. growing at depths of 1–50 m on Davies Reef, Great Barrier Reef, Australia. (After Chalker *et al.*, 1984).

Concomitant changes in chlorophyll content of corals might be anticipated in response to changing irradiance levels. Chlorophyll concentrations per zooxanthella fluctuate with season in temperate sea anemones (Dykens and Shick, 1984) and in inter tidal reef corals (Brown *et al.*, 1995, and unpublished data), values varying inversely with mean solar radiation. The higher chlorophyll content of zooxanthellae in temperate anemones during winter and tropical corals during the wet season presumably compensates for the lower solar radiation values at these times. In temperate sea anemones disproportionate increases in chlorophyll c_2 during winter months were described as a further adaptation to reduced irradiance (Dykens and Shick, 1984). A similar pattern was observed during the wet season in intertidal reef corals.

Changes in chlorophyll content per zooxanthella may also occur in corals in the field over a period of days (Brown *et al.*, unpublished data) with lower chlorophyll levels and reduced chlorophyll c_2 values associated with exposure to high radiant flux. Such physiological plasticity is characteristic of photosynthetic organisms that have evolved in a highly dynamic solar radiation environment.

5.4. Responses to Ultraviolet Radiation

Shallow water reef-building organisms have frequently been described as being exposed to high levels of UVR (Jokiel and York, 1982; Dunlap and Chalker, 1986; Drollet *et al.*, 1993; Gleason, 1993a; Gleason and Wellington, 1993; Kinzie, 1993). While this may be the case for clear open ocean waters, many reefs exist in turbid or coloured water where short-wavelength solar radiation is very rapidly attenuated under water (Jerlov, 1950).

Nevertheless, even in these environments shallow water tropical marine organisms are exposed to the penetration of both UVA (320–400 nm) and UVB (280–320 nm). Damaging effects of UVR on survival, growth and physiology of marine biota are well documented (Calkins, 1982). For marine invertebrates and algae the effects include damage to DNA and proteins, oxidation of membrane lipids and reduction of photosynthesis (Shick *et al.*, 1991). The relative responses of an organism to a particular spectral range may be summarized by an action spectrum. This plot is a measure of spectrally weighted functions for specific biological effects. A set of action spectra for generalized plant responses, for DNA damage and for the expression of photosynthesis with respect to UVR is shown in Figure 11a (after Stordal *et al.*, 1982). Although action spectra have been prepared for the photosynthetic (Halldal, 1968; Scott and Jitts, 1977) (Figure 11b) and absorption responses (Dustan, 1979; Leletkin *et al.*, 1980; Jokiel and York, 1982; Schlichter and Fricke, 1990) of zooxanthellae isolated from corals, we can most confidently predict effects only over the wavelengths 400–700 nm and not in the UV part of the spectrum.

Unequivocal evidence of damage to corals from UV exposure is difficult to find from published work (see Brown *et al.*, 1994b) although Kinzie (1993) demonstrated reductions in areal chlorophyll concentrations in *Montipora verrucosa* as a result of elevated solar UV (280–380 nm) while Gleason (1993a) showed reduced rates of zooxanthellae division and linear skeletal extension in *Porites astreoides* exposed to UVR in a field manipulation. More experimental evidence is available for the damaging synergistic effects of UVR and other factors such as PAR, and increased temperature on tropical symbiotic cnidarians (Coles and Jokiel, 1978; Lesser *et al.*, 1990; Glynn *et al.*, 1993).

When encountering high and potentially damaging levels of UVR the coral's response may be avoidance, damage repair or tolerance. Avoidance may be effected by retraction of the tentacles and tissues as observed in

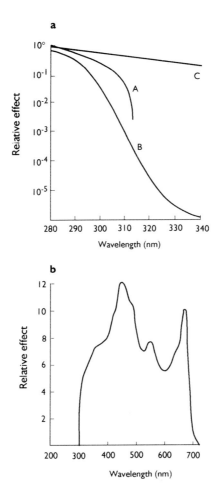

Figure 11 Action spectra for: (a) generalized plant responses (A), DNA damage (B) and photoinhibition (C) (after Stordal *et al.*, 1982); and (b) photosynthetic responses for isolated zooxanthellae from the coral *Favia* (after Halldal, 1968).

both corals and anemones living in environments subject to high irradiance. These responses allow shading and protection of zooxanthellae (Dykens and Shick, 1984; Brown *et al.*, 1994b; Stochaj *et al.*, 1994).

Alternatively, corals and anemones may use screening as a protection. In this way mucus (Drollet *et al.*, 1993), pigments sited in the host tissues (Kawaguti, 1944; Shibata, 1969) and UV-absorbing compounds found in both host and symbionts (Shick *et al.*, 1991, 1995) may all play a role. Most work has been carried out on UV-absorbing compounds—the mycosporine amino acids (MAAs). A suite of such compounds has now been isolated in both corals (Dunlap and Chalker, 1986; Dunlap et al., 1986; Gleason, 1993a; Shick et al., 1995) and anemones (Stochaj et al., 1994) each having characteristic absorbance peaks over the range 310–334 nm. While the presence of MAAS and UV protection have been linked in a number of studies on tropical marine organisms, it is by no means clear that this is either their primary or even secondary function (Karentz and McEuen, 1991).

A protective role for these compounds has been inferred from their UV-absorbing properties (Dunlap et al., 1986), their decrease in concentration with increasing depth (Dunlap et al., 1986; Scelfo, 1986) or after artificial screening from UV (Jokiel and York, 1982; Scelfo, 1986; Kinzie, 1993). In addition, the increase in concentration of these compounds on transplanting corals from deep to shallow water (Scelfo, 1986; Gleason, 1993a) has also been attributed to the effects of increased UVR. However, not all coelenterates show such compensatory changes (Shick et al., 1991; Stochaj et al., 1994), and both handling stress (Scelfo, 1986) and increased temperature (Sebastian et al., 1994) also appear to induce production of some MAAs (Scelfo, 1986). Results of transplantation experiments must therefore be viewed with some caution. Also, it is clear that the mucus of some corals contains considerable quantities of MAAs (Drollet et al., 1993), with the highest concentrations noted in the first 2 min of secretion and decreasing thereafter. If UV-absorbing compounds are generally found in coral mucus then considerable variability in recorded concentrations may result from inter- and intraspecific differences in mucus production, and from handling stresses when excessive amounts of mucus may be produced. Preliminary results on seasonal production of UV-absorbing compounds in corals from shallow water reefs in Thailand suggest that highest concentrations of these compounds are not found during the dry season when solar radiation is maximal but during the wet season (Brown, unpublished). This unexpected result may indicate that the actual concentrations of MAAs measured at any one time are less important than their turnover. The time taken to synthesize these compounds on transplantation, or during experiments, is variable ranging from 13 d (Scelfo, 1986) to over 100 d (Shick et al., 1991).

Many issues remain to be clarified regarding the UV photophysiology of corals. These include a better understanding of the underwater UV environment over a variety of different conditions, the damage action spectra for zooxanthellae and coral symbiotic unit, and the role and regulation of the UV-absorbing compounds.

In terms of damage repair little is known about the processes involved in either corals or their symbiotic algae. However, repair systems for UV-induced DNA damage have been well documented in bacteria (Jagger,

1985) where they may involve both recombination and excision repair. Evidence for a repair system induced by exposure to UV over the wavelengths 320–400 nm, has also been described in actively growing cultures of *Escherichia coli* (Peters and Jagger, 1981), a process which involved induction of several new proteins. Repair proteins induced by exposure to UV (290–400 nm) and white light (incandescent lamps) have also been described in *Amoeba proteus* (Chatterjee and Bhattacharjee, 1976) and the cyanobacterium *Anacystis nidulans* (Bhattacharjee, 1977). In assessing biological consequences of UV radiation from sunlight, interactions with longer wavelengths must also be considered. Such interactions include both photoreactivation (or photoenzymatic repair) and also photoprotective effects (see Tyrell, 1986, for review). Recent work by Karentz *et al.* (1991) has shown that photoreactivation is likely a predominant pathway for DNA repair in antarctic diatoms exposed to UVR. Furthermore, Karentz *et al.* (1991) interpret increased zooxanthellae division under enhanced UV (Jokiel and York, 1982) as possibly attributable to increased production of DNA replication enzymes that are required for excision repair. It is likely that zooxanthellae and their coral hosts have diverse capabilities for sustaining and repairing UV-induced damage to DNA and that they are not defenceless against this environmental stress. The scope of these adaptations remains to be explored, however.

In relation to tolerance, Jokiel and York (1982) were the first to suggest that different genetic strains of zooxanthellae may have different tolerances to UVR. Certainly work on different strains of free-living dinoflagellates has shown marked differences in UV sensitivity, the responses being correlated with previous irradiance history of the organism (Sebastian *et al.*, 1994). Jokiel and York (1982) argued that a wide range of potential pigment and symbiont clone combinations might provide considerable flexibility in the tolerances achieved by these symbioses. Such a conclusion is also reached by Gleason (1993a), who demonstrated different adaptations to UVR in two morphs of the same Caribbean coral species *Porites astreoides*. Whether these adaptations involve genetic differences in either host and/or symbiotic algae is not known, although Gleason states that the two coral morphs are genetically distinct phenotypes (Potts and Garthwaite, 1991).

6. ADAPTATION OF REEF CORALS TO DIFFERENT WATER FLOW REGIMES

The profound effects of water motion on coral community structure are well documented (Rosen, 1975; Geister, 1977; Done, 1983), as are the effects of water flow on intraspecific variations in coral colony morphology (Veron and Pichon, 1976; Done, 1983). With regard to the latter, numerous investigators have used reciprocal transplant techniques where colonies of a specific morphology are transferred from their natural habitats into reef environments of differing physical and chemical characteristics known to harbour resident populations of the same species with a significantly different morphology. In the majority of these studies, where transplants have been in place for periods of 1–3 years, significant changes in colony morphology (ranging from growth form changes to spacing of corallites) have been noted (Maragos, 1972; Dustan, 1979; Foster, 1979, 1980; Hudson, 1981a, b; Graus and MacIntyre, 1982; Oliver et al., 1983; Potts, 1984b; Willis, 1985). Such studies clearly highlight the morphological phenotypic plasticity of corals but they do not allow the effects of single factors, such as water motion, to be isolated from the milieu of interacting environmental influences which impinge on corals.

Recently, a number of manipulative experiments have allowed the identification of specific adaptations of corals to water flow. It is therefore worth briefly reviewing such responses because they reveal once more the high degree of phenotypic plasticity (behavioural, biochemical and morphological) exhibited by reef corals in response to a physical variable. Indeed, it has been argued that exposure of corals to a range of hydrodynamic conditions effects a similar scale of phenotypic physiological responses as those noted in fluctuating light environments (Patterson et al., 1991).

Sessile organisms like corals risk being dislodged or damaged by high water flow yet they depend on such a flow for supply of nutrients and essential chemical elements and to disperse wastes and gametes. Corals may either avoid rapid flow and show adaptations which compensate for lack of external transport or face rapid flow and show adaptations which reduce the drag forces on the organism (Wainwright and Koehl, 1976). For example, flow over many branching corals is generally at a high Reynolds number. In exposed locations corals might be expected to have most of their surface area parallel to the direction of flow, thus minimizing drag. This is certainly the case for the Caribbean branching coral *Acropora palmata* (Chamberlain and Graus, 1975) and the Indo-Pacific coral *Acropora palifera* (Done, 1983). Experiments have also shown that branching corals undergo morphological modifications (affecting branching patterns, porosity of the colony and relative branch size) in relation to water flow to accommodate physiological requirements (Chamberlain and Graus, 1975; Graus, 1977).

It is very important to appreciate that the water flow pattern over the coral colony may differ from that over the polyps. Expanded polyps possess diffusive boundary layers (a laminar layer of slow-moving fluid near the body surface) of considerable depth which may limit exchange of nutrients and waste products (Patterson, 1992; Shashar *et al.*, 1993). At high water flows more rapid exchanges are allowed through thinner diffusive boundary layers, potentially allowing higher rates of respiration, photosynthesis, calcification and growth (Jokiel, 1978; Dennison and Barnes, 1988; Patterson and Sebens, 1989; Shashar *et al.*, 1993).

The degree to which behavioural responses of corals (i.e. expansion/contraction) may enhance the exchange of nutrients and other elements by altering the thickness of the boundary layer has been addressed by Patterson (1992). In expanded colonies the extended tentacles, by increasing the polyp's surface : volume ratio and exposed surface area, can enhance diffusion rates across the boundary layer. At the same time, behavioural patterns and tentacle morphologies may also play an important role not only in solute exchange but also in efficient food capture in different water flows. Colonies living in strongly unidirectional currents often exhibit a vertical planar morphology associated with capture of food particles from a downstream wake (Wainwright and Koehl, 1976; Wainwright *et al.*, 1976; Patterson, 1980) but many flattened and rounded corals also live in unidirectional currents. Some, with relatively short tentacles such as *Agaricia agaricites*, benefit from enhanced downstream capture in faster flows (Helmuth and Sebens, 1993) whereas others, such as *Meandrina meandrites*, may make use of long tentacles and horizontal morphology to enhance relative upstream retention of food particles (Johnson and Sebens, 1993).

As well as these marked interspecific behavioural and morphological adaptations, corals also show considerable phenotypic plasticity in their growth form under different wave energy conditions. *Pocillopora damicornis* from protected low flow and exposed high flow habitats in Hawaii, where light environments were similar, differed markedly in their morphology (Lesser *et al.*, 1994). Colonies from high flow had significantly thicker branches with small interbranch spacing while low flow colonies exhibited thinner branches with greater interbranch spacing. Earlier transplants of this species between high flow and low flow habitats in Hawaii (Maragos, 1972) resulted in the alteration of growth morphologies appropriate to the new hydrodynamic setting. Of interest is recent work showing that colonies from both habitats have similar Reynolds numbers and similar productivities (Lesser *et al.*, 1994). These authors suggest that the morphological plasticity of the skeleton provides a mechanism that minimizes diffusional boundary layer thickness and maximizes supply of carbon to the zooxanthellae under different flow regimes.

Physiological and biochemical plasticity are also important when a coral of a fixed morphology is exposed to different flow regimes. Acclimation of high flow *Pocillopora damicornis* to different flow regimes for 14 days resulted in alteration of physiological performance in a manner which compensated, at least in part, for the constraints imposed by a fixed morphology. An increase in photosynthesis with increasing flow was mirrored by increased activity of antioxidant enzymes while increased activity of other enzymes under low flow conditions augmented the supply of carbon to sites of assimilation (Lesser *et al.*, 1994). This biochemical plasticity under low flow conditions did not fully compensate for the thicker diffusive boundary layers encountered and the resultant low dissolved inorganic carbon availability. Nevertheless, the experiment showed that the coral has some scope to adjust physiological processes to meet changed environmental conditions in the short term.

7. DISCUSSION

From the foregoing sections it is clear that corals are well adapted to the changing and often extreme environments of the reef by virtue of an armoury of phenotypic responses at organism, cellular and molecular levels (Table 1). The physical fluctuations that they experience may be unpredictable (hurricanes, storms, ENSO events) or predictable (seasonal changes in temperature, irradiance and water flow). Clearly, survival of the unpredictable physical challenge depends upon the severity of the event but there is an increasing number of case histories to suggest that the short-term (weeks–months) exposure of corals to increased sea water temperatures (Brown *et al.*, 1993; Gleason, 1993b) and high irradiance (Brown *et al.*, 1994b) can be tolerated by some colonies of even the most susceptible species. Although little work has been carried out on adaptations of corals to predictable seasonal influences, evidence is beginning to accrue of seasonal physiological adaptation to altered temperature and irradiance (Chalker *et al.*, 1984; Al-Sofyani and Davies, 1992).

The phenotypic responses documented in Table 1 are all relatively short-term and indicate that many of these defences and repair mechanisms operate within a very short time span of detection of physical change. One factor that is likely to be very significant in overall tolerance of physical fluctuations is the energetic costs of protection and damage repair (Schäfer, 1994). Protein synthesis and turnover are metabolically costly. Protein synthesis and degradation are central to the majority of responses shown in Table 1, whether they be induction of stress proteins, repair proteins or protein metabolism associated with acclimatization responses. Metabolic costs of protein production are not available for corals but protein synthesis in marine mussels accounts for 16% of basal metabolic rate (Hawkins, 1985). Such figures do not account for the energy costs

Table 1 Selected phenotypic responses to altered temperature, irradiance and water motion regimes by corals and symbiotic anemones.

Time-scale	Temperature	Irradiance	Water flow
Seconds–minutes		Organism—behavioral contraction/expansion[1]	Organism—behavioral contraction/expansion[2]
Minutes–hours	Molecular—induction of stress proteins[3]	Molecular—induction of repair proteins[4]	
Hours–days	Cellular—altered enzyme activities[5]; Organism—degradation of zooxanthellae and release[6,7]	Cellular—altered antioxidant enzyme activities[5]; alterations in pigment concentrations[7]	Cellular—altered antioxidant enzyme activities, also altered activities of enzymes RUBISCO and carbonic anhydrase[12]
Days–weeks	Acclimation[8] of photosynthetic/respiration rates	Cellular—altered concentrations of MAAs[13]; Organism—degradation of zooxanthellae and release[7]; Acclimation of photosynthetic/respiration rates[11]; Growth form changes[14]	Organism—acclimation of photosynthetic/respiration rates[12,13]
Weeks–months	Organism—seasonal acclimatization of photosynthetic/respiration rates[9]	Organism—seasonal acclimatization of photosynthetic/respiration rates[15]; Growth form changes[9]	Organism—growth form changes[16]

MAAs, Mycosporine amino acids; RUBISCO, ribulose 1,5-bisphosphate carboxylase.

1 Brown et al. (1994b).
2 Johnson and Sebens (1993).
3 Sharp et al. (1993).
4 Karentz et al. (1991).
5 Lesser et al. (1990).
6 Glynn and D'Croz (1990).
7 Brown et al. (in press).
8 Jokiel and Coles (1977).
9 Al-Sofyani and Davies (1993).
10 Dykens and Shick (1984).
11 Harland and Davies (1994).
12 Lesser et al. (1994).
13 Jokiel (1978).
14 Hidaka and Shirasaka (1992).
15 Willis (1985).
16 Maragos (1972).

of protein degradation—a process which is also energy demanding. An ability to adjust the concentration and composition of intracellular proteins must be especially important to shallow water corals living in a potentially hostile and physically changing environment. For these organisms, added to the energetics costs already described, will be the additional costs of UVB protection and possible production of repair proteins. The synthesis of UVB screening capacity inherent in $100 \, \text{mol} \, \text{m}^{-3}$ palythine, costs 19% of the total energy needed for synthesis of a microalgal cell (Raven, 1991), and there would be additional costs if damage should ensue to this target and any surrounding tissues.

What, then, are the likely costs for corals? We know relatively little about the nitrogen budget of corals (Bythell, 1988) let alone protein turnover under stressed and non-stressed conditions but these processes and the energetic costs involved for both host and zooxanthellae are central to our understanding of short-term adaptive responses of corals to environmental change. We might suspect that given such a changeable environment, in time and space, protein turnover in shallow water corals would be relatively high. Experiments with marine bivalves have measured rates of protein synthesis and breakdown and components of the energy budget for organisms subjected to temperature elevation (Hawkins et al., 1987). Individuals expressing high rates of protein turnover and higher metabolic rates were more sensitive to temperature and more susceptible to ultimately lethal temperature than individuals with lower rates of protein turnover. While it might have been supposed that a high inherent rate of protein synthesis would confer a fitness advantage under temperature stress, this was not the case. On the contrary, by elevating metabolic demand the result was a significant disadvantage compared with individuals with a lower and more efficient rate of protein synthesis.

Further physiological studies on bivalves have shown that the most heterozygous individuals exhibit low intensities of protein turnover and reduced energetics costs of metabolism (Hawkins et al., 1986, 1994). A too intense or too prolonged period of stress may affect the energetics balance in an organism with energy production no longer sufficient for maintenance requirements (Sibly and Calow, 1989). As a result, higher energetics efficiency in heterozygotes has been shown to increase resistance to stress (Rodhouse and Gaffney, 1984; Hawkins et al., 1989; Koehn and Bayne, 1989; Scott and Koehn, 1990; Borsa et al., 1992). This work is still in its infancy and the genetic mechanisms by which maintenance requirements vary with heterozygosity remain speculative. Phenotypic effects of heterozygosity may be locus-specific with a particular contribution by loci encoding enzymes involved in protein catabolism and glycolysis (Koehn et al., 1988). An alternative theory is that the enzyme loci may be considered as neutral markers of genes or part of the genome which determine these phenotypic effects (Zouros *et al.*, 1988; Zouros and Mallet, 1989; Zouros, 1990). Whatever the mechanism that links genome composition and physiological adaptation, the relevance of this work to corals is clear. Indeed, recent work by Edmunds (1994) has suggested that the high intraspecific variability in coral bleaching may be due to the distribution of bleaching susceptible clones and/or the expression of different phenotypes by adjacent genets.

Levinton (1973) was one of the first to suggest that heterozygosity was selected in a fluctuating environment. For coral reefs, Potts and Garthwaite (1991) have argued that physical variations within the reef mosaic have acted within each generation of corals to select for individuals which show morphological, physiological and ecological variations. Such physical fluctuations, and the fact that there is extensive back-crossing in organisms with a long generation time, result in shallow water coral populations showing a large amount of genetic variation (Potts, 1984a, b; Potts and Garthwaite, 1991). For *Porites* from the Pacific the case for intraspecific genetic variation was supported by electrophoretic data from 14 enzyme loci. The level of heterozygosity observed was higher than that reported for most other invertebrates (Potts and Garthwaite, 1991). Apart from such inherent genetic variation, the possibility that a single colony does not always constitute a single genotype has also been raised (Hughes *et al.*, 1992), highlighting further the significance of individual variability and the bearing it may have on future adaptational potential.

Do such results account for the high intraspecific variations in physiological tolerances to elevated sea water temperatures described in corals and discussed earlier with specific reference to *Porites* (Tudhope *et al.*, 1993)? And do those coral–algal associations which do not bleach in response to increased temperature and irradiance show higher heterozygosity than those which do? Furthermore, are such tolerances the consequence of lower metabolic expenditure and more efficient protein turnover by either host and/or zooxanthellae? These and many other questions regarding responses of corals to environmental stress remain unanswered. Future research into the scope of adaptation of reef corals to environmental change will demand much closer collaboration between geneticists and physiologists than has been seen to date if any headway is to be made.

8. SUMMARY

1. Coral reefs are not stable communities living in benign environments but ecosystems which are subject to frequent physical disturbances on time-scales which vary from minutes to years.

2. Throughout their 260 million year history, corals have responded to these physical fluctuations by a variety of phenotypic and genetic adaptations.

3. The major influences on phenotypic and genotypic adaptive processes are likely to be the products of fluctuating abiotic (and biotic) factors, which have varied in both geological and ecological time, and the heterogeneity of the coral reef environment.

4. The survival of most coral taxa in the Caribbean and Indo-Pacific over the last 2–3 million years has been attributed, at least in part, to their extensive intraspecific genetic variation which has permitted rapid local adaptation to environmental conditions encountered at a particular site in any one generation.

5. Corals are capable of rapid phenotypic responses to predictable changes on a day-to-day and seasonal basis, the best example of this being their remarkable photoadaptation ability.

6. Physiological adaptations to physical variability include fluxes in photosynthetic pigments and protective carotenoids in the zooxanthellae, production of stress proteins, defensive enzymes and possibly specific amino acids which may act as sun screens in both coral host and zooxanthellae. The energy costs of such defences for the coral symbiosis are likely to be significant but are largely unknown.

7. The phenomenon of coral bleaching has been suggested as adaptive by potentially allowing the repopulation of corals by different types of algae which may confer stress resistance, although evidence is still sought to support this claim. It is also possible that bleaching may be a damage-limitation process which permits corals to withstand stressful periods while harbouring a reduced zooxanthellae population.

8. On a geological time-scale, corals have shown themselves capable of colonizing extreme environments such as the Arabian Gulf where rigorous physical conditions have no doubt demanded the evolution of particular genetic strains of both coral hosts and their zooxanthellae.

9. This review has focused primarily on short-term phenotypic physiological responses with the aim of highlighting the scope that corals have for responding to environmental stress, for it is these responses that may be significant for present generations of long-lived corals in the next 40–50 years. Longer-term influences on the genotype through regional differentiation and speciation are clearly also critically important in predicting future responses of reef corals not only to climate change but also to other stresses. In this regard, our understanding of the genetics of coral host and algae and their evolutionary potential is even more limited than our appreciation of short-term phenotypic responses. The prediction of the physiological responses of reef corals

to future environmental change will demand not only a better understanding of the defence mechanisms at work and their energetic costs but also the phenotypic and genetic bases of any observed tolerances in individual corals.

ACKNOWLEDGEMENTS

I would like to thank Professor A. Clarke, Dr P. S. Davies, Dr B. Rosen and Ms J. Hawkridge for their helpful comments on the manuscript. I should also like to thank the Royal Society, the Natural Environment Research Council, the Natural Resources and Environment Department of the Overseas Development Administration and the Phuket Marine Biological Centre, Thailand, for their support of my research which has contributed significantly in the synthesis of this review. I am particularly grateful to Mr R. P. Dunne for his contribution to the review as a whole and in particular to the discussion of adaptations of reef corals to fluctuating levels of solar radiation.

In the compilation of Figure 10 the author gratefully acknowledges copyright permission of Elsevier Science Publishers. This figure was reprinted from Z. Dubinsky (1990), "Ecosystems of the World: Coral Reefs", p. 113, with kind permission from Elsevier Science, Sara Burerhartstraat 25, 1055 KV Amsterdam, The Netherlands.

REFERENCES

Abe, N. (1939). On the expansion and contraction of the polyp of a reef coral, *Caulastrea furcata* Dana. *Palao Tropical Biological Station Studies* **1**, 651–669.

Adams, C. G., Lee, D. E. and Rosen, B. R. (1990). Conflicting isotopic and biotic evidence for tropical sea-surface temperatures during the Tertiary. *Palaeogeography, Palaeoclimatology and Palaeoecology* **77**, 289–314.

Allison, N., Tudhope, A. W. and Fallick, A. E. (1996). A study of the factors influencing the stable carbon and oxygen isotopic composition of *Porites lutea* coral skeletons from Phuket, South Thailand. *Coral Reefs* **15**, 43–57.

Alongi, D. M. (1987). Intertidal zonation and seasonality of meiobenthos in tropical mangrove estuaries. *Marine Biology* **95**, 447–458.

Al-Sofyani, A. and Davies, P. S. (1992). Seasonal variation in production and respiration of Red Sea corals. *Proceedings of the 7th International Coral Reef Symposium* **1**, 351–357.

Anderson, D. M. and Webb, R. S. (1994). Ice-age tropics revisited. *Nature, London* **367**, 23–24.

Anderson, J. J. and Sapulette, D. (1981). Deep water renewal in inner Ambon

Bay, Ambon, Indonesia. *Proceedings of the 4th International Coral Reef Symposium* **1**, 369–374.

Andrews, J. C. and Pickard, G. L. (1990). The physical oceanography of coral-reef systems. *In* "Ecosystems of the World: Coral Reefs" (Z. Dubinsky, ed.), vol. 25, pp. 11–48. Elsevier, New York, USA.

Andrews, J. C., Dunlap, W. C. and Bellamy, N. F. (1984). Stratification in a small lagoon in the Great Barrier Reef. *Australian Journal of Marine and Freshwater Research* **35**, 273.

Asada, K. and Takahashi, M. (1987). Production and scavenging of active oxygen in photosynthesis. *In* "Photoinhibition" (D. J. Kyles, C. B. Osmond and C. J. Arntzen, eds), pp. 228–287. Elsevier, Amsterdam, Netherlands.

Ayre, D. J. and Willis, B. L. (1988). Population structure in the coral *Pavona cactus*: clonal genotypes show little phenotypic plasticity. *Marine Biology* **99**, 495–506.

Ayukai, T. (1993). Temporal variability of the nutrient environment on Davies Reef in the central Great Barrier Reef, Australia. *Pacific Science* **47**, 171–179.

Babcock, R. C., Bull, G. D., Harrison, P. L., Heyward, A. J., Oliver, J. K., Wallace, C. C. and Willis, B. L. (1986). Synchronous spawnings of 105 scleractinian coral species on the Great Barrier Reef (Australia). *Marine Biology* **90**, 379–394.

Babcock, R. C., Wills, B. L. and Simson, C. J. (1994). Mass spawning of corals on a high latitude coral reef. *Coral Reefs* **13**, 161–169.

Ball, M. M., Shinn, E. A. and Stockman, K. W. (1967). The geological effects of Hurricane Donna in south Florida. *Journal of Geology* **75**, 583–597.

Barnes, D. J. and Chalker, B. E. (1990). Calcification and photosynthesis in reef-building corals and algae. *In* "Ecosystems of the World: Coral Reefs" (Z. Dubinsky, ed.), vol. 25, pp. 109–131. Elsevier, New York, USA.

Bayne, B. (1975). Aspects of physiological condition in *Mytilus edulis* L. with respect to the effects of oxygen tension and salinity. *In* "Proceedings of the 9th European Marine Biology Symposium" (H. Barnes, ed), pp. 213–238. University Press, Aberdeen, UK.

Bayne, B. L. (1985). Responses to environmental stress: tolerance, resistance and adaptation. *In* "Marine Biology of Polar Regions and Effects of Stress on Marine Organisms" (J. S. Gray and M. E. Christianen, eds), pp. 331–349. John Wiley, New York, USA.

Bergmann, W., Creighton, S. M. and Stokes, W. M. (1956). Contributions to the study of marine products. XL. Waxes and triglycerides of sea anemones. *Journal of Organic Chemistry* **21**, 721–728.

Berner, T., Achituv, Y., Dubinsky, Z. and Benayahu, Y. (1987). Pattern of distribution and adaptation to different irradiance levels of zooxanthellae in the soft coral *Litophyton arboreum* (Octocorallia, Alcyonacea). *Symbiosis* **3**, 23–39.

Bhattacharjee, S. K. (1977). Unstable protein mediated ultraviolet light resistance in *Anacystis nidulans*. *Nature, London* **269**, 82–83.

Bjerknes, J. (1969). Atmospheric teleconnections from the equatorial Pacific. *Monthly Weather Reviews* **97**, 163–172.

Black, N. A., Voellmy, R. and Szmant, A. M. (1995). Heat shock protein induction in *Montastrea faveolata* and *Aiptasia pallida* exposed to elevated temperatures. *Biological Bulletin, Marine Biological Laboratory, Woods Hole* **188**, 234–240.

Blanquet, R. S., Nevenzel, J. C. and Benson, A. A. (1979). Acetate incorporation into the lipids of the anemone *Anthopleura elangantissima* and its associated zooxanthellae. *Marine Biology* **54**, 185–194.

Borsa, P., Jousselin, Y. and Delay, B. (1992). Relationships between allozymic heterozygosity, body size and survival to natural anoxic stress in the palourde *Ruditapes decussatus* L. (Bivalvia: Veneridae). *Journal of Experimental Marine Biology and Ecology* **155**, 169–181.

Bosch, T. G. C., Krylow, S. M., Bode, H. R. and Steele, R. E. (1988). Thermotolerance and synthesis of heat shock proteins; these responses are present in *Hydra attenuata* but absent in *Hydra oligactis*. *Proceedings of the National Academy of Sciences of the United States of America* **85**, 7927–7931.

Bradshaw, A. D. (1965). Evolutionary significance of phenotypic plasticity in plants. *Advances in Genetics* **13**, 115–155.

Bradshaw, A. D., and Hardwick, K. (1989). Evolution and stress—genotypic and phenotypic components. *Biological Journal of the Linnean Society* **37**, 137–155.

Brakel, W. H. (1979). Small-scale spatial variation in light available to coral reef benthos: quantum irradiance measurements from a Jamaican reef. *Bulletin of Marine Science* **29**, 406–413.

Bray, N., Hautala, S. and Pariwono, J. (1996) (in press). Large scale sea level, wind and thermocline variations in the Indonesian flow through region. *Proceedings of IOC-Westpac 3rd International Scientific Symposium, Bali, Indonesia.*

Brown, B. E. (1987). Worldwide death of corals—natural cyclical events or man-made pollution. *Marine Pollution Bulletin* **18**, 9–13.

Brown, B. E. (1996) (in press). Coral bleaching: causes and consequences. *Coral Reefs.*

Brown, B. E. and Dunne, R. P. (1980). Environmental controls of patch reef growth and development, Anegada, British Virgin Islands. *Marine Biology* **56**, 85–96.

Brown, B. E. and Howard, L. S. (1985). Assessing the effects of stress on reef corals. *Advances in Marine Biology* **22**, 1–63.

Brown, B. E., and Suharsono. (1990). Damage and recovery of coral reefs affected by El Niño-related seawater-warming in the Thousand Islands, Indonesia. *Coral Reefs* **8**, 163–170.

Brown, B. E., Le Tissier, M. D. A., Howard, L. S., Charuchinda, M. and Jackson, J. A. (1986). Asynchronous deposition of dense skeletal bands in *Porites lutea* Edwards and Haine. *Marine Biology* **93**, 83–89.

Brown, B. E., Le Tissier, M. D. A., Dunne, R. P. and Scoffin, T. P. (1993). Natural and anthropogenic disturbances on intertidal reefs of SE Phuket, Thailand 1979–1992. *In* "Proceedings of the Colloquium on Global Aspects of Coral Reefs: Health, Hazards and History" (R. N. Ginsberg, ed.) 420 pp. Rosenstiel School of Marine and Atmospheric Science, University of Miami, Miami, USA

Brown, B. E., Dunne, R. P., Scoffin, T. P. and Le Tissier, M. D. A. (1994a). Solar damage in intertidal corals. *Marine Ecology Progress Series* **105**, 219–230.

Brown, B. E., Le Tissier, M. D. A. and Dunne, R. P. (1994b). Tissue retraction in the scleractinian coral *Coeloseris mayeri*, its effect upon pigmentation, and preliminary implications for heat balance. *Marine Ecology Progress Series* **105**, 209–218.

Brown, B. E., Le Tissier, M. D. A. and Bythell, J. C. (1995). Mechanisms of bleaching deduced from histological studies of reef corals sampled during a natural bleaching event. *Marine Biology* **122**, 655–663.

Budd, A. F., Stemann, T. A. and Stewart, R. H. (1992). Eocene Caribbean reef corals—a unique fauna from the Gatuncillo formation of Panama. *Journal of Paleontology* **66**, 570–594.

Budd, A. F., Johnson, K. G. and Stemann, T. A. (1993). Plio–Pleistocene extinctions and the origin of the modern Caribbean reef-coral fauna. *In* "Proceedings of the Colloquium on Global Aspects of Coral Reefs: Health, Hazards and History" (R. N. Ginsberg, ed.). Rosenstiel School of Marine and Atmospheric Science, University of Miami, 420 pp.

Buddemeier, R. W. and Fautin, D. G. (1993). Coral bleaching as an adaptive mechanism—a testable hypothesis. *Bioscience* **43**, 320–326.

Buddemeier, R. W. and Hopley, D. (1988). Turn-ons and turn-offs: causes and mechanisms of the initiation and termination of coral reef growth. *Proceedings of the 6th International Coral Reef Symposium* **1**, 253–261.

Buddemeier, R. W. and Kinzie, R. A. (1976). Coral growth. *Oceanography and Marine Biology, Annual Review* **14**, 183–225.

Budyko, M. I. (1974). "Climate and Life." Academic Press, New York, USA.

Burns, J. R. (1975). Seasonal changes in the respiration of pumpkin seed, *Lepomis gibbosus*, correlated with temperature, daylength and stage of reproductive development. *Physiological Zoology* **48**, 142–149.

Bythell, J. C. (1988). A total nitrogen and carbon budget for the elkhorn coral *Acropora palmata* (Lamarck). *Proceedings of the 6th International Coral Reef Symposium* **2**, 535–540.

Bythell, J. C., Sharp, V. A., Miller, D. and Brown, B. E. (1995). A novel environmentally-regulated 33 kDa protein from tropical and temperate cnidarian zooxanthellae. *Journal of Thermal Biology* **20**, 15–22.

Calkins, J. (1982). "The Role of Solar Ultraviolet Radiation in Marine Ecosystems." NATO Conference Series IV. Marine Sciences, vol. 7. Plenum Press, New York and London.

Cane, M. A. (1983). Oceanographic events during El Niño. *Science, N.Y.* **222**, 1189–1194.

Cane, M. A. (1986). El Niño. *Earth and Planetary Science Letters* **14**, 43–70.

Chalker, B. E. (1977). Daily variation in the calcification capacity of *Acropora cervicornis*. *Proceedings of the 3rd International Coral Reef Symposium* **2**, 418–423.

Chalker, B. E. (1983). Calcification by corals and other animals on the reef. *In* "Perspectives on Coral Reefs" (D. J. Barnes, ed.), pp. 29–45. Australian Institute of Marine Science, Brian Clouston, Manuka.

Chalker, B. E. and Taylor, D. L. (1978). Rhythmic variation in calcification and photosynthesis associated with the coral *Acropora cervicornis* (Lamarck). *Proceedings of the Royal Society of London, B* **201**, 179–189.

Chalker, B. E., Dunlap, W. C. and Oliver, J. K. (1983). Bathymetric adaptations of reef-building corals at Davies Reef, Great Barrier Reef, Australia. II. Light saturation curves for photosynthesis and respiration. *Journal of Experimental Marine Biology and Ecology* **15**, 51–56.

Chalker, B. E., Cox, T. and Dunlap, W. C. (1984). Seasonal changes in primary production and photoadaptation by the reef-building coral *Acropora granulosa*. *In* "Marine Phytoplankton and Productivity" (O. Holm-Hansen, L. Bolis and R. Giles, eds), pp. 73–87. Springer, Berlin, Germany.

Chalker, B. E., Carr, K. and Gill, E. (1985). Measurement of primary production and calcification *in situ* on coral reefs using electrode techniques. *Proceedings of the 5th International Coral Reef Congress* **6**, 167–172.

Chamberlain, J. A. and Graus, R. R. (1975). Water flow and hydromechanical adaptations of branched reef corals. *Bulletin of Marine Science* **25**, 112–125.

Chappell, J. (1981). Relative and average sea level changes, and endo-, epi- and exogenic processes on the earth. *In* "Sea Level, Ice and Climatic Change." International Association of Hydrology, Scientific Publications No. 131.

Chappell, J. (1983). Sea-level changes and coral reef growth. Iin "Perspectives on Coral Reefs" (D. J. Barnes, ed.), pp. 46–55. Australian Institute of Marine Science, Brian Clouston, Manuka.

Chappell, T. G., Welch, W. J., Schlossman, D. M., Palter, K. B., Schlesinger, M. J. and Rothman, J. E. (1986). Uncoating ATPase is a member of the 70 kilodalton family of stress proteins. *Cell* **45**, 3–13.

Chatterjee, S. and Bhattacharjee, S. K. (1976). Differential response of cells grown in the light and dark to near-ultraviolet light. *Nature, London* **259** 676–677.

Chevalier, J. P. (1971). Les scléractiniares del la Mélanésie française (Nouvelle Calédonie, Isles Chesterfield, Isles Loyauté, Nouvelles Hebrides). *In* "Expedition Française récifs Coralliens Nouvelle Calédonie", pp. 5–307. Fondation Singer-Polignac, Paris, France.

Chirico, W. J., Waters, M. G. and Blobel, G. (1988). 70 kDa heat shock related proteins stimulate protein translocation into microsomes. *Nature, London* **332**, 805–810.

Clark, J. A., Farrell, W. E. and Peltier, W. R. (1978). Global changes in post-glacial sea level: a numerical calculation. *Quaternary Research* **9**, 265–287.

Clark, K. B. and Jensen, K. R. (1982). Effects of temperature on carbon fixation and carbon budget partitioning in the zooxanthellal symbioses of *Aiptasia pallida*. *Journal of Experimental Marine Biology and Ecology* **64**, 215–230.

Clarke, A. (1993). Temperature and extinction in the sea: a physiologist's view. *Palaeobiology* **19**, 498–517.

Clausen, C. D. and Roth, A. A. (1975). Effect of temperature and temperature adaptation on calcification rate in the hermatypic coral *Pocillopora damicornis*. *Marine Biology* **33**, 93–100.

CLIMAP project members. (1981). Seasonal reconstruction of the earth's surface at the last glacial maximum. Map and Chart Series of the Geological Society of America, MC-36.

Cloud, P. E. (1952). Preliminary report on the geology and marine environment of Onotoa Atoll, Gilbert Islands. *Atoll Research Bulletin* **12**, 1–73.

Coles, S. L. (1975). A comparison of effects of elevated temperature versus temperature fluctuations on reef corals at Kahe Point, Oahu. *Pacific Science* **29**, 15–18.

Coles, S. L. (1985). The effects of elevated temperature on reef coral planula settlement as related to power station entrainment. *Proceedings of the 5th International Coral Reef Congress* **4**, 171–176.

Coles, S. L. and Fadlallah, Y. H. (1991). Reef coral survival and mortality at low temperatures in the Arabian Gulf: New species-specific lower temperature limits. *Coral Reef* **9**, 231–237.

Coles, S. L. and Jokiel, P. L. (1977). Effects of temperature on photosynthesis and respiration in hermatypic corals. *Marine Biology* **43**, 209–216.

Coles, S. L. and Jokiel, P. L. (1978). Synergistic effects of temperature, salinity and light on the hermatypic coral *Montipora verrucosa* (Lamarck). *Marine Biology* **49**, 187–195.

Coles, S. L., Jokiel, P. L. and Lewis, C. R. (1976). Thermal tolerance in tropical versus subtropical Pacific reef corals. *Pacific Science* **30**, 159–166.

Colinvaux, P. (1987). Amazon diversity in light of the paleoecological record. *Quaternary Science Reviews* **6**, 93–114.

Cook, C. B., Logan, A., Ward, J., Luckhurst, B. and Berg, C. J. Jr (1990). Elevated temperatures and bleaching on a high latitude coral reef: the 1988 Bermuda event (North Atlantic Ocean). *Coral Reefs* **9**, 45–49.

Copper, P. (1994). Ancient reef ecosystem expansion and collapse. *Coral Reefs* **13**, 3–11.

Cossins, A. R. and Bowler, K. (eds) (1987). "Temperature Biology of Animals." Chapman and Hall, London, UK.

Crowley, T. J. (1994). Pleistocene temperature changes. *Nature, London* **371**, 664.

Cubit, J. D., Caffey, H. M., Thompson, R. C. and Windsor, D. M. (1989). Meteorology and hydrography of a shoaling reef flat on the Caribbean coast of Panama. *Coral Reefs* **8**, 59–66.

Cullen, J. L. (1981). Microfossil evidence for changing salinity patterns in the Bay of Bengal over the last 20 000 years. *Palaeogeography, Palaeoclimatology and Palaeoecology* **35**, 315–356.

Cutchis, P. (1982). A formula for comparing annual damaging ultraviolet (DUV) radiation doses at tropical and mid-latitude sites. *In* "NATO Conference Series. IV. Marine Sciences", vol. 7: "The Role of Solar Ultraviolet Radiation in Marine Ecosystems" (J. Calkins, ed.), pp. 213–228. Plenum Press, New York, USA.

Darwin, C. R. (1881) In a letter to Professor C. G. Semper dated 19th July 1881. Source: *The Correspondence of Charles Darwin*, University of Cambridge, UK.

Davies, P. S. (1991). Effect of daylight variations on the energy budgets of shallow-water corals. *Marine Biology* **108**, 137–144.

Davies, P. S. (1992). Endosymbiosis in marine cnidarians. *In* "Plant–Animal Interactions in the Marine Benthos" (D. M. John, S. J. Hawkins and J. H. Price, eds), vol. 46, pp. 511–540. Systematics Association Special Volume No. 46. Clarendon Press, Oxford, UK.

Dennison, W. C. and Barnes, D. J. (1988). Effect of water motion on coral photosynthesis and calcification. *Journal of Experimental Marine Biology and Ecology* **115**, 67–77.

Done, T. J. (1983). Coral zonation: its nature and significance. *In* "Perspectives on Coral Reefs" (D. J. Barnes, ed.), pp. 107–147. Australian Institute of Marine Science, Brian Clouston, Manuka.

Done, T. J. (1992). Phase shifts in coral reef communities and their ecological significance. *Hydrobiologia* **247**, 121–132.

Downing, N. (1985). Coral reef communities in an extreme environment: the Northwestern Arabian Gulf. *Proceedings of the 5th International Coral Reef Congress* **6**, 343–348.

Drollet, J. H., Glaziou, P. and Martin, P. M. V. (1993). A study of mucus from the solitary coral *Fungia fungites* (Scleractinia: Fungiidae) in relation to photobiological UV adaptation. *Marine Biology* **115**, 263–266.

Dubinsky, Z., Falkowski, P. G., Porter, I. W. and Muscatine, L. (1984). Absorption and utilisation of radiant energy by light- and shade-adapted colonies of hermatypic coral *Stylophora pistillata*. *Proceedings of the Royal Society of London, B* **222**, 203–214.

Dunlap, W. C. and Chalker, B. E. (1986). Identification and quantitation of near-UV absorbing compounds (S-320) in a hermatypic scleractinian. *Coral Reefs* **5**, 155–160.

Dunlap, W. C., Chalker, B. E. and Oliver, J. K. (1986). Bathymetric adaptations of reef-building corals at Davies Reef, Great Barrier Reef, Australia. III. UV-B absorbing compounds. *Journal of Experimental Marine Biology and Ecology* **104**, 239–248.

Dunne, R. P. (1994a). "Environmental Data Handbook for North Male Atoll, Maldives." Overseas Development Administration, London, UK.

Dunne, R. P. (1994b). "Environmental Data Handbook for Pari Island, Indonesia." Overseas Development Administration, London, UK.

Dunne, R. P. (1994c). "Environmental Data Handbook for Phuket Marine Biological Centre, Thailand." Overseas Development Administration, London, UK.

Dunne, R. P. (1994d). Radiation and coral bleaching. *Nature, London* **368**, 697.

Dustan, P. (1979). Distribution of zooxanthellae and photosynthetic chloroplast pigments of the reef-building coral *Montastrea annularis* Ellis and Solander in relation to depth on a West Indian coral reef. *Bulletin of Marine Science* **29**, 79–95.

Dustan, P. (1982). Depth-dependent photoadaptation by zooxanthellae of the reef coral *Montastrea annularis*. *Marine Biology* **68**, 253–264.

Dykens, J. A. and Shick, J. M. (1984). Photobiology of the symbiotic sea anemone, *Anthopleura elegantissima*: defenses against photodynamic effects, and seasonal photoacclimatization. *Biological Bulletin, Marine Biological Laboratory, Woods Hole* **167**, 683–697.

Eakin, C. M. (1992). Post El Niño Panamanian reefs: less accretion, more erosion and damselfish protection. *Proceedings of the 7th International Coral Reef Symposium* **1**, 387–396.

Edmondson, C. H. (1946). Behaviour of coral planulae under altered saline and thermal conditions. *Occasional Papers of the Bernice Bishop Museum* **18**, 283–304.

Edmunds, P. J. (1994). Evidence that reef-wide patterns of coral bleaching may be the result of the distribution of bleaching-susceptible clones. *Marine Biology* **121**, 137–142.

Edmunds, P. J. and Davies, P. S. (1986). An energy budget for *Porites porites* (Scleractinia). *Marine Biology* **92**, 339–347.

Edmunds, P. J. and Davies, P. S. (1989). An energy budget for *Porites porites* (Scleractinia), growing in a stressed environment. *Coral Reefs* **8**, 37–44.

Endean, R., Stephenson, W. and Kenny, R. (1956). The ecology and distribution of intertidal organisms on certain islands off the Queensland coast. *Australian Journal of Marine and Freshwater Research* **7**, 317–342.

Fabricius, K. (1995). Nutrition and community regulation in tropical reef-inhibiting soft corals (Coelenterata: Octocorallia). PhD Thesis, University of Munich, Germany.

Falkowski, P. G., Jokiel, P. L. and Kinzie, R. A. (1990). Irradiance and Corals. *In* "Ecosystems of the World: Coral Reefs" (Z. Dubinsky, ed.), vol. 25, pp. 89–107. Elsevier, New York, USA.

Farrow, G. E. and Brander, K. M. (1971). Tidal studies on Aldabra. *Philosophical Transactions of the Royal Society of London, B* **260**, 92–121.

Fennessey, M. J. and Shukla, J. (1988). Impact of the 1982/3 and 1986/7 Pacific SST anomalies on time–mean prediction with the Glas GCM. *World Climate Report Programme* **15**, 26–44.

Fishelson, L. (1973). Ecological and biological phenomena influencing coral-species composition on the reef tables at Eilat (Gulf of 'Aqaba, Red Sea). *Marine Biology* **19**, 183–196.

Fitt, W. K. (1985). Effect of different strains of the zooxanthella *Symbiodinium microadriaticum* on growth and survival of their coelenterate and molluscan hosts. *Proceedings of the 5th International Coral Reef Congress* **6**, 131–136.

Fitt, W. K., Pardy, R. L. and Littler, M. M. (1982). Photosynthesis, respiration, and contribution to community productivity of the symbiotic sea anemone *Anthopleura elegantissima* (Brandt, 1835). *Journal of Experimental Marine Biology and Ecology* **61**, 213–232.

Fitt, W. K., Spero, H. J., Halas, J., White, M. W. and Porter, J. W. (1993). Recovery of the coral *Montastrea annularis* in the Florida Keys after the 1987 Caribbean "bleaching event". *Coral Reefs* **12**, 57–64.

Foster, A. B. (1979). Phenotypic plasticity in the reef corals *Montastrea annularis* (Ellis and Solander) and *Siderastrea siderea* (Ellis and Solander). *Journal of Experimental Marine Biology and Ecology* **39**, 25–34.

Foster, A. B. (1980). Environmental variation in skeletal morphology within the Caribbean reef corals *Montastrea annularis* and *Siderastrea siderea*. *Bulletin of Marine Science* **30**, 678–709.

Frost, S. H. (1977). Miocene to Holocene evolution of Caribbean Province building corals. *Proceedings of the 3rd International Coral Reef Symposium* **2**, 353–359.

Gardiner, J. S. (1903). "The Fauna and Geography of the Maldive and Lacadive Archipelagoes." University of Cambridge, Cambridge, UK.

Gates, R. D., Baghdasarian, G. and Muscatine, L. (1992). Temperature stress causes host cell detachment in symbiotic cnidarians: implications for coral bleaching. *Biological Bulletin, Marine Biological Laboratory, Woods Hole* **182**, 324–332.

Gates, W. L., Rowntree, P. R. and Zeng, Q. C. (1990). Validation of climate models. *In* "Climate Change. The IPCC Scientific Assessment" (J. T. Houghton, G. J. Jenkins and J. J. Ephraums, eds), pp. 99–130. Cambridge University Press, Cambridge, UK.

Gattuso, J. P. and Jaubert, J. (1984). Preliminary data concerning the *in situ* effects of light on metabolism, growth and classification of the hermatypic coral *Stylophora pistillata*. *Comptes Rendus de l'Academie des Sciences Serie III Sciences de la Vie* **299**, 585–590.

Geister, J. (1977). The influence of wave exposure on the ecological zonation of Caribbean coral reefs. *Proceedings of the 3rd International Coral Reef Symposium* **1**, 23–29.

Geister, J. (1978). Occurrence of *Pocillopora* in late Pleistocene coral reefs. *Mémoires de la Bureau de Récherches Geologigues et Minières* **89**, 378–388.

Gleason, D. F. (1993a). Differential-effects of ultraviolet-radiation on green and brown morphs of the Caribbean corals *Porites astreoides*. *Limnology and Oceanography* **38**, 1452–1463.

Gleason, D. F. and Wellington, G. M. (1993). Ultraviolet radiation and coral bleaching. *Nature, London* **365**, 836–838.

Gleason, M. G. (1993b). Effects of disturbance on coral communities: bleaching in Moorea, French Polynesia. *Coral Reefs* **12**, 193–201.

Glider, W. V., Phipps, D. W. Jr and Pardy, R. L. (1980). Localisation of symbiotic dinoflagellate cells within tentacle tissue of *Aiptasia pallida* (Coelenterata, Anthozoa). *Transactions of the American Microscopical Society* **99**, 426–438.

Glynn, P. W. (1968). Mass mortalities of echinoids and other reef flat organisms coincident with midday, low water exposures in Puerto Rico. *Marine Biology* **1**, 226–243.

Glynn, P. W. (1973). Ecology of a Caribbean coral reef: the *Porites* reef-flat biotope. Part 1. Meteorology and hydrography. *Marine Biology* **20**, 297–318.

Glynn, P. W. (1984). Widespread coral mortality and the 1982/83 El Niño warming event. *Environmental Conservation* **11**, 133–146.

Glynn, P. W. (1988). El Niño warming, coral mortality and reef framework destruction by echinoid bioerosion in the eastern Pacific. *Galaxea* **7**, 129–160.

Glynn, P. W. (1993). Coral-reef bleaching—ecological perspectives. *Coral Reefs* **12**, 1–17.

Glynn, P. W. and Colgan, M. W. (1992). Sporadic disturbances in fluctuating coral-reef environments—El Niño and coral-reef development in the eastern Pacific. *American Zoologist* **32**, 707–718.

Glynn, P. W. and D'Croz, L. (1990). Experimental evidence for high temperature stress as the cause of El Niño-coincident coral mortality. *Coral Reefs* **8**, 181–191.

Glynn, P. W. and Stewart, R. H. (1973). Distribution of coral reefs in the Pearl Islands (Gulf of Panama) in relation to thermal conditions. *Limnology and Oceanography* **18**, 367–379.

Glynn, P. W., Perez, M. and Gilchrist, S. L. (1985a). Lipid decline in stressed corals and their crustacean symbionts. *Biological Bulletin, Marine Biological Laboratory, Woods Hole* **168**, 276–284.

Glynn, P. W., Peters, E. C. and Muscatine, L. (1985b). Coral tissue microstructure and necrosis: relation to catastrophic coral mortality in Panama. *Diseases of Aquatic Organisms* **1**, 29–38.

Glynn, P. W., Imai, R., Sakai, K., Nakano, Y. and Yamazato, K. (1993). Environmental responses of Okinawan (Ryuku Islands, Japan) reef corals to high sea temperature and UV radiation. *Proceedings of the 7th International Coral Reef Symposium* **1**, 27–37.

Goreau, T. J. and Macfarlane, A. H. (1990). Reduced growth rate of *Montastrea annularis* following the 1987 to 1988 coral-bleaching event. *Coral Reefs* **8**, 211–216.

Graus, R. R. (1977). Investigation of coral growth adaptation using computer modelling. *Proceedings of the 3rd International Coral Reef Symposium* **2**, 463–469.

Graus, R. R. and MacIntyre, I. G. (1982). Variation in growth forms of the reef coral *Montastrea annularis* (Ellis and Solander): a quantitative evaluation of growth response to light distribution using computer simulation. *In* "The Atlantic Barrier Reef Ecosystem at Carrie Bow Bay, Belize" (K. Rutzler and I. G. MacIntyre, eds), pp. 441–464. Smithsonian Institution, Washington, DC, USA.

Graves, J. E. and Somero, G. N. (1982). Electrophoretic and functional enzymic evolution in four species of eastern Pacific barracudas from different thermal environments. *Evolution* **36**, 97–106.

Gregory, S. (1989a). The changing frequency of drought in India, 1871–1985. *Journal of Geography* **155**, 322–334.

Gregory, S. (1989b). Macro-regional definition and characteristics of Indian summer monsoonal rainfall. *International Journal of Climate* **9**, 465–483.

Grigg, R. W. (1994). Science management of the worlds fragile coral-reefs. *Coral Reefs* **13**, 1.

Grime, J. P. (1989). The stress debate: symptom of impending synthesis? *Biological Journal of the Linnean Society* **37**, 3–17.

Guilderson, T. P., Fairbanks, R. G. and Rubenstone, J. L. (1994). Tropical temperature-variations since 20 000 years ago—modulating interhemispheric climate-change. *Science, NY* **263**, 663–665.

Halldal, P. (1968). Photosynthetic capacities and photosynthetic action spectra of endozoic algae of the massive coral *Favia*. *Biological Bulletin, Marine Biological Laboratory, Woods Hole* **134**, 411–424.

Harland, A. D. and Davies, P. S. (1994). Time-course of photoadaptation in the symbiotic sea anemone *Anemonia viridis*. *Marine Biology* **119**, 45–51.

Harland, A. D., Fixter, L. M., Davies, P. S. and Anderson, R. A. (1991). Distribution of lipids between the zooxanthellae and animal compartment in the symbiotic sea anemone *Anemonia viridis*: wax esters, triglycerides and fatty acids. *Marine Biology* **110**, 13–19.

Harmelin-Vivien, M. L. (1985). Hurricane effects on coral reefs: introduction. *Proceedings of the 5th International Coral Reef Congress* **3**, 315.

Harmelin-Vivien, M. L. and Laboute, P. (1986). Catastrophic impact of hurricanes on atoll outer reef slopes in the Tuamoto (French Polynesia). *Coral Reefs* **5**, 59–68.

Harrison, P. L. and Wallace, C. C. (1990). Reproduction, dispersal and recruitment of scleractinian corals. *In* "Ecosystems of the World: Coral Reefs" (Z. Dubinsky, ed.), vol. 25, pp. 133–207. Elsevier, New York, USA.

Hawkins, A. J. S. (1985). Relationships between the synthesis and breakdown of protein, dietary absorption and turnovers of nitrogen and carbon in the blue mussel, *Mytilus edulis* L. *Oecologia* **66**, 42–49.

Hawkins, A. J. S. (1991). Protein turnover: a functional appraisal. *Functional Ecology* **5**, 222–233.

Hawkins, A. J. S., Bayne, B. L. and Day, A. J. (1986). Protein turnover, physiological energetics and heterozygosity in the blue mussel *Mytilus edulis*: the basis of variable age specific growth. *Proceedings of the Royal Society of London, B* **229**, 161–176.

Hawkins, A. J. S., Wilson, I. A. and Bayne, B. L. (1987). Thermal responses reflect protein turnover in *Mytilus edulis* L. *Functional Ecology* **1**, 339–351.

Hawkins, A. J. S., Bayne, B. L., Day, A. J., Rusin, J. and Worrall, C. M. (1989). Genotype-dependent interrelations between energy metabolism, protein metabolism and fitness. *In* "Reproduction, Genetics and Distributions of Marine Organisms" (J. S. Ryland and P. A. Tyler, eds), Proceedings of the 23rd European Marine Biology Symposium, pp. 283–292. Olsen and Olsen, Fredensborg.

Hawkins, A. J. S., Day, A. J., Gerard, A., Naciri, Y., Ledu, C., Bayne, B. L. and Heral, M. (1994). A genetic and metabolic basis for faster growth among triploids induced by blocking meiosis I but not meiosis II in the larviparous European flat oyster, *Ostrea edulis* L. *Journal of Experimental Marine Biology and Ecology* **184**, 21–40.

Hayashibara, T., Shimoike, K., Kimura, T., Hosaka, S., Heyward, A., Harrison, P., Kudo, K. and Omori, M. (1993). Patterns of coral spawning at Akajima Island, Okinawa, Japan. *Marine Ecology Progress Series* **101**, 253–262.

Helmuth, B. and Sebens, K. (1993). The influence of colony morphology and orientation to flow on particle capture by the scleractinian coral *Agaricia agaricites* (Linnaeus). *Journal of Experimental Marine Biology and Ecology* **165**, 251–278.

Hemmingsen, S. M., Woolford, C., van der Vies, S. M., Tilly, K., Dennis, D. T., Georgopoulos, C. P., Hendria, R. W. and Ellis, R. J. (1988). Homologous plant and bacterial proteins chaperone oligomeric protein assembly. *Nature, London* **33**, 330–334.

Hidaka, M. and Shirasaka, S. (1992). Mechanism of phototropism in young corallites of the coral *Galaxea fascicularis* (L.). *Journal of Experimental Marine Biology and Ecology* **157**, 69–77.

Hochachka, P. W. and Somero, G. N. (1984). "Biochemical Adaptation." Princeton University Press, Princeton, USA.

Hoegh-Guldberg, O. and Smith, G. J. (1989). The effect of sudden changes in temperature, light and salinity on the population density and export of zooxanthellae from the reef corals *Stylophora pistillata* Esper and *Seriatopora hystrix* Dana. *Journal of Experimental Marine Biology and Ecology* **129**, 279–304.

Hoffman, A. A. and Parsons, P. A. (1991). "Evolutionary Genetics and Environmental Stress." Oxford University Press, Oxford, UK.

Hoffman, R. J. (1981). Evolutionary genetics of *Metridium senile* 1 Kinetic differences in phosphoglucose isomerase allozymes. *Biochemistry and Genetics* **19**, 129–144.

Hoffman, R. J. (1983). Temperature modulation of the kinetics of phosphoglucose isomerase genetic variants from the sea anemone *Metridium senile*. *Journal of Experimental Zoology* **227**, 361–370.

Hoffman, R. J. (1986). Variation in contributions of asexual reproduction to the genetic structure of populations of the sea anemone *Metridium senile*. *Evolution* **40**, 357–365.

Hollaran, M. K. and Witteman, G. J. (1986). Diurnal periodicity in planual release by the reef coral *Pocillopora damicornis*. In "Coral Reef Population Biology" (P. R. Jokiel, R. H. Richmond and R. A. Rogers, eds), vol. 37, pp. 161–169. Hawaii Institute Marine Biology Technical Report, Hawaii.

Hudson, J. H. (1981a). Growth rates in *Montastrea annularis*; a record of environmental change in the Key Largo coral reef marine sanctuary, Florida. *Bulletin of Marine Science* **31**, 444–459.

Hudson, J. H. (1981b). Response of *Montastrea annularis* to environmental change in the Florida Keys. *Proceedings of the 4th International Coral Reef Symposium* **2**, 233–240.

Hughes, T. P., Ayre, D. and Connell, J. H. (1992). The evolutionary ecology of corals. *Trends in Ecology and Evolution* **7**, 292–295.

Hutson, W. H. and Prell, W. L. (1980). A paleoecological transfer function, F1–2, for Indian Ocean planktonic foraminifera. *Journal of Paleontology* **54**, 381–399.

Iglesias-Prieto, R. and Trench R. K. (1994). Acclimation and adaptation to irradiance in symbiotic dinoflagellates. 1. Responses of the photosynthetic unit to changes in photon flux density. *Marine Ecology Progress Series* **113**, 163–175.

Iglesias-Prieto, R., Matta, J. L., Robins, W. A. and Trench, R. K. (1992). Photosynthetic response to elevated-temperature in the symbiotic dinoflagellate *Symbiodinium microadriaticum* in culture. *Proceedings of the National Academy of Sciences of the USA* **89**, 302–305.

Jackson, J. B. C. (1991). Adaptation of diversity of reef corals. *Bioscience* **41**, 475–482.

Jackson, J. B. C. (1992). Pleistocene perspectives on coral-reef community structure. *American Zoologist* **32**, 719–731.

Jackson, J. B. C. and Hughes, T. P. (1985). Adaptive strategies of coral-reef invertebrates. *American Scientist* **73**, 265–274.

Jagger, J. (ed.) (1985). "Solar-UV Actions on Living Cells." Praeger, New York, USA.

Jaubert, J. M. and Vasseur, P. (1974). Light measurements: duration aspect and the distribution of benthic organisms in an Indian Ocean coral reef (Tuléar, Madagascar). *Proceedings of the 2nd International Coral Reef Symposium* **2**, 127–142.

Jeffrey, S. W. (1980). Algal pigment systems. In "Productivity in the Sea" (E. P. Falkowski, ed.), pp. 33–58. Plenum Press, New York, USA.

Jeffrey, S. W. and Haxo, F. T. (1968). Photosynthetic pigments of symbiotic dinoflagellates (zooxanthellae) from corals and clams. *Biological Bulletin, Marine Biological Laboratory, Woods Hole* **135**, 149–165.

Jeffrey, S. W. and Vesk, M. (1977). Effect of blue-green light on the photosynthetic pigments and chloroplast structure in the marine diatom *Stephanopyxis turris*. *Journal of Phycology* **13**, 271–279.

Jerlov, N. G. (1950). Ultra-violet radiation in the sea. *Nature, London* **4211**, 111–112.

Jerlov, N. G. (1968). "Optical Oceanography." Elsevier, Amsterdam, Netherlands.

Jerlov, N. G. (1976). "Marine Optics." Elsevier, Oxford, UK.

Johnson, A. S. and Sebens, K. P. (1993). Consequences of a flattened morphology: effects of flow on feeding rates of the scleractinian coral *Meandrina meandrites*. *Marine Ecology Progress Series* **99**, 99–114.

Johnston, I. A. (1983). Cellular responses to an altered body temperature: the role of alterations in the expression of protein isoforms. In "Cellular Acclimatisation to Environmental Change" (A. R. Cossins and P. Sheterline, eds), pp. 121–143. Cambridge University Press, Cambridge, UK.

Jokiel, P. J. (1978). Effects of water motion on reef corals. *Journal of Experimental Marine Biology and Ecology* **35**, 87–97.

Jokiel, P. L. and Coles, S. L. (1974). Effects of heated effluent on hermatypic corals at Kahe Point, Oahu. *Pacific Science* **28**, 1–18.

Jokiel, P. L. and Coles, S. L. (1977). Effects of temperature on the mortality and growth of Hawaiian reef corals. *Marine Biology* **43**, 201–208.

Jokiel, P. L. and Coles, S. L. (1990). Response of Hawaiian and other Indo-Pacific reef corals to elevated temperature. *Coral Reefs* **8**, 155–162.

Jokiel, P. L. and Morrissey, J. I. (1986). Influence of size on primary production in the reef coral *Pocillopora damicornis* and the macroalga *Acanthophora spicifera*. *Marine Biology* **91**, 15–26.

Jokiel, P. L. and York, R. H. Jr (1982). Solar ultraviolet photobiology of the reef coral *Pocillopora damicornis* and symbiotic zooxanthellae. *Bulletin of Marine Science* **32**, 301–315.

Kaiser, P., Schlichter, D. and Fricke, H. W. (1993). Influence of light on algal symbionts and the deep water coral *Leptoseris fragilis*. *Marine Biology* **117**, 45–52.

Karentz, D. and McEuen, F. S. (1991). Survey of mycosporine-like amino acid compounds in Antarctic marine organisms: potential protection from ultraviolet exposure. *Marine Biology* **108**, 157–166.

Karentz, D., Cleaver, J. E. and Mitchell, D. L. (1991). Cell survival characteristics and molecular responses of Antarctic phytoplankton to ultraviolet-B radiation. *Journal of Phycology* **27**, 326–341.

Kawaguti, S. (1944). On the physiology of reef corals. VI. Study on the pigments. *Palao Tropical Biological Station Studies* **II**, 617–672.

Kawaguti, S. (1954). Effects of light and ammonium on the expansion of polyps in the reef corals. *Biological Journal of Okayama University* **2**, 45–50.

Kellog, R. B. and Patton, J. S. (1983). Lipid droplets, medium of energy exchange in the symbiotic anemone *Condylactis gigantea*: a model coral polyp. *Marine Biology* **75**, 137–149.

Kenny, R. (1974). Inshore surface sea temperatures at Townsville. *Australian Journal of Marine and Freshwater Research* **25**, 1–5.

Kerr, R. A. (1991). Could the sun be warming the climate? *Science, NY* **254**, 652–653.

Kershaw, R. (1988). The effect of a sea surface temperature anomaly on a prediction of the onset of the south-west monsoon over India. *Quarterly Journal of the Royal Meteorological Society* **114**, 325–345.

Kinsey, D. W. (1985). Metabolism, calcification and carbon production. *Proceedings of the 5th International Coral Reef Congress* **4**, 505–526.

Kinsey, D. W. and Domm, A. (1974). Effects of fertilisation on a coral reef environment—primary production studies. *Proceedings of the 2nd International Coral Reef Symposium* **1**, 49–66.

Kinzie, R. A. and Chee, G. S. (1979). The effect of different zooxanthellae on the growth of experimentally reinfected hosts. *Biological Bulletin, Marine Biological Laboratory, Woods Hole* **156**, 315–327.

Kinzie, R. A., Jokiel, P. L. and York, R. (1984). Effects of light of altered spectral composition on coral zooxanthellae associations and on zooxanthellae *in vitro*. *Marine Biology* **78**, 239–248.

Kinzie, R. A. III (1993). Effects of ambient levels of solar ultraviolet radiation on zooxanthellae and photosynthesis of the reef coral *Montipora verrucosa*. *Marine Biology* **116**, 319–327.

Kinzie, R. A. III and Hunter, T. (1987). Effect of light quality on photosynthesis of the reef coral *Montipora verrucosa*. *Marine Biology* **94**, 95–100.

Kirk, J. T. O. (1983). "Light and Photosynthesis in Aquatic Ecosystems." Cambridge University Press, Cambridge, UK.

Kirk, J. T. O. (1992). The nature and measurement of the light environment in the ocean. *In* "Primary Productivity and Biogeochemical Cycles in the Seas" (P. G. Falkowski and A. D. Woodhead, eds) pp. 9–29. Plenum Press, New York, USA.

Knowlton, N. and Jackson, J. B. C. (1994). New taxonomy and niche partitioning on coral reefs: Jack of all trades or master of some? *Trends in Ecology and Evolution* **9**, 7–9.

Koehn, R. K. and Bayne, B. L. (1989). Towards a physiological and genetical understanding of the energetics of the stress response. *Biological Journal of the Linnean Society* **37**, 157–171.

Koehn, R. K., Diehl, W. J. and Scott, T. M. (1988). The differential contribution by individual enzymes of glycolysis and protein catabolism to the relationship between heterozygosity and growth rate in the coot clam *Mulinia lateralis*. *Genetics* **118**, 121–130.

Krause, G. H. (1988). Photoinhibition of photosynthesis. An evaluation of damaging and protective mechanisms. *Physiologia Plantarum* **74**, 566–574.

Kyle, D. J. (1987). The biochemical basis for photoinhibition of photosystem II. *In* "Photoinhibition" (D. J. Kyle, C. B. Osmond and C. J. Arntzen, eds), pp. 197–226. Elsevier, Amsterdam, Netherlands.

Kyle, D. J., Ohad, I. and Arntzen, C. J. (1984). Membrane protein damage and repair; selective loss of the quinone-protein function in chloroplast membranes. *Proceedings of the National Academy of Sciences of the USA* **81**, 4070–4074.

Lasker, H. R. (1976). Intraspecific variability of zooplankton feeding in the hermatypic coral *Montastrea cavernosa*. *In* "Coelenterate Ecology and Behaviour" (G. O. Mackie, ed.), pp. 101–109. Plenum Press, New York, USA.

Lasker, H. R. (1979). Light dependent activity patterns among reef corals: *Montastrea cavernosa*. *Biological Bulletin, Marine Biological Laboratory, Woods Hole* **156**, 196–211.

Lasker, H. R. (1981). Phenotypic variation in the coral *Montastrea cavernosa* and its effects on colony energetics. *Biological Bulletin, Marine Biological Laboratory, Woods Hole* **160**, 292–302.

Leletkin, V. A., Zvalinskii, V. I. and Titlyanov, E. A. (1980). Photosynthesis of zooxanthellae in corals of different depths. *Physiology of Plants (USSR)* **27**, 863–870. [Eng. transl. of *Fizologii Rastenii*.]

Lesser, M. P. (1989). Photobiology of natural populations of zooxanthellae from the sea anemone *Aiptasia pallida*: assessment of the host's role in protection against ultraviolet radiation. *Cytometry* **10**, 653–658.

Lesser, M. P. and Shick, J. M. (1989). Effects of irradiance and ultraviolet radiation on photoadaptation in the zooxanthellae of *Aiptasia pallida*: primary production, photoinhibition, and enzymatic defenses against oxygen toxicity. *Marine Biology* **102**, 243–255.

Lesser, M. P., Stochaj, W. R., Tapley, D. W. and Shick, J. M. (1990). Bleaching in coral reef anthozoans: effects of irradiance, ultraviolet radiation, and temperature on the activities of protective enzymes against active oxygen. *Coral Reefs* **8**, 225–232.

Lesser, M. P., Weiss, V. M., Patterson, M. R. and Jokiel, P. L. (1994). Effects of morphology and water motion on carbon delivery and productivity in the reef coral *Pocillopora damicornis* (Linnaeus): diffusion barriers, inorganic carbon limitation, and biochemical plasticity. *Journal of Experimental Marine Biology and Ecology* **178**, 153–179.

Le Tissier, M. D. A. and Brown, B. E. (1996) Dynamics of cellular mechanisms of solar bleaching in the intertidal reef coral *Goniastrea aspera* at Ko Phuket, Thailand. *Marine Ecology Progress Series* **136**, 235–44.

Levinton, J. (1973). Genetic variation in a gradient of environmental variability: marine Bivalvia (Mollusca). *Science, NY* **180**, 75–76.

Lichtenthaler, H. K. and Meier, D. (1984). Regulation of chloroplast photomorphogenesis by light intensity and light quality. *In* "Chloroplast Biogenesis" (R. J. Ellis, ed.), pp. 261–281. Cambridge University Press, Cambridge, UK.

Lough, J. M. (1994). Climate variation and El Niño Southern Oscillation events on the Great Barrier Reef: 1958 to 1987. *Coral Reefs* **13**, 181–195.

Ludlow, M. M. (1987). Light stress at high temperature. *In* "Photoinhibition" (D. J. Kyle, C. B. Osmond and C. J. Arntzen, eds), pp. 89–110. Elsevier, Amsterdam, Netherlands.

McBride, J. L. (1987). The Australian summer monsoon. In "Monsoon Meteorology", pp. 203–232. Oxford University Press, Oxford, UK.

McClanahan, T. R. (1988). Seasonality in East Africa's coastal waters. *Marine Ecology Progress Series* **44**, 191–199.

McFarland, W. N. (1991). The visual world of coral reef fishes. In "The Ecology of Fishes on Coral Reefs" (P. F. Sale, ed.), pp. 16–38. Academic Press, New York, USA.

McIntyre, A., Ruddiman, W. F., Karling, K. and Mix, A. C. (1989). Surface water response of the equatorial Atlantic Ocean to orbital forcing. *Paleoceanography* **4**, 19–55.

Maragos, J. E. (1972). A study of the ecology of Hawaiian coral reefs. PhD Thesis, University of Hawaii, Honolulu.

Marcus, J. and Thorhaug, A. (1981). Pacific versus Atlantic responses of the subtropical hermatypic coral *Porites* spp. to temperature and salinity effects. *Proceedings of the 4th International Coral Reef Symposium* **2**, 15–20.

Markell, D. A., Trench, R. K. and Iglesias-Prieto, R. (1992). Macromolecules associated with the cell walls of symbiotic dinoflagellates. *Symbiosis* **12**, 19–31.

Marsh, J. A., Ross, R. M. and Zolan, W. J. (1981). Water circulation on two Guam reef flats. *Proceedings of the 4th International Coral Reef Symposium* **1**, 355–360.

Mayer, A. G. (1914). The effects of temperature on tropical marine animals. *Carnegie Institute of Washington Publications* **183**, 1–24.

Mayer, A. G. (1918). Ecology of the Murray Island coral reef. *Papers of the Marine Biology Dept Carnegie Institute* **9**, 1–48.

Mazel, C. H. (1995). Spectral measurements of fluorescence emission in Caribbean cnidarians. *Marine Ecology Progress Series* **120**, 185–191.

Meehl, G. A. (1994). Coupled land–ocean–atmosphere processes and south Asian monsoon variability. *Science, NY* **266**, 263–267.

Mesolella, K. J. (1967). Zonation of uplifted Pleistocene coral reefs on Barbados, West Indies. *Science, NY* **156**, 638–640.

Mesolella, K. J., Sealy, H. A. and Matthews, R. K. (1970). Facies geometrics within Pleistocene reefs of Barbados, West Indies. *American Association of Petroleum Geologists Bulletin* **54**, 1899–1917.

Meyers, P. A. (1979). Polyunsaturated fatty acids in corals: indicators of nutritional sources. *Marine Biology Letters* **1**, 69–75.

Moore, H. B. (1972). Aspects of stress in the tropical marine environment. *Advances in Marine Biology* **10**, 217–269.

Mopper, K. and Zhou, X. (1990). Hydroxyl radical photoproduction in the sea and its potential impact on marine processes. *Science, NY* **250**, 661–664.

Morris, B., Barnes, J., Brown, F. and Markham, J. (1977). The Bermuda marine environment. Bermuda Biological Station Special Publication, no. 15, Bermuda, 120 pp.

Muller-Parker, G. (1984). Photosynthesis-irradiance responses and photosynthetic periodicity in the sea anemone *Aiptasia pulchella* and its zooxanthellae. *Marine Biology* **82**, 225–232.

Muller-Parker, G. (1987). Seasonal variation in light-shade adaptation of natural populations of the symbiotic sea anemone *Aiptasia pulchella* (Carlgren, 1943) in Hawaii. *Journal of Experimental Marine Biology and Ecology* **112**, 165–183.

Munk, W. M., Ewing, G. C. and Revelle, R. R. (1949). Diffusion in Bikini lagoon. *Transactions of the American Geophysical Union* **30**, 159–166.

Murata, N. and Yamaya, J. (1984). Temperature-dependent phase behaviour of phosphatidyglycerols from chilling-sensitive and chilling-resistant plants. *Plant Physiology* **74**, 1016–1024.

Muscatine, L., Grossman, D. and Doino, J. (1991). Release of symbiotic algae by tropical sea-anemones and corals after cold shock. *Marine Ecology Progress Series* **77**, 233–243.

Newell, N. D. (1971). An outline history of tropical organic reefs. *American Museum Novitates* **2465**, 1–37.

Ogden, J. C., Porter, J. W., Smith, N. P., Szmant, A. M., Jaap, W. C. and Forcucci, D. (1994). A long-term interdisciplinary study of the Florida Keys seascape. *Bulletin of Marine Science* **54**, 1059–1071.

Ohad, I., Koike, H., Shochat, S. and Inoue, Y. (1988). Changes in the properties of reaction centre II during the initial stages of photoinhibition as revealed by thermoluminescence measurements. *Biochimica Biophysica Acta* **933**, 288–298.

Ohlhorst, S. L. and Liddell, W. D. (1988). The effect of substrate microtopography on reef community structure 60–120 m. *Proceedings of the 6th International Coral Reef Symposium* **3**, 355–360.

Oliver, J. K., Chalker, B. E. and Dunlap, W. C. (1983). Bathymetric adaptations of reef building corals at Davies Reef, Great Barrier Reef, Australia. I. Long-term growth responses of *Acropora formosa* (Dana 1846). *Journal of Experimental Marine Biology and Ecology* **73**, 11–35.

Orr, A. P. (1933a). Physical and chemical conditions in the sea in the neighbourhood of the Great Barrier Reef. *Scientific Reports of the Great Barrier Reef Expedition 1928–1929* **2**, 37–86.

Orr, A. P. (1933b). Variations in some physical and chemical conditions on and near Low Isles reef. *Scientific Reports of the Great Barrier Reef Expedition 1928–1929* **2**, 87–98.

Osmond, C. B. (1994). What is photoinhibition? Some insights from comparisons of shade and sun plants. *In* "Photoinhibition of Photosynthesis" (N. R. Baker and J. R. Bowyer, eds), pp. 1–24. Bios, Oxford, UK.

Patterson, M. R. (1980). Hydromechanical adaptations in *Alcyonium siderium* (octocorallia). *In* "Biofluid Mechanics" (D. J. Schneck, ed.), vol. 2, pp. 183–201. Plenum Press, New York, USA.

Patterson, M. R. (1992). A chemical-engineering view of cnidarian symbioses. *American Zoologist* **32**, 566–582.

Patterson, M. R. and Sebens, K. P. (1989). Forced convection modulates gas exchange in cnidarians. *Proceedings of the National Academy of Sciences of the USA* **86**, 8833–8836.

Patterson, M. R., Sebens, K. P. and Olson, R. R. (1991). *In situ* measurements of flow effects on primary production and dark respiration in reef corals. *Limnology and Oceanography* **36**, 936–948.

Patton, J. S., Abraham, S. and Benson, A. A. (1977). Lipogenesis in the intact coral *Pocillopora capitata* and its isolated zooxanthellae: evidence for a light-driven carbon cycle between symbiont and host. *Marine Biology* **44**, 235–247.

Paulay, G. (1990). Effects of late Cenozoic sea-level fluctuations on the bivalve faunas of tropical oceanic islands. *Paleobiology* **16**, 415–434.

Pearse, V. B. (1974). Modification of sea anemone behavior by symbiotic zooxanthellae: expansion and contraction. *Biological Bulletin, Marine Biological Laboratory, Woods Hole* **147**, 641–651.

Pelham, H. B. R. (1986). Speculation on the functions of the major heat shock and glucose regulated proteins. *Cell* **46**, 959–961.

Peters, J. and Jagger, J. (1981). Inducible repair of near-UV radiation lethal damage in *E. coli*. *Nature, London* **289**, 194–195.

Piazena, H. and Häder, D. P. (1994). Penetration of solar UV irradiation in coastal lagoons of the southern Baltic Sea and its effect on phytoplankton communities. *Photochemistry and Photobiology* **60**, 463–469.

Pickard, G. L. (1986). Effects of wind and tide on upper layer currents at Davies Reef, during MECOR (July–August 1984). *Australian Journal of Marine and Freshwater Research* **37**, 545–565.

Porter, J. W. (1976). Autotrophy, heterotrophy and resource partitioning in Caribbean reef-building corals. *American Naturalist* **110**, 731–742.

Porter, J. W. (1980). Primary productivity in the sea: reef corals *in situ*. *In* "Primary Productivity in the Sea" (P. G. Falkowski, ed.), pp. 403–410. Plenum Press, New York, USA.

Porter, J. W. (1985). The maritime weather of Jamaica: its effects on annual carbon budgets of the massive reef-building coral *Montastrea annularis*. *Proceedings of the 5th International Coral Reef Congress* **6**, 363–379.

Potts, D. C. (1983). Evolutionary disequilibrium among Indo-Pacific corals. *Bulletin of Marine Science* **33**, 619–632.

Potts, D. C. (1984a). Generation times and the quaternary evolution of reef building corals. *Palaeobiology* **10**, 48–58.

Potts, D. C. (1984b). Natural selection in experimental populations of reef building corals (Scleractinia). *Evolution* **38**, 1059–1078.

Potts, D. C. and Garthwaite, R. L. (1991). Evolution of reef-building corals during periods of rapid global change. *In* "The Unity of Evolutionary Biology", Proceedings of the Fourth International Congress of Systematic and Evolutionary Biology, University of Maryland, College Park, USA, July 1990 (E. C. Dudley, ed.), pp. 170–177. Dioscorides Press, Portland, Oregon, USA.

Potts, D. C. and Swart, P. K. (1984). Water temperature as an indicator of environmental variability on a coral reef. *Limnology and Oceanography* **29**, 504–516.

Prell, W. L. (1984). Monsoonal climate of the Arabian Sea during the late Quaternary: a response to changing solar radiation. *In* "Milankovitch and Climate" (A. Berger, J. Imbrie, J. Hays, G. Kukla and B. Saltzman, eds), pp. 349–366. Reidel, Hingham, MA, USA.

Prell, W. L. and Kutzbach, J. E. (1987). Monsoon variability over the past 150 000 years. *Journal of Geophysical Research* **92**, 8411–8425.

Prezelin, B. B. and Sweeney, B. M. (1978). Photoadaptation of photosynthesis in *Gonyaulax polyedra*. *Marine Biology* **48**, 27–35.

Prezelin, B. B., Meeson, B. W. and Sweeney, B. M. (1977). Characterization of photosynthetic rhythms in marine dinoflagellates. I. Pigmentation, photosynthetic capacity and respiration. *Plant Physiology* **60**, 384–387.

Prosser, C. L. (ed.) (1973). "Comparative Animal Physiology." Saunders, Philadelphia, USA.

Pugh, D. T. and Rayner, R. F. (1981). The tidal regimes of three Indian Ocean atolls and some ecological implications. *Estuarine Coastal and Shelf Science* **13**, 389–407.

Quinn, W. H. (1992). A study of Southern Oscillation-related climate activity for AD622–1900 incorporating Nile River flood data. *In* "El Niño: Historical and Palaeoclimatic Aspects of the Southern Oscillation" (H. Diaz and V. Markgraf, eds), pp. 119–149.

Quinn, W. H., Neal, V. T. and Antunez de Mayolo, S. E. (1987). El Niño occurrences over the past four and a half centuries. *Journal of Geophysics* **92**, 14 449–14 462.

Rasmusson, E. M. (1985). El Niño and variation in climate. *American Scientist* **73**, 168–177.

Raup, D. M. and Boyajian, G. E. (1988). Patterns of generic extinction in the fossil record. *Paleobiology* **14**, 109–125.

Raven, J. A. (1989). Fight or flight: the economics of repair and avoidance of photoinhibition of photosynthesis. *Functional Ecology* **3**, 5–19.

Raven, J. A. (1991). Responses of aquatic photosynthetic organisms to increased solar UVB. *Photochemistry and Photobiology* **9**, 239–244.

Raven, J. A. (1994). The cost of photoinhibition to plant communities. *In* "Photoinhibition of Photosynthesis" (N. R. Baker and J. R. Bowyer, eds), pp. 449–464. Bios, Oxford, UK.

Raven, J. A., Johnston, A. M., Parsons, R. and Kübler, J. (1994). The influence of natural and experimental high oxygen concentrations on oxygen-evolving phototrophs. *Biological Reviews of the Cambridge Philosophical Society* **69**, 61–94.

Richardson, K., Beardall, J. and Raven, J. A. (1983). Adaptation of unicellular algae to irradiance: an analysis of strategies. *The New Phytologist* **93**, 157–191.

Richmond, R. H. (1985). Reversible metamorphosis in coral planula larvae. *Marine Ecology Progress Series* **22**, 181–186.

Roberts, H. H. and Suhayda, J. N. (1983). Wave–current interactions on a shallow reef (Nicaragua, Central America). *Coral Reefs* **1**, 209–214.

Roberts, H. H., Murray, S. P. and Suhayda, J. N. (1977). Physical processes on a fore-reef shelf environment. *Proceedings of the 3rd International Coral Reef Symposium* **2**, 507–516.

Roberts, J. L. (1964). Metabolic responses of freshwater sunfish to seasonal photoperiods and temperatures. *Helgolander Wissenschaftliche Meeresunter-suchungen* **9**, 459–473.

Rodhouse, P. G. and Gaffney, P. M. (1984). Effect of heterozygosity on metabolism during starvation in the American oyster *Crassostrea virginica*. *Marine Biology* **80**, 179–187.

Roos, P. J. (1967). Growth and occurrence of the reef coral *Porites astreoides* Lamarck in relation to submarine radiance distribution. PhD Thesis, University of Amsterdam, Elinkwijk, Utrecht, Netherlands.

Rosen, B. R. (1975). The distribution of reef corals. *Report of the Underwater Association* **1**, 2–16.

Rosen, B. R. (1981). The tropical high diversity enigma—the corals'-eye view. *In* "The Evolving Biosphere" (P. H. Greenwood, ed.), pp. 103–129. Cambridge University Press, Cambridge, UK.

Rosen, B. R. (1984). Reef coral biogeography and climate through the late Cainozoic: just islands in the sun or a critical pattern of islands? *In* "Fossils and Climate" (P. Brenchley, ed.), pp. 201–262. John Wiley, New York, USA.

Rosen, B. R. and Turnsek, D. (1989). Extinction patterns and biogeography of scleractinian corals across the Cretaceous/Tertiary boundary. *Memoirs of the Association of Australian Palaeontologists* **8**, 355–370.

Rowan, R. and Knowlton, N. (1995). Intraspecific diversity and ecological zonation in coral algal symbiosis. *Proceedings of the National Academy of Sciences of the USA* **92**, 2850–2853.

Sammarco, P. (1982). Polyp bail-out: an escape response to environmental stress and a new means of reproduction in corals. *Marine Ecology Progress Series* **10**, 57–65.

Sanders, B. M., Hope, C., Pascoe, V. M. and Martin, L. S. (1991). Characterisation of the stress protein response in two species of *Collisella* limpets with temperature tolerances. *Physiological Zoology* **64**, 1471–1489.

Satapoomin, U. (1993). Responses of corals and coral reefs to the 1991 coral reef bleaching event in the Andaman Sea, Thailand. MSc Thesis, Chulalongkorn University, Bangkok, Thailand.

Scelfo, G. (1986). Relationship between solar radiation and pigmentation of the coral *Montipora verrucosa* and its zooxanthellae. *In* "Coral Reef Population Biology" (P. L. Jokiel, R. H. Richmond and R. A. Rogers, eds), pp. 440–451. Hawaii Institute of Marine Biology, Honolulu.

Schäfer, C. (1994). Controlling the effects of excessive light energy fluxes: dissipative mechanisms, repair processes and long-term acclimation. *In* "Flux Control in Biological Systems from Enzymes to Populations and Ecosystems" (E. D. Schulze, ed.), pp. 37–54. Academic Press, New York, USA.

Schlichter, D. and Fricke, H. W. (1990). Coral host improves photosynthesis of endosymbiotic algae. *Naturwissenschaften* **77**, 447–450.

Schoenberg, D. A. and Trench, R. K. (1980). Genetic variation in *Symbiodinium (= Gymnodinium) microadriaticum* Freudenthal, and specificity in its symbiosis with marine invertebrates. III. Specificity and infectivity of *Symbiodinium microadriaticum*. *Proceedings of the Royal Society of London, B* **207**, 445–460.

Scoffin, T. P. (1993). The geological effects of hurricanes on coral reefs and the interpretation of storm deposits. *Coral Reefs* **12**, 203–221.

Scott, B. D. and Jitts, H. R. (1977). Photosynthesis of phytoplankton and zooxanthellae on a coral reef. *Marine Biology* **41**, 307.

Scott, T. M. and Koehn, R. K. (1990). The effect of environmental stress on the relationship of heterozygosity to growth rate in the coot clam, *Mulinia lateralis* (Say). *Journal of Experimental Marine Biology and Ecology* **135**, 109–116.

Sebastian, C., Scheuerlein, R. and Häder, D.-P. (1994). Effects of solar and artificial ultraviolet radiation on pigment composition and photosynthesis in three *Prorocentrum* strains. *Journal of Experimental Marine Biology and Ecology* **182**, 251–263.

Sebens, K. P. and de Riemer, K. (1977). Diel cycles of expansion and contraction in coral reef anthozoans. *Marine Biology* **43**, 247–256.

Sebens, K. P. and Done, T. J. (1993). Water flow, growth form and distribution of scleractinian corals: Davies Reef (GBR), Australia. *Proceedings of the 7th International Coral Reef Symposium* **1**, 557–568.

Sepkoski, J. J. (1986). Global bioevents and the question of periodicity. *Lecture Notes in Earth Sciences* **8**, 46–61.

Sharp, V. A. (1995). Temperature tolerance and heat shock protein production by temperate and tropical cnidarians. PhD Dissertation, University of Newcastle upon Tyne, UK.

Sharp, V. A., Miller, D., Bythell, J. C. and Brown, B. E. (1994). Expression of low molecular weight HSP 70 related polypeptides from the symbiotic sea anemone *Anemonia viridis* Forskall in response to heat shock. *Journal of Experimental Marine Biology and Ecology* **179**, 179–193.

Shashar, N., Cohen, Y. and Loya, Y. (1993). Extreme diel fluctuations of oxygen in diffusive boundary layers surrounding stony corals. *Biological Bulletin, Marine Biological Laboratory, Woods Hole* **185**, 455–461.

Sheehan, P. M. (1985). Reefs are not so different—they follow the evolutionary pattern of level-bottom communities. *Geology* **13**, 46–49.

Shen, G. T., Cole, J. E., Lea, D. W., Linn, L. J., McConnaughey, T. A. and Fairbanks, R. G. (1992). Surface ocean variability at Galapagos from 1936 to 1982: calibration of geochemical tracers in corals. *Paleoceanography* **7**, 563–588.

Sheppard, C. R. C. (1981). Illumination and the coral community beneath tabular *Acropora* species. *Marine Biology* **64**, 53–58.

Shibata, K. (1969). Pigments and a UV-absorbing substance in corals and a blue-green alga living in the Great Barrier Reef. *Plant and Cell Physiology* **10**, 325–335.

Shick, J. M. (1991). "A Functional Biology of Sea Anemones." Chapman and Hall, London, UK.

Shick, J. M. and Dykens, J. A. (1984). Photobiology of the symbiotic sea anemone, *Anthopleura elangantissima*: photosynthesis, respiration, and behavior under intertidal conditions. *Biological Bulletin, Marine Biological Laboratory, Woods Hole* **166**, 608–619.

Shick, J. M., Lesser, M. P. and Stochaj, W. R. (1991). Ultraviolet radiation and photooxidative stress in zooxanthellate anthozoa: the sea anemone *Phyllodiscus semoni* and the octocoral *Clavularia* sp. *Symbiosis* **10**, 145–173.

Shick, J. M., Lesser, M. P., Dunlap, W. C., Stochaj, W. R., Chalker, B. E. and Wu Won, J. (1995). Depth-dependent responses to solar ultraviolet radiation and oxidative stress in the zooxanthellate coral *Acropora microphthalma*. *Marine Biology* **122**, 41–51.

Shine, K. P., Derwent, R. G., Wuebbles, D. J. and Morcrettte, J. J. (1990). Radiative forcing of climate. *In* "Climate Change; the IPCC Scientific Assessment" (J. T. Houghton, G. J. Jenkins and J. J. Ephraums, eds), pp. 41–68. Cambridge University Press, Cambridge, UK.

Sibly, R. M. and Calow, P. (1989). A life-cycle theory of responses to stress. *Biological Journal of the Linnean Society* **37**, 101–116.

Smith, S. V. and Buddemeier, R. W. (1992). Global change and coral-reef ecosystems. *Annual Review of Ecology and Systematics* **23**, 89–118.

Somero, G. N. (1978). Temperature adaptation of enzymes: biological optimization through structure–function compromises. *Annual Review of Ecology and Systematics* **9**, 1–29.

Somero, G. N. (1986). Protein adaptation and biogeography: threshold effects on molecular evolution. *Trends in Ecology and Evolution* **1**, 124–127.

Spotila, J. R., Standora, E. A., Easton, D. P. and Rutledge, P. S. (1989). Bioenergetics, behaviour and resource partitioning in stressed habitats: biophysical and molecular approaches. *Physiological Zoology* **62**, 253–285.

Steen, R. G. and Muscatine, L. (1987). Low temperature evokes rapid exocytosis of symbiotic algae by a sea anemone. *Biological Bulletin, Marine Biological Laboratory, Woods Hole* **172**, 246–263.

Stochaj, W. R., Dunlap, W. C. and Shick, J. M. (1994). Two new UV-absorbing mycosporine-like amino acids from the sea anemone *Anthopleura elegantissima* and the effects of zooxanthellae and spectral irradiance on chemical composition and content. *Marine Biology* **118**, 149–156.

Stoddart, D. R. (1969). Ecology and morphology of recent coral reefs. *Biological Reviews of the Cambridge Philosophical Society* **44**, 433–498.

Stordal, F., Hov, Ý. and Isaksen, I. S. A. (1982). The effect of perturbation of the total ozone column due to CFC on the spectral distribution of UV fluxes and the damaging UV doses at the ocean surface: a model study. In "NATO Conference Series. IV. Marine Sciences", vol. 7: "The Role of Solar Ultraviolet Radiation in Marine Ecosystems" (J. Calkins, ed.), pp. 93–120. Plenum Press, New York, USA.

Suharsono, Pipe, R. K. and Brown, B. E. (1993). Cellular and ultrastructural-changes in the endoderm of the temperate sea-anemone Anemonia viridis as a result of increased temperature. Marine Biology 116, 311–318.

Sverdrup, H. U., Johnson, M. W. and Fleming, R. H. (1942). "The Oceans, their Physics, Chemistry and General Biology." Prentice-Hall, Englewood Cliffs, NJ, USA.

Szmant, A. M. and Gassman, N. J. (1990). The effects of prolonged "bleaching" on the tissue biomass and reproduction of the reef coral Montastrea annularis. Coral Reefs 8, 217–224.

Taylor, J. D. (1978). Faunal response to the instability of reef habitats: Pleistocene molluscan assemblages of Aldabra atoll. Palaeontology 21, 1–30.

Thomson, D. J. (1995). The seasons, global temperature and precession. Science, NY 268, 59–68.

Titlyanov, E. A., Shaposhnikova, M. G. and Zvalinsky, V. I. (1980). Photosynthesis and adaptation of corals to irradiance. 1. Contents and native state of photosynthetic pigments in symbiotic macroalgae. Photosynthetica 14, 413–421.

Tourre, Y. M. and White, W. B. (1996) (in press). ENSO signals in global upper ocean temperature. Journal of Physical Oceanography.

Tribble, G. W. and Randall, R. H. (1986). Description of the high-latitude shallow water coral communities of Miyake-jima, Japan. Coral Reefs 4, 151–159.

Tudhope, A. W. (1993). Extracting high-resolution climatic records from coral skeletons. Geoscientist 4, 17–20.

Tudhope, A. W., Allison, N., Le Tissier, M. D. A. and Scoffin, T. P. S. (1993). Growth characteristics and susceptibility to bleaching in massive Porites corals, South Thailand. Proceedings of the 7th International Coral Reef Symposium 1, 64–69.

Tudhope, S., Chilcott, C., Fallick, T., Jebb, M. and Shimmield, G. (1994). Southern oscillation-related variations in rainfall recorded in the stable oxygen isotopic composition of living and fossil massive corals in Papua New Guinea. Mineralogical Magazine 58, 914–915.

Tyrell, R. M. (1986). Repair of genetic damage induced by UVB (290–32 nm) radiation. In "Stratospheric Ozone Reduction and Plant Life" (R. C. Worrest and M. M. Caldwell, eds), pp. 139–149. Springer, Berlin, Germany.

van Campo, E., Duplessy, J. C. and Rossignol-Strick, M. (1982). Climatic conditions deduced from a 150-kyr oxygen isotope-pollen record from the Arabian Sea. Nature, London 296, 56–59.

Vernberg, F. J. (1962). Comparative physiology: latitudinal effects on physiological properties of animal populations. Annual Reviews in Physiology 24, 517–546.

Vernberg, F. J. and Costlow, J. D. (1966). Studies on the physiological variation between tropical and temperate zone fiddler crabs of genus Uca. IV. Oxygen consumption of larvae and young crabs reared in the laboratory. Physiological Zoology 39, 36–52.

Veron, J. E. N. (1995). "Corals in Space and Time: the Biogeography and Evolution of the Scleractinia." Comstock/Cornell, London, UK.

Veron, J. E. N. and Kelley, R. (1988). "Species Stability in Reef Corals of Papua New Guinea and the Indo Pacific." Association of Australian Palaeontologists, Sydney.

Veron, J. E. N. and Pichon, M. (1976). "Scleractinia of Eastern Australia." Australian Government Publishing Service, Canberra, Australia.

Vesk, M. and Jeffrey, S. W. (1977). Effect of blue-green light on photosynthetic pigments and chloroplast structure in unicellular marine algae from six classes. *Journal of Phycology* **13**, 280–288.

Wainwright, S. A. and Koehl, M. A. R. (1976). The nature of flow and reaction of benthic cnidaria to it. *In* "Coelenterate Ecology and Behavior" (G. O. Mackie, ed.), pp. 5–21. Plenum Press, New York, USA.

Wainwright, S. A., Biggs, W. D., Currey, J. D. and Gosline, J. M. (eds) (1976). "Mechanical Design in Organisms." Edward Arnold, London, UK.

Wallace, A. R. (1878). "Tropical Nature and Other Essays." Macmillan, London, UK.

Walsh, P. J. (1985). Temperature adaptation in sea anemones: physiological and biochemical variability in geographically separate populations of *Metridium senile*. *Marine Biology* **62**, 25–34.

Walsh, P. J. and Somero, G. N. (1981). Temperature adaptation in sea anemones; physiological and biochemical variability in geographically separate populations of Metridium senile. *Marine Biology* **62**, 25–34.

Weil, E. (1993). Genetic and morphological variation in Caribbean and eastern Pacific *Porites* (Anthozoa, Scleractinia) preliminary results. *Proceedings of the 7th International Coral Reef Symposium* **2**, 643–656.

Wellington, G. M. and Glynn, P. W. (1983). Environmental influences on skeletal banding in Eastern Pacific (Panama) corals. *Coral Reefs* **1**, 215–222.

Wells, J. W. (1952). The coral reefs of Arno Atoll, Marshall Islands. *Atoll Research Bulletin* **9**, 14.

Wiens, H. J. (1962). "Atoll Environment and Ecology." Yale University Press, New Haven, USA.

Wilkinson, C. R. and Buddemeier, R. W. (1994). Global climate change and coral reefs: Implications for people and reefs. Report of the UNEP-IOC-ASPEI-IUCN Global Task Team on the implications of climate change on coral reefs. IUCN, Gland, Switzerland.

Willis, B. L. (1985). Phenotypic plasticity versus phenotypic stability in the reef corals *Turbinaria mesenterina* and *Pavona cactus*. *Proceedings of the 5th International Coral Reef Congress* **4**, 107–112.

Willis, B. L. and Ayre, D. J. (1985). Asexual reproduction and genetic determination of growth form in the coral *Pavona cactus*: biochemical genetic and immunogenic evidence. *Oecologia* **65**, 516–525.

Wolanski, E. and Hamner, W. L. (1988). Topographically controlled fronts in the ocean and their biological influence. *Science, NY* **241**, 177–181.

Wolanski, E. and Pickard, G. L. (1985). Long-term observations of currents on the central Great Barrier Reef continental shelf. *Coral Reefs* **4**, 47–57.

Wolter, K. and Hastenrath, S. (1989). Annual cycle and long-term trends of circulation and climate variability over the tropical oceans. *Journal of Climate* **2**, 1329–1351.

Woodley, J. D. (1992). The incidence of hurricanes on the north coast of Jamaica since 1870—are the classic reef descriptions atypical. *Hydrobiologia* **247**, 133–138.

Wyrtki, K. (1975). El Niño—the dynamic response of the equatorial Pacific Ocean to atmospheric forcing. *Journal of Physical Oceanography* **5**, 572–584.

Wyrtki, K. (1979). The response of sea surface topography to the 1976 El Niño. *Journal of Physical Oceanography* **9**, 1223–1231.

Wyrtki, K. (1985). Sea level fluctuations in the Pacific during the 1982–3 El Niño. *Geophysical Research Letters* **12**, 125–128.

Yap, H. T., Montebon, A. R. F. and Dizon, R. M. (1994). Energy flow and seasonality in a tropical coral reef flat. *Marine Ecology Progress Series* **103**, 35–43.

Yeh, T. C., Tao, S. Y. and Li, M. C. (1959). The abrupt change of circulation over the Northern Hemisphere during June and October. *In* "The Atmosphere and the Sea in Motion", pp. 249–267.

Zouros, E. (1990). Heterozygosity and growth in marine bivalves: response to Koehn's remarks. *Evolution* **44**, 216–218.

Zouros, E. and Mallet, A. L. (1989). Genetic explanations of the growth/heterozygosity correlation in marine molluscs. *In* "Reproduction, Genetics and Distributions of Marine Organisms" (J. S. Ryland and P. A. Tyler, eds), pp. 317–324. Olsen and Olsen, Fredensborg, Denmark.

Zouros, E., Romero-Dorey, M. and Mallet, A. L. (1988). Heterozygosity and growth in marine bivalves: further data and possible explanations. *Evolution* **42**, 1332–1341.

Harmful or Exceptional Phytoplankton Blooms in the Marine Ecosystem

K. Richardson

Danish Institute for Fisheries Research, Charlottenlund Castle, DK-2920 Charlottenlund, Denmark

ADVANCES IN MARINE BIOLOGY VOL. 31
ISBN 0-12-026131-6

1. INTRODUCTION

A phytoplankton (also called "microalgal" or "algal") bloom is the rapid growth of one or more species which leads to an increase in biomass of the species. In recent years, there has been considerable popular and scientific attention directed towards "exceptional" (also called "noxious" or "nuisance") and "harmful" phytoplankton blooms. Often, in the case of exceptional/harmful blooms, it is a single species that comes to dominate the phytoplankton community (i.e. the blooms are "monospecific"). However, when toxic algae are involved in a harmful bloom, the mere presence of the toxic alga in concentrations sufficient to elicit effects is often enough to cause the scientific community and public at large to refer to a "bloom" of that particular species. In other words, one refers to "blooms" of toxic phytoplankton on the basis of the effects observed and not necessarily because of a large biomass.

2. DEFINITIONS

Exceptional phytoplankton blooms have been defined as "those which are noticeable, particularly to the general public, directly or indirectly through their *effects* such as visible discoloration of the water, foam production, fish or invertebrate mortality or toxicity to humans" (ICES, 1984). Here, it is important to note that the list of potential "noticeable effects" of exceptional algal blooms contains effects that are caused by very different processes and which affect humans in very different ways. Some bloom-forming phytoplankton are directly harmful to humans and thus form a subset of exceptional blooms referred to as "harmful".

Harmful phytoplankton may contain potent neurotoxins that, when they become concentrated in, for example, filter-feeding bivalves, can pose a serious public health threat. Indeed, it has been estimated (Hallegraeff, 1993) that, on a global scale, approximately 300 people die annually as a result of eating shellfish contaminated with toxic phytoplankton. Other phytoplankton species contain toxins that can induce sublethal responses in humans (diarrhoea, eye/skin irritation, breathing difficulties, etc.). Some phycotoxins appear to be carcinogenic. This, for example, is the case for some blue-green algae (cyanobacteria), which are primarily found in fresh water but which invade and can bloom in brackish waters (Falconer, 1991; Carmichael, 1992).

Other phytoplankton (*Phaeocystis* spp.) are considered to be harmful because they exude protein-rich compounds that can be whipped into a stiff foam by wave action and which, under certain hydrographic

Figure 1   Phaeocystis foam accumulation on the beach—eastern North Sea. Photograph courtesy of Helene Munk Sørensen.

conditions, can accumulate along beaches and deter (paying) beach guests (Figure 1). "Red tides", which have nothing to do with tidal action *per se* but are simply accumulations of phytoplankton so dense that the water appears to be coloured (red, green, brown, orange, etc.), can also discourage coastal visitors. Still other blooms attract public attention and earn their designation "harmful" because their occurrence in large numbers can pose an economic threat to fin- or shellfish aquaculture (see, e.g., Shumway, 1992). Mortalities of caged fish or wild fish and shellfish can occur as a result of phytoplankton blooms for a number of reasons. In some cases, there can be mechanical interaction between the phytoplankter and the gills which leads to gill damage and, ultimately, suffocation of the fish. Diatoms are often implicated in such events (Bell, 1961; Taylor *et al.*, 1985; Farrington, 1988; Rensel, 1993; Kent *et al.*, 1995; Tester and Mahoney, 1995). In some cases, anoxia and/or bacterial infection in combination with the damaged gill tissue leads to mortality of the fish (Jones and Rhodes, 1994; Tester and Mahoney, 1995). Hypoxia or anoxia resulting from the respiration or decay of dense blooms of phytoplankton can also, on its own, lead to fish or shellfish kills—especially of caged fish that are unable to swim from the affected area (Steimle and Sindermann, 1978; Taylor *et al.*, 1985). Finally, some types of phytoplankton blooms cause fish mortality through the production of toxins. A number of toxins produced by phytoplankton that affect humans also affect fish (White, 1977; Gosselin *et al.*, 1989; Riley *et al.*, 1989; Robineau *et al.*, 1991). In addition, however, some bloom-forming phytoplankton species that have not been shown to be toxic to humans produce toxins that affect fish or other marine organisms (Moshiri *et al.*, 1978; Granmo *et al.*, 1988; Change *et al.*, 1990; Black *et al.*, 1991; Aune *et al.*, 1992; Heinig and Campbell, 1992; Eilertsen and Raa, 1994).

Thus the responses elicited by harmful phytoplankton blooms are very different. The common feature that such blooms share is that they attract public attention and that they often have public health or economic implications. No common physiological, phylogenetic or structural feature has yet been identified that distinguishes "harmful" phytoplankton species from non-harmful and the scientific basis for treating harmful phytoplankton blooms as a distinct subset of algal blooms is not obvious. Rao and Pan (1995) conducted a literature survey in which they compared photosynthetic and respiratory characteristics of some toxic and non-toxic phytoplankters in an attempt to identify physiological differences between them. They concluded that there were no differences in the characteristics examined between the toxic and non-toxic dinoflagellates studied. However, they did find that photon efficiency and maximum rates of photosynthesis were lower in the diatom *Pseudo-nitzschia pungens* f. *multiseries* when domoic acid, the toxin with which this species is associated, was produced. This, they suggest, may have been a result of the fact that domoic acid may be a product of physiologically stressed cells while at least one of the dinoflagellates considered (*Alexandrium tamarensis*) is known to produce toxin during all phases of the life cycle. Such studies should serve to remind us that even within the subset of "harmful" phytoplankton that produce toxins, the differences are great and it may not be wise to expect similar responses or behaviour from these organisms.

A number of workers have argued that there has been an increase in harmful blooms in recent years (Anderson, 1989; Smayda, 1990; Hallegraeff, 1993) and it is often argued that the apparent increase in the occurrence of "harmful" blooms is linked to eutrophication.[1] Indeed, in some areas—especially those with limited water exchange such as fjords,

---

[1] Eutrophication as defined by Nixon (1995) = "an increase in the rate of supply of organic matter to an ecosystem". Eutrophication can occur via natural processes or as a result of human activities (cultural eutrophication). In practice, since the supply of organic matter to marine ecosystems is almost entirely the result of primary production and an increase in mineral nutrients can stimulate primary production, we equate cultural eutrophication with an increase in nutrients to a given water body resulting from human activities. When the term "eutrophication" is used in this review, it is cultural eutrophication that is referred to.

estuaries and inland seas—there does seem to be good evidence for a stimulation of the number of algal blooms occurring by eutrophication. However, the relationship between the occurrence of harmful phytoplankton blooms and environmental conditions is complicated and anthropogenic perturbation of the environment is certainly not a prerequisite for all harmful algal blooms. Thus, the occurrence of a harmful bloom may or may not have as one of its underlying causes a change in human activities or behaviour.

In Section 3 of this review, a discussion of our current state of knowledge with respect to harmful or exceptional phytoplankton blooms is presented. This section draws heavily on other recent reviews dealing with aspects of the problem of harmful phytoplankton blooms and, to the authors of these reviews, I am grateful. In the present review, particular attention is given to consideration of whether or not anthropogenic alteration of the environment may, directly or indirectly, selectively stimulate harmful phytoplankton species and the types of data required in order to be able to identify possible trends in the occurrence of harmful blooms.

Section 4 deals with the occurrence of exceptional/harmful blooms in the context of the seasonal cycles of phytoplankton bloom distribution in different latitudinal regions. In the last substantive section of the review (Section 5), the dynamics relating to the formation, maintenance and decay of exceptional/harmful blooms are discussed.

3. "HARMFUL" OR "EXCEPTIONAL" PHYTOPLANKTON BLOOMS

3.1. History

Given that exceptional/harmful algal blooms were defined above as phytoplankton blooms where the effects are noticeable and/or harmful to humans, the real history of these blooms can only begin with the records left behind by man. Indeed, it is often cited that the first written record of a harmful algal bloom (albeit in fresh water) appears in the Bible (Exodus 7: 20–21). However, there is fossil evidence that harmful algal blooms were occurring long before this. Noe-Nygaard et al. (1987) have suggested, on the basis of the distribution of dinoflagellate cysts and bivalve shells in fossil sediments taken from the island of Bornholm in the Baltic Sea, that toxic dinoflagellates caused mass mortalities of bivalves on several occasions dating back to about 130 million years ago (Figure 2).

Dale et al. (1993) have also studied the distribution of fossil dinoflagellate cysts and present evidence suggesting bloom formation by the toxic

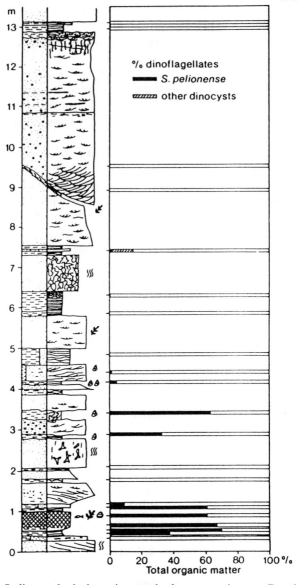

Figure 2 Sedimentological section made from a region on Bornholm, Denmark, showing the monospecific occurrences of the dinoflagellate *Sentusidium pelionense* in beds with mass occurences of the bivalve, *Neomiodon angulata* (indicated in the drawings to the left of the bar diagram). The relative proportion of dinoflagellate cysts in relation to the total organic matter is indicated. It is not possible from such a section to establish cause and effect between the dinoflagellate blooms and mass mortalities of *Neomiodon*. However, the repeated coincidence of these events suggests that the bivalve mortalities may have been caused by the dinoflagellate blooms. (After Noe-Nygaard *et al.*, 1987.)

dinoflagellate *Gymnodinium catenatum* in the Kattegat–Skagerrak long before anthropogenic activities can have influenced these waters. The species first appeared in the region about 6000 years ago and achieved a "minor peak in production" about 4500 years ago. After that, it occurred in relatively low numbers until about 2000 years ago. During the next 1500 years, periodic blooms of the species took place. In the sediment records for the last 300 years, these workers found no evidence of this dinoflagellate. However, in a postscript to another paper (Dale and Nordberg, 1993), they indicate that living cysts of *G. catenatum* have recently been isolated from Kattegat sediments. Dale and Nordberg (1993) believe that the most important regulating factor with respect to these ancient phytoplankton blooms was climate change (the blooms appear to be associated with periods of relatively warm water) and its resultant influence on hydrographic conditions.

These forays into the fossil record emphasize two important points with respect to exceptional algal blooms: (1) the occurrence of such blooms is not only a recent phenomenon (i.e. anthropogenic perturbation of the environment is not a necessary prerequisite for all exceptional blooms) and (2) phytoplankton species are not necessarily permanent residents of a given water body. Their relative abundance within an area can vary dramatically over a relatively short time span.

The fact that harmful blooms can occur in pristine waters is further illustrated by the chronicles of a number of European explorers who report encounters with poisoned shellfish upon arriving on the shores of North America. In Canada, for example, there is still a cove known as "Poison Cove" so named by Captain George Vancouver in 1793 after an ill-fated dinner on the local bivalves (as cited in Hallegraeff, 1993).

It is not only reports of algae that are directly toxic to humans that appear in the historical literature—Pouchet described blooms of what must have been *Phaeocystis* in Norwegian waters occurring in 1882. He also recorded observing this same plankton organism at the Faeroe Islands in 1890 (cited in Moestrup, 1994). *Phaeocystis* blooms concentrated enough to clog fishing nets were recorded along the New Zealand coast (here called "Tasman Bay Slime") as early as the 1860s (Hurley, 1982). Wyatt (1980) has suggested that a massive seal mortality event recorded off south-west Africa by Benjamin Morrell in the early 1800s may have been the result of a toxic phytoplankton bloom.

3.2. Causative Organisms

Sournia (1995) has recently conducted a survey of known marine phytoplankton species and estimated numbers of species within each

Table 1 Census of the total known number of species, number of species implicated in exceptional/harmful blooms and toxic species in the world's ocean for each class of the phytoplankton flora. (From Sournia, 1995.)

Class	Number of phytoplankton species	Number of exceptional/ harmful bloom species	Number of toxic species
Chlorarachniophyceae	1	0	0
Chlorophyceae	107–122	5–6	0
Chrysophyceae	96–126	6	1
Cryptophyceae	57–73	5–8	0
Cyanophyceae	7–10	3–4	1–2
Diatomoph. Centrales	870–999	30–65	1–2
Diatomoph. Pennales	300	15–18	3–4
Dictyochophyceae	1–3	1–2	0
Dinophyceae	1514–1880	93–127	45–57
Euglenophyceae	36–37	6–8	1
Eustigmatophyceae	3	0	0
Prasinophyceae	103–136	5	0
Prymnesiophyceae	244–303	8–9	4–5
Raphidophyceae	11–12	7–9	4–6
Rhodophyceae	6	0	0
Tribophyceae	9–13	0	0
Total	3365–4024	184–267	60–78

phytoplankton class, how many of these have been implicated in the formation of exceptional or harmful blooms[2] and how many have been identified as being toxic. In his census (Table 1), the dinoflagellates (Dinophyceae) comprise the class which is numerically largest and it is also within this class that we find the greatest numbers of both exceptional/harmful bloom-forming and toxic species: Sournia estimates that 5.5–6.7% of the known phytoplankton species in the world's oceans (i.e. about 200 species) have been identified as causing exceptional/harmful blooms. About half of these species are dinoflagellates. Diatoms were the second most important algal class in Sournia's analysis in terms of causing exceptional/harmful blooms.

Not all exceptional/harmful blooms are toxic, of course, and Sournia has estimated that only 1.8–1.9% of the world's phytoplankton flora has so far been identified as toxic. However, he also makes the point that this value may be an underestimate as several new toxic species have become

[2] Sournia actually identifies organisms implicated in "red tides". However, his use of the term red tide corresponds to the definition of exceptional/harmful used here.

Table 2 Census of the number of species, exceptional/harmful bloom forming, and toxic species in the world's ocean for each order of the class *Dinophyceae* (dinoflagellates). (From Sournia, 1995.)

Order	Number of species	Number of exceptional/ harmful bloom species	Number of toxic species
Actiniscales	8–11	0	0
Brachydiniales	7–8	0	0
Desmomonadales	6	0–1	0
Dinococcales	4	0	0
Dinophysales	240–382	3–4	7–11
Dinotrichales	3	0	0
Ebriales	3	0–1	0
Gymnodiniales	512–529	31–52	9–14
Noctilucales	15–19	1	0–1
Oxyrrhinales	2	1	0
Peridiniales	656–788	46–53	21–22
Prorocentrales	30–83	11–13	7–8
Protaspidales	4–6	0–1	0
Pyrocystales	7–17	0	0
Doubtful dinoflagellates	15–17	0	0
Total	1514–1880	93–127	45–57

known in recent years. Approximately 75% of the species that have been identified as being toxic belong to the Dinophyceae and most of these belong to the orders Peridiniales, Gymnodiniales and Dinophysales (Table 2). Four genera (*Alexandrium*, *Dinophysis*, *Gymnodinium* and *Prorocentrum*) dominate in terms of causing toxic blooms.

Moestrup (1994) has reviewed bloom formation by phytoplankton belonging to the *Prymnesiophyceae* and concluded that only species of *Chrysochromulina*, *Prymnesium* and *Phaeocystis* are known to form exceptional/harmful blooms. A number of species from both *Chrysochromulina* and *Prymnesium* have been implicated in fish kills in both fresh and marine waters. However, the mechanism of their toxic effect is, in most cases, not well documented. Like Sournia (1995), Moestrup (1994) concludes that there is a good chance that more species will be discovered as being potentially toxic in the coming years.

Considerable attention has been given to "brown tides" occurring in recent years in Narragansett Bay and around Long Island on the east coast of the USA. The causative organism here is the chrysophyte *Aureococcus anophagefferens* (Nuzzi and Waters, 1989; Smayda and Villareal, 1989). This organism has been shown to reduce or stop filter feeding in some shellfish (Tracey, 1988; Gallager et al., 1989; Gainey and Shumway, 1991). In addition, adverse effects of blooms of this organism have been observed in field studies on a number of different organisms (Cosper et al., 1987; Durbin and Durbin, 1989; Smayda and Fofonoff, 1989) although the mechanism of the interaction between A. anophagefferens and the various affected components of the ecosystem is not, in all cases, well documented. The southern Texas coast has also recently been plagued by "brown tides". Here, the causative organism is also a chrysophyte but somewhat larger than A. anophagefferens. Buskey and Stockwell (1993) have demonstrated in their field studies that micro- and mesozooplankton populations are, apparently, reduced during these blooms.

While it is clearly a relatively small percentage of the world's total phytoplankton population that has been identified as having the potential to develop exceptional/harmful blooms, Sournia (1995) makes the point that there is still a large number of species involved representing great taxonomic diversity. In view of this diversity he argues that there is "no hope of defining a single algal type or target organism for use in understanding, modeling or protection against" exceptional/harmful algal blooms.

3.3. Toxic Algal Blooms

The analytical chemistry surrounding the description and identification of the toxins associated with algal blooms is complex. In addition, the technical capabilities in terms of toxin identification have been evolving rapidly in recent years. It is not the purpose of the present review to consider the more chemical aspects of the toxins associated with algal blooms. The more chemically inclined reader is referred to the following recent works for a status of knowledge concerning the chemistry of algal toxins (WHO, 1984; Falconer, 1993; Premazzi and Volterra, 1993). For the purposes of this review, it is sufficient to consider types of "toxicity events" rather than deal with the individual toxins themselves.

3.3.1. Amnesic Shellfish Poisoning (ASP)

Amnesic shellfish poisoning gets its name from the fact that one of the symptoms of the poisoning is loss of memory. ASP was first recognized in 1987 on Prince Edward Island on the Canadian east coast when a very serious incident occurred which caused several human fatalities and over a hundred cases of acute poisoning following the consumption of blue mussels (Bates et al., 1989). The cause of this poisoning was traced to a bloom dominated by the pennate diatom, *Pseudo-nitzschia multiseries* (formerly *Pseudonitzschia pungens f. multiseries*) which produces the neurotoxin domoic acid (Bates *et al.*, 1989, 1995).

After the discovery of ASP, many countries began routine monitoring for the occurrence of domoic acid in shellfish products. Already in 1988, unacceptably high concentrations of domoic acid were found in soft-shelled clams, *Mya arenaria*, and blue mussels, *Mytilus edulis*, in the south-western Bay of Fundy (Canadian coast). Shellfish harvesting areas were closed and no incidents of poisoning were reported at this time (Haya *et al.*, 1991; Martin *et al.*, 1993).

Coastal fisheries along the North American west coast were closed in the autumn of 1991 due to the detection of domoic acid (Drum *et al.*, 1993; Langlois *et al.*, 1993; Villac *et al.*, 1993). There was also a report of poisonings by domoic acid at this time from Santa Cruz, California (Work *et al.*, 1993). In this poisoning event, no human victims were reported; however, mortalities of pelicans and cormorants were observed. Remnants of frustules from the diatom, *Pseudo-nitzschia australis* (= *Nitzschia pseudoseriata*) and high levels of domoic acid were found in the stomachs of the dead birds. A bloom of *Pseudo-nitzschia australis* was occurring at the time of the mortalities. Thus, it was possible to confirm that domoic acid production was associated with this diatom. Work *et al.* (1993) also demonstrated the presence of domoic acid in anchovies from the area of the bloom, thus identifying finfish as potential vectors of ASP.

Lundholm *et al.* (1994) have shown that the diatom *Pseudo-nitzschia seriata* also produces domoic acid at levels similar to those observed with *Pseudo-nitzschia multiseries*. This diatom is common in colder waters of the Northern Hemisphere. There is also an indication in the literature that several other common diatom species may produce domoic acid (e.g. *Fragilaria* sp.: Pauley *et al.*, 1993; *Amphora coffeaeformis*: Maranda *et al.*, 1990) so it would seem likely that other species may soon be confirmed as domoic acid producers. Thus, although ASP has only recently been recognized, the fact that the causative organisms have a wide geographic distribution (Douglas *et al.*, 1993; Hallegraeff, 1993) and that commercial finfish have the potential to harbour and transmit domoic acid poisoning suggests that there is a serious risk of ASP incidents occurring in many parts of the world's oceans.

3.3.2. *Paralytic Shellfish Poisoning (PSP)*

In contrast to ASP, the existence of PSP has been recognized for many centuries. It has even been claimed that the Indians along the Pacific west coast of North America practised a form of public health control by not

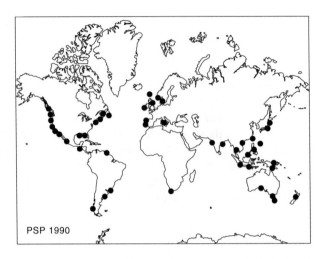

Figure 3 Known global distribution of paralytic shellfish poisoning (PSP) in 1990. (After Hallegraeff, 1993.)

harvesting shellfish in periods when there was bioluminescence in the water (many of the dinoflagellates responsible for PSP are bioluminescent; Dale and Yentsch, 1978). This form of shellfish poisoning induces muscular paralysis and, in severe cases, can lead to death through paralysis of the respiratory system. PSP is caused by one or more of about 18 different toxins which include saxitoxins, neosaxitoxins and gonyautoxins (Frémy, 1991). These toxins are often referred to collectively as "PSP toxins". The combination of toxins found in contaminated shellfish seems to depend upon the type of toxin produced by the causative phytoplankton organism as well as the storage conditions and the metabolism of the compounds in the shellfish (Kirschbaum *et al.*, 1995).

   Hallegraeff (1993) reviewed the distribution of areas known in 1990 to have been affected by PSP events at one time or another (Figure 3). From his census it would appear that, with the exception of the African and parts of the South American coasts, PSP is a worldwide problem. To what degree the lack of reports of PSP in Africa and South America indicate a true absence of the potential for PSP incidents in these regions is not yet clear. Recent anecdotal evidence (Baddyr, 1992) suggests that PSP incidents have occurred along the Moroccan coast during the 1980s and 1990s. Thus it seems likely that PSP is even more widespread than Figure 3 would suggest.

   Shellfish are the usual vectors of PSP to human consumers. However, crustaceans can also accumulate PSP toxins (Desbiens and Cembella, 1995). Lobsters harbour the toxins primarily in the hepatopancreas rather than in the meat—a fact that caused the Department of Health and Welfare in Canada to recommend that no more than two lobster "tomalley" be consumed at any one meal (Todd *et al.*, 1993). Lobster larvae do not appear to be sensitive to PSP toxins but a number of fish larvae (mackerel: Robineau *et al.*, 1991; capelin and herring: Gosselin *et al.*, 1989) have been shown to be vulnerable to these toxins. PSP toxins have been found in zooplankton and the guts of dead or diseased fish in the vicinity of blooms (White, 1977). Occasionally, some PSP toxins have also been found in the muscle of affected fish and there is at least circumstantial evidence that PSP toxins may be implicated in mortalities of marine mammals (Hofman, 1989; Anderson and White, 1992).

Sublethal effects of PSP toxins on marine food webs have also been recorded. A number of studies have indicated reduced grazing rates by copepods on PSP-containing phytoplankton (Ives, 1985, 1987; Huntley *et al.*, 1986; Turriff *et al.*, 1995). In an elegant study, Hansen (1989) examined the effect of the presence of the PSP producers, *Alexandrium tamarensis* and *A. fundyense*, on a tintinnid, *Favella ehrenbergii*. In this case, it was not the ingestion of the alga that affected the grazer but, rather, the presence of algal exudates in the medium. These exudates appear to affect the cell membrane and induce ciliary reversals which cause the organism to swim backwards. A similar response has been observed when *F. ehrenbergii* is presented with *Alexandrium ostenfeldii* (Hansen *et al.*, 1992).

3.3.3. *Neurotoxic Shellfish Poisoning (NSP)*

Blooms of the dinoflagellate *Ptychodiscus breve* (formerly *Gymnodinium breve*) have been associated with human poisonings which are characterized by neurological symptoms but no paralysis. Here again, it appears that a group of toxins are responsible for these intoxications and these are referred to collectively as "brevetoxins". Brevetoxins are potent polyether neurotoxins (Baden, 1989; Trainer *et al.*, 1990). Two types of human poisonings are recorded: one causing paraesthesia, alternating hot and cold sensations, nausea, diarrhoea and ataxia; the other form is characterized by upper respiratory distress and/or eye irritation (WHO, 1984). NSP is a serious problem along the south-eastern coast of North America and in the Gulf of Mexico. However, the causative organism has also been recorded in other parts of the world. It was recorded for the first time in 1993 in New Zealand waters concomitant with an outbreak of respiratory problems in the local population (Smith *et al.*, 1993).

In addition to causing human poisonings, brevetoxins affect marine organisms (Riley *et al.*, 1989) and have been implicated in a number of

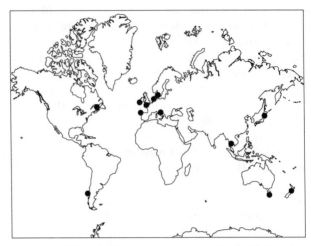

Figure 4 Known global distribution of diarrhoeic shellfish poisoning (DSP). (After Hallegraeff, 1993.)

mortality events for marine mammals. (Hofman, 1989; O'Shea *et al*., 1991; Anderson and White, 1992). Recent studies (Shimuzu *et al*., 1995) have suggested that toxin production by *P. breve* is increased when the organism is metabolizing heterotrophically.

3.3.4. *Diarrhoeic Shellfish Poisoning (DSP)*

DSP was first identified in Japan in 1976 (Yasumoto *et al*., 1978). Incidents of DSP had, however, almost certainly occurred earlier when they were probably mistaken for being caused by bacterial gastrointestinal infections. The symptoms of DSP include stomach pain, nausea, vomiting and diarrhoea (WHO, 1984). No fatalities directly associated with DSP have ever been recorded. However, Suganuma *et al*. (1988) have reported that some of the toxins associated with DSP may promote the development of stomach tumours. Thus prolonged or chronic exposure to DSP toxins may have long-term negative effects on public health. The toxins implicated here are okadaic acid and derivatives and polyether lactones (WHO, 1984). A number of dinoflagellate species (Hallegraeff, 1993), especially from the genus *Dinophysis*, have been identified as producing these toxins. Again, these organisms are broadly distributed on a global scale and confirmed DSP incidents have been reported from all continents except Africa and Antarctica (Figure 4). As in the case of PSP, there is evidence that DSP may be geographically more widespread than the scientific literature would suggest. Mendez (1992), for example, reports in a newsletter on algal blooms that DSP occurred along the Uruguayan coast during January 1992.

3.3.5. Ciguatera

Ciguatera poisoning has been known for centuries in subtropical and tropical areas—Captain Cook was stricken in New Caledonia in 1774—and is a growing problem in some areas. During the period 1960–84, over 24 000 cases were reported in French Polynesia alone (Hallegraeff, 1993). Ciguatera is caused by benthic dinoflagellates such as *Gambierdiscus toxicus* and, possibly, *Ostreopsis siamensis*, *Coolia monotis*, *Prorocentrum lima* and related species (Hallegraeff, 1993). The toxins are transported through the food chain and usually reach humans through the consumption of finfish.

There are chemical similarities between ciguatoxin and brevetoxins and the toxic effects of both of these types of toxin are, apparently, caused by changes in sodium ion influx in the affected organism (Baden, 1989, 1995; Trainer *et al.*, 1990). Many of the symptoms associated with ciguatera poisoning also resemble NSP and, in severe cases, death can result from circulatory collapse or respiratory failure (WHO, 1984). Ciguatoxins have also been implicated in marine mammal kills (Hofman, 1989).

3.3.6. Cyanobacterial Toxins

Toxic cyanobacteria (blue-green algae) blooms are most often associated with fresh waters. However, it has recently been shown that blooms of the cyanobacterium, *Trichodesmium thiebautii* occurring in the open ocean of the US Virgin Islands can be toxic (Hawser and Codd, 1992). Guo and Tester (1994) conducted toxicity tests with *Trichodesmium* sp. cells (taken from a 1992 bloom off the North American coast) on the copepod, *Acartia tonsa*. These workers demonstrated that intact cells were not toxic to *A. tonsa* but that homogenized cells were. Thus, the authors suggest that the alga contained one or more intracellular biotoxins. Brackish waters can also harbour toxic cyanobacteria blooms. Some Australian estuaries and inlets (Huber and Hamel, 1985; Lenanton *et al.*, 1985; Blackburn and Jones, 1995) and the Baltic Sea (Kononen, 1992) are examples of brackish areas where the toxic cyanobacterium, *Nodularia spumigena*, frequently forms blooms.

Other toxic cyanobacteria have also been reported to occur in the Baltic but these have apparently not been associated with bloom formation (Kononen, 1992). Nodularin, the toxin associated with *Nodularia spumigena*, is (like some of the toxins associated with the most common freshwater blooms of toxic cyanobacteria) hepatotoxic and may promote tumours with chronic exposure (Falconer, 1991). There are a number of incidents reported in the literature of animal poisonings/mortalities with associated liver damage which have been seen in connection with blooms of *N. spumigena* (Kononen, 1992).

Some marine regions (e.g. Hawaii, USA and Okinowa, Japan) have experienced blooms of the filamentous cyanobacterium, *Lyngbya majuscula*, which has been documented to produce toxins that induce dermatitis in animals. Thus, swimmers exposed to blooms of this organism have suffered attacks of dermatitis (WHO, 1984).

While toxic cyanobacteria blooms comprise a relatively small proportion of the phytoplankton blooms occurring in the world's oceans, the fact that they occur in brackish waters means that they are often found near the mouths of rivers and in bays and inlets affected by freshwater runoff. Such regions are often heavily used by the local human population. Given the recent discovery that some cyanobacterial toxins may stimulate tumour formation (Falconer, 1991; Carmichael, 1992), it is likely that there will be increasing awareness directed towards this type of phytoplankton bloom in coming years.

3.3.7. Other Types of Toxic Blooms

In addition to the toxic bloom events that directly threaten public health, there occur each year phytoplankton blooms that elicit mortality in wild or cultured marine animals without exhibiting any adverse effects for humans. Much publicity is given to such blooms, especially when cultured stocks (i.e. aquaculture activities) are threatened, and there has been a tendency to consider the occurrence of such blooms as a recent phenomenon. In fact, reports of mortalities of wild marine organisms in association with algal blooms have appeared in the scientific literature for at least the last century:

> During the last two months the inhabitants in Rhode Island witnessed the following remarkable phenomenon. The water of a considerable portion of the Bay became thick and red, emitting an odor almost intolerable to those living near by. The situation became alarming when, on the 9th and 10th of September, thousands of dead fish, crabs and shrimps were found strewn along the shores or even piled up in the windrows . . .
> During the last of August, throughout September and a part of October streaks of red or 'chocolate' water were observed from near Quonset Point and Providence Island. (Mead, 1898)

As identified in Section 2, the mortalities resulting from such blooms can have different causes or combination of causes including suffocation due to anoxia and/or gill damage and/or bacterial infection in weakened animals. However, in other cases the causative phytoplankton are known toxin producers or evoke responses in the affected organisms that suggest that toxin production is implicated.

Toxic blooms of Prymnesiophytes are frequently reported to be a problem for aquaculture facilities. For example, a massive bloom of *Chrysochromulina polylepis* in the Skagerrak/Kattegat during May–June 1988, caused havoc with salmon-raising facilities along the Swedish and Norwegian coasts (insurance companies in Norway paid out the equivalent of approximately 10 million US$ in connection with the bloom (Moestrup, 1994). This bloom also affected wild stocks of a number of different genera (Olsgard, 1993) and, in addition to causing mortalities of mature animals, it has been shown that the presence of the phytoplankton inhibits the reproduction process in the ascidian, *Ciona intestinalis* and the mussel, *Mytilus edulis* (Granmo et al., 1988). Thus, Granmo et al. (1988) suggested that the phytoplankton bloom may have had long-term effects in the affected area by reducing larval settlement and recruitment for these and possibly other species. Other plankton organisms were also shown to be affected by this *Chysochromulina* bloom (Nielsen et al., 1990). *Chrysochromulina leadbeateri* caused mortalities in caged salmon along the northern Norwegian coast in 1991 and two *Prymnesium* species (*P. patelliferum* and *P. parvum*) have also caused mortalities in aquaculture facilities for salmon and rainbow trout (Aune et al., 1992; Meldahl et al., 1995).

The raphidophyte, *Heterosigma akashiwo*, presents a serious problem for aquaculture in many parts of the world including Japan, North America, Asia and New Zealand (Haigh and Taylor, 1990; Change et al., 1990; Black et al., 1991; MacKenzie, 1991; Qi-Yusao et al., 1993; Honjo, 1994). Affected salmon show signs of gill and intestinal pathology and death has been attributed to impairment of the gills' respiratory and osmoregulatory functions (Change et al., 1990). It has been suggested that the toxic effects elicited by this organism on salmon may be brought about by the formation of toxic concentrations of oxygen radical and hydrogen peroxide (Yang and Albright, 1994).

A number of dinoflagellates are also dreaded by aquaculture operators for their toxic effects although the mechanism by which these dinoflagellates cause fish mortalities is not always well documented. An example here is *Gyrodinium aureolum* which has been recognized as a fish killer for many years (Tangen, 1977). Gill histopathology in affected fish has been reported to be similar to that observed in fish exposed to *Heterosigma* (Change et al., 1990). However, there has been considerable discussion in the scientific literature as to whether or not this organism is toxic and the nature of the toxin(s) that it may produce (see discussion in Bullock *et al.*, 1985). It is now generally accepted that toxin production by *G. aureolum* is, at least in some cases, involved in mortalities caused by this organism (Boalch, 1983; Bullock *et al.*, 1985; Heinig and Campbell, 1992).

It has also been suggested (Jenkinson, 1989, 1993) that *Gymnodinium aureolum* (and some other bloom-forming flagellates) may alter sea water characteristics through the production of extracellular organic material. This extracellular material should increase the viscosity of the medium surrounding the fish so that the energy expended in filtering water through the gills exceeds that which can be supported by the oxygen uptake. Support for the hypothesis that certain phytoplankton blooms can alter sea water characteristics is found in measurements made in the North Sea (Jenkinson and Biddanda, 1995). There is also some experimental evidence (Jenkinson, unpublished data) that the presence of certain phytoplankton in the medium decreases the rate at which fish are capable of filtering water over the gills. Thus, the mechanism(s) by which phytoplankton blooms may lead to fish mortalities are not always straightforward and, in some cases, several factors may be involved.

A new toxic dinoflagellate (*Pfiesteria piscicida*; Steidinger *et al.* 1996 (dubbed the "phantom" dinoflagellate) has recently been described (Burkholder *et al.*, 1992). This dinoflagellate requires live finfish (or their excrement) in order to excyst and to release its potent neurotoxin. It appears to be highly lethal to both fin- and shellfish in laboratory studies and it is believed that this organism may be responsible for major fish kills. "Blooms" of this dinoflagellate appear to be very short-lived in that the organism appears in relatively large numbers during a fish kill (i.e. while the fish are dying) but within hours of the death of the fish, the dinoflagellate appears to encyst and settle into the sediments. Fish kills associated with this dinoflagellate usually last less than 3 d (Burkholder *et al.*, 1995).

3.4. Role of Bacteria in Toxin Production

As noted by Sournia (1995; see Table 1), the greatest number, by far, of identified toxic species are found within the Dinophyceae. Two provocative papers presented (Kodama, 1990; Sousa-Silva, 1990) suggested that bacteria (either intra- or extracellular) may be implicated in the production of dinoflagellate toxicity. In the wake of these papers, considerable activity has been directed towards elucidating the potential role of bacteria in the production of phytoplankton toxicity. Although it now seems clear that not all phytoplankton toxicity is the direct result of bacterial activity, some bacteria have been identified as being capable of autonomous "phyco"toxin production (Doucette and Trick, 1995; and references contained in Doucette, 1995). In addition, there is growing evidence that bacteria or some unidentified algal–bacterial interaction(s) may play a potential role in many types of toxicity events.

Bates *et al.* (1995) examined the effects of bacteria on domoic acid production by the diatom, *Pseudo-nitzschia multiseries*. They observed a dramatic increase in domoic acid production in non-axenic cultures as compared to axenic ones (Figure 5). Furthermore, they demonstrated that toxin production was stimulated when bacteria were reintroduced to the axenic cultures. This stimulation occurred with the introduction of different types of bacteria indicating that the response observed was not specific to the presence of a single bacterium. Thus, while the presence of bacteria is not essential for the production of domoic acid by *P. multiseries*, some sort of extracellular bacterial interaction with the diatom may enhance the production of toxin. Further research is clearly needed to elucidate the nature of the relationship between bacteria and algal toxin production.

3.5. Role of Nutrient Availability in Toxin Production

As in the case of the potential influence of bacteria on toxin production, the role of nutritional status of toxin producers on the rate of toxin production is still not clear. Given the many different types of toxins produced by phytoplankton and the varying metabolic pathways involved in this toxin production, it seems likely that nutrient availability will affect different toxin producers in different ways. Certainly, the studies that have been conducted thus far on the influence of nutritional status on toxin production do not suggest a consistent relationship between nutrient availability and toxin production.

Many studies have been carried out on changes in toxin concentration throughout the life cycle or under changing nutrient/environmental conditions (Boyer *et al.*, 1987; Ogata *et al.*, 1987; Anderson *et al.*, 1990; Reguera and Oshimo, 1990; Bates *et al.*, 1991; Aikman *et al.*, 1993). Most, but not all, of the studies examining rates of toxin production or content throughout the toxin-producing organism's growth cycle suggest that toxin production varies as a function of life cycle stages. It should be noted, however, that the majority of these studies have been carried out on batch cultures. In addition, most have been limited to a description of the toxin responses observed and no attempt has been made to describe the underlying physiological mechanisms.

Anderson *et al.* (1990) examined different PSP-producing dinoflagellates

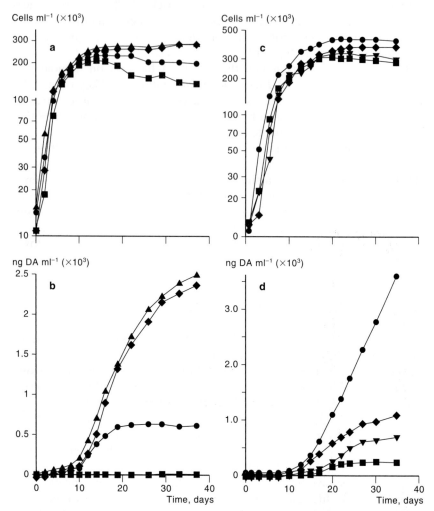

Figure 5 Cell growth and domoic acid (DA) production by two different strains of *Pseudo-nitzschia multiseries*: (a) and (c) illustrate cell numbers (based on optical density) in strains "POM" and "KP-14", respectively. ■, Axenic and ●, non-axenic cultures; ▲, reintroduced with strain BO-2 (originally isolated from a *P. multiseries* culture); ◆, introduced bacterial strain BD-1 (originally isolated from a *P. multiseries* culture); ▼, introduced bacterial strain CH-1 (originally isolated from a *Chaetoceros* sp. culture). (b) and (d) are domoic acid (DA) measured as concentration in the whole culture. (After Bates *et al.*, 1995.)

Table 3 Alexandrium fundyense. Net toxin production rates (R_{tox}) in batch culture during different growth stages. Intervals define early exponential growth and late exponential/early stationary phase growth. (From Anderson *et al.*, 1990.)

Treatment	Interval (d)	R_{tox}
Control	0–4	73.5
	4–6	30.0
Low PO_4^{3-}	0–5	88.6
	5–9	160.6
Low NO_3^-	0–4	69.7
	4–5	11.7
High salinity	0–4	58.3
	4–10	31.2
Low temperature	0–11	106.8
	11–19	29.6

(*Alexandrium fundyense*, *A. tamarense* and *A.* sp.) in both batch and semi-continuous culture. They identified differences in toxin production that were directly related to changes in growth rate associated with the various stages of the life cycle ("growth stage variability"). In addition, they were able to identify changes in toxin production related to "environmental" factors (i.e. temperature) and "nutrient" stress (Table 3). In particular, they found that phosphate limitation dramatically stimulated toxin production in *A. fundyense*. These workers suggested that the observed increase in toxic production in phosphate-stressed cells may be related to a build-up of arginine in the cells.

Other studies have also suggested an increase in toxicity for some phytoplankton species when grown under phosphate-limiting conditions (Edvardsen *et al.*, 1990; Aure and Rey, 1992) and demonstrated increased haemolytic activity of extracts from the prymnesiophyte, *Chrysochromulina polylepis*, when the organism was phosphate-stressed. This observation caused these and other workers (Maestrini and Granéli, 1991) to suggest that the high N : P ratio, which often is recorded in association with eutrophication and which was observed in the Skagerrak prior to the massive bloom of *C. polylepis* in 1988, may have stimulated the toxicity. It is worth noting here that this organism has been a component of the plankton community in the region both before and after 1988 but toxic effects have only been recorded in 1988 when an unusually high N : P ratio was reported (Maestrini and Granéli, 1991).

There is a developing awareness that the relative availability of different nutrients may also affect rates of toxin production. This may help in the search for an explanation for the physiological mechanisms behind toxin production and in our understanding of the relationship between environmental conditions and toxic events. Flynn and Flynn (1995) point out that toxins are secondary metabolites, the production of which is dependent upon a cell's physiological condition which, in turn, is a function of nutrient status. Production of secondary metabolites may increase either at "upshock" (e.g. nutrient refeeding of starved cells) or "downshock" (e.g. nutrient deprivation). These workers also argue that the unnaturally high nitrogen concentrations relative to carbon and phosphorus found in most laboratory culture media may complicate interpretation of results of studies designed to describe the interaction between toxin synthesis and nutrient availability.

3.6. Do Human Activities Stimulate Blooms/Select for Harmful Species?

3.6.1. *Has there been an Increase in Harmful Algal Blooms?*

Several authors have argued that there has been a global increase in harmful phytoplankton blooms in recent decades (Anderson, 1989; Smayda, 1990; Hallegraeff, 1993) and considerable concern has been expressed with respect to the potential role that human activities may have played in this increase. Before examining the potential influence that human activities may have had on the occurrence of harmful algal blooms, the evidence for the supposed increase in such algal blooms will be considered.

Quantitative demonstration of an increase in harmful algal blooms is difficult for several reasons. First, there is a degree of subjectivity associated with the definition of harmful blooms (see Section 2). The utilization of coastal resources has been expanding in recent decades. Thus, what is "harmful" today in a region heavily exploited for aquaculture might not even have been noticed in the period prior to the development of the aquaculture industry.

Environmental monitoring programmes have also been expanding during the past decades. Many of these programmes include comprehensive monitoring of chlorophyll distributions, primary production and/or mapping of the distributions of algal species. Thus, more effort is expended in looking for algal blooms than was the case, for example, 10 or 20 years ago. This increased effort in looking for blooms may also contribute to the fact that more are apparently found.

These considerations make it difficult to deduce on the basis of an increased number of reportings that there has been a real increase in the number of potentially harmful blooms occurring on a global basis. They

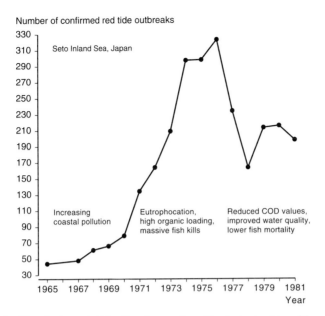

Figure 6 Trends in red tide (i.e "exceptional" phytoplankton blooms) from 1965 to 1981 in relation to stages of environmental changes in the Seto Inland Sea. (After Prakash, 1987.)

also illustrate the problems in establishing a quantitative database that can be used to assess the development of phytoplankton blooms and possible relationships between their frequency and human activities. Nevertheless, it does seem clear that in some semi-enclosed marine areas there has been a real increase in the number of harmful algal blooms.

3.6.2. *Influence of Cultural Eutrophication on the Frequency of Algal Blooms*

Perhaps the most frequently cited examples of areas exhibiting an increased bloom frequency are the Seto Inland Sea, Japan (Prakash, 1987, Figure 6) and Hong Kong Harbour (Lam and Ho, 1989). For the Seto Inland Sea, there is also evidence that the observed increase is directly related to human activities in that a decrease in the number of exceptional/harmful blooms ("red tides") occurred following a reduction in chemical oxygen demand (COD) in effluents (Prakash, 1987). Thus, there would appear to be a direct link in this region between eutrophication and the frequency of algal blooms.

Similar suggestions of increases in exceptional or harmful algal blooms caused by eutrophication have been made for other marine areas (especially those with relatively slow flushing times). The Baltic Sea, the coastal North Sea and the Black Sea have all been identified as regions where such an increase in algal blooms may have occurred (references cited in Smayda, 1990). However, while it intuitively seems likely that cultural eutrophication may have influenced phytoplankton growth and species distribution in these areas, quantitative evidence of an absolute increase in harmful algal blooms for most of these areas is lacking.

Here, it is worth considering what types of data might be useful in identifying possible trends in the occurrence of exceptional/harmful algal blooms related to cultural eutrophication. The difficulties in using reports of harmful algal blooms *per se* in order to address this question have been discussed above. Other types of data that are routinely collected which pertain to phytoplankton and might, therefore, be able to provide information concerning the incidences of harmful blooms include primary production, chlorophyll (as a proxy for biomass) distributions and algal species abundance.

An International Council for the Exploration of the Seas (ICES) Working Group dealing with phytoplankton collated in 1991 a list of data time series available that might be useful for quantifying potential changes in the frequency of harmful blooms (Table 4). While this list is not meant to include all long-term data sets pertaining to phytoplankton, it does illustrate the types of data available and the typical lengths/geographic coverages of such data sets. Note that, for the most part, these data series are no more than a few decades in length. In addition, sampling frequency has not been constant throughout the duration of some of these time series.

Species abundance data potentially offer the opportunity of quantifying changes in the relative frequency of occurrence of exceptional/harmful bloom species. However, the abundance of individual species varies dramatically interannually and identification of an increase in a particular area requires that a change in the normally occurring abundance pattern can be demonstrated (Figure 7). Increases in the occurrence of individual harmful species have been identified in some regions (e.g. *Phaeocystis* at the Marsdiep (Dutch Wadden Sea): Cadée and Hegemann, 1986). However, for the most part, the time series upon which these increases have been demonstrated are not, biologically speaking, very long. Until longer data series are available, it is difficult to ascertain with any degree of certainty that the increases observed are true increases and not simply an expression of the naturally occurring changes in species abundance.

Table 4 Examples of available time series of data pertaining to phytoplankton and harmful blooms. (From ICES, 1991.)

Country	Area	Data	First year
Norway	Oslofjord	PSP	1962
	Rest of coast	PSP	1982
	All	Mortalities	1966
		Gyrodinium	1981
		Dinophysis	1984
		DSP	1984
Sweden	West coast	PSP	1982
		Phytoplankton	1989
		DSP	1984
Finland	All	Mortalities	1984
	Gulf of Finland	Phytoplankton	1968
Germany	German Bight	Phytoplankton	1962
UK	NE England	PSP	1968
France	All	PSP	1984
		Mortalities	1976
		Phytoplankton	1984
		Gyrodinium	1980
		DSP	1983
Spain	Galicia	PSP	1976
		Phytoplankton	1977
		Dinophysis	1985
		DSP	1982
Portugal	All	PSP	1986
		Phytoplankton	1987
		DSP	1987
USA	Maine	PSP	1958
	Massachusetts	PSP	1972
	California	PSP	1962
	Washington	PSP	1978
	Oregon	PSP	1980
	Florida	NSP	1978
	Florida	Mortalities	1980
	Washington	Mortalities	1989
	Narragansett Bay	Phytoplankton	1958
Canada	East coast	PSP	1942
	West coast	PSP	1950
HELCOM	Baltic Sea	Phytoplankton	1979

PSP, Paralytic shellfish poisoning; DSP, diarrhoeic shellfish poisoning; NSP, neurotoxic shellfish poisoning.

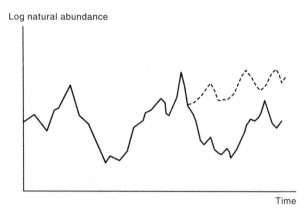

Figure 7 Hypothetical time series of mean annual abundance of a single phytoplankton species over a period of a few decades. Solid line: no overall increase in abundance of species over time. If the last third of the time series were described by the broken line, then species abundance would have increased during the period of study. The challenge in identifying whether or not there has been an increase in harmful algal blooms in recent years is to identify such changes from the naturally occurring interannual variability in phytoplankton species abundance. This requires relatively long time series that are, for the most part, lacking. (After Wyatt, 1995.)

3.6.3. *Harmful Phytoplankton Species as a Subset of the Total Phytoplankton Community*

If we assume that exceptional/harmful phytoplankton represent a relatively constant percentage of the total phytoplankton population, then another way in which we can investigate potential changes in the occurrence of exceptional/harmful events is to look for changes in the total phytoplankton abundance in a given area. Choosing this approach allows us to use phytoplankton biomass data (i.e. chlorophyll determined either by direct sampling or through remote sensing methods such as satellite-borne sensors) and/or primary production data. An advantage in using this type of data is that interannual variability in the primary production/algal biomass recorded is generally much less than that observed in connection with annual abundance of individual species.

This can be demonstrated by data presented in Heilmann *et al.* (1994). These workers examined biological and hydrographic characteristics of the water column during May 1988 (when the *Chrysochromulina polylepis* bloom discussed in Section 2.3.7 was underway) with similar characteristics observed in this region during the month of May in six different years distributed on either side of the bloom (Table 5). During none of the other

Table 5 Interannual variations (mean ± SD) in characteristics derived from chlorophyll *a* (μg l^{-1}) at the surface (2 m), integrated water column chlorophyll *a* (mgm) and primary production (mg C m^{-2} d^{-1}) measurements for May cruises in the period 1987–93. (After Heilmann *et al.*, 1994.)

Year	Surface chlorophyll	Surf. pigm./ max. pigm.*	Integrated chlorophyll	Primary production
1987	2.15 ± 0.51	0.60 ± 0.16	65 ± 7.2	908 ± 412
1988	0.58 ± 0.25	0.39 ± 0.28	19 ± 4.5	591 ± 242
1989	1.30 ± 1.04	0.79 ± 0.12	47 ± 29	—
1990	0.77 ± 0.24	0.57 ± 0.28	28 ± 9.6	699 ± 334
1991	1.29 ± 0.59	0.36 ± 0.19	29 ± 10	697 ± 355
1992	1.16 ± 0.44	0.55 ± 0.22	31 ± 4	838 ± 255
1993	0.38 ± 0.10	0.17 ± 0.09	41 ± 17	575 ± 232

* Surf. pigm./max. pigm. is concentration of chlorophyll *a* in surface (2.5 m) waters/greatest concentration of chlorophyll measured in the water column. When this value is 1, then the highest chlorophyll concentrations are found at the surface or chlorophyll is homogeneously distributed throughout the surface and the rest (or part) of the water column. The smaller this ratio, the larger the amount of chlorophyll found in a subsurface peak relative to the surface chlorophyll concentration.

Sampling in May 1988 occurred during the *Chrysochromulina polylepis* bloom in the Skagerrak–Kattegat. Note that there is no significant difference in the total primary production occurring during the *C. polylepis* bloom and in the other May studies.

six years was *C. polylepis* identified as a major component of the phytoplankton community. Despite the apparently very great differences in abundance of *C. polylepis*, no significant differences could be identified in the primary production occurring at this time during the different years. The reduced variability seen in primary production rates relative to abundances of individual species may make it possible to identify trends with shorter time series from data.

Unfortunately, there are not many long time series pertaining to marine phytoplankton primary production. This is largely due to the fact that the most sensitive method available for routine determination of primary production (^{14}C incorporation method) was first developed in the 1950s (Steemann Nielsen, 1952). As this method requires the use of radioactive material, it was some time before it found its way into routine use. Another problem with the ^{14}C incorporation method is that most laboratories have "adapted" the original method to fit their own needs and facilities. Thus, it is not certain that exactly the same measurements are being conducted by different laboratories. Indeed, there have been shown to be large interlaboratory differences in the estimates obtained for primary production even when made on the same water sample (Richardson, 1991). Nevertheless, some primary production and/or chlorophyll data do suggest that changes have occurred in the total phytoplankton biomass or activity in certain regions. Richardson and Heilmann (1995) have considered the data concerning primary production in the Kattegat. For this region, there is no continuous long-term data set. However, this was a region extensively studied by the developer of the ^{14}C method, Steemann Nielsen (1964). Thus, there are primary production measurements from the 1950s and the period 1954–60 was especially well studied (approximately fortnightly measurements made throughout these years). One of the stations considered by Steemann Nielsen is located in the middle of an area which has been extensively studied in the late 1980s to early 1990s (Heilmann et al., 1994).

Richardson and Heilmann (1995) examined, and as far as possible corrected for, the differences between the methods used during the two different periods in order to compare the magnitude of the primary production occurring in this region during the two studies. They concluded that a real increase (at least a doubling) in pelagic primary production had occurred in the Kattegat between the 1950s and the late 1980s to early 1990s. Furthermore, their analysis suggested that there was no significant difference in the measurements made during the winter (i.e. November–February) in the two studies, when light is usually limiting for phytoplankton growth (Figure 8). Considerable differences between the two studies were noted in primary production measurements made during most of the rest of the year. As the period March–October corresponds well to the period in which phytoplankton are nutrient-limited in the surface waters of the Kattegat (Richardson and Christoffersen, 1991), Richardson and Heilmann argue that the observed difference in pelagic primary production between the 1950s and the 1980 to 90s is a result of increased nutrient availability. Richardson (1996) examined the various natural and anthropogenically affected processes that might contribute to a change in nutrient availability for phytoplankton in the Kattegat and concluded that the observed increase in primary production in this region was most likely the result of eutrophication.

There are many examples of freshwater systems that respond to increased nutrient loading by increasing the total phytoplankton biomass production. The Kattegat is a marine region with limited water exchange with the open ocean and heavily influenced by freshwater input from land. Thus, it is perhaps not surprising that eutrophication effects seem to be evident here. Other studies carried out in marine waters heavily influenced by land processes have also indicated an increase in phytoplankton biomass and/or production which may be linked to eutrophication (i.e. Adriatic Sea: Justic et al., 1987; Wadden Sea

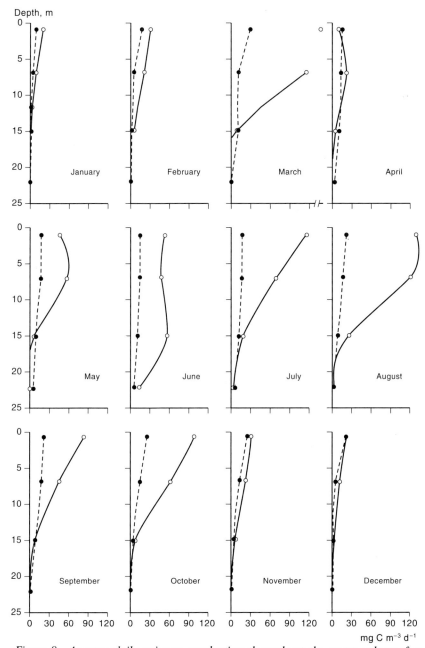

Figure 8   Average daily primary production throughout the water column for each month of the year in the period 1954–60 (●) and in the 1980s to early 1990s (○). (After Richardson and Heilmann, 1995.)

(southern North Sea): Cadée, 1986; Helgoland (south-eastern North Sea): Berg and Radach, 1985).

There does, then, appear to be evidence of an increase in total phytoplankton biomass/primary production in some marine areas which may be related to eutrophication. If exceptional/harmful bloom-forming phytoplankton species represent a constant proportion of the total phytoplankton community, then anthropogenically induced nutrient enrichment leading to an increase in total phytoplankton production will also lead to an increase in the production of these species. There is, of course, no way at present of determining whether or not the biomass of harmful species represents a relatively constant proportion of total phytoplankton biomass. However, there is no intuitive reason to suggest that the proportion of harmful phytoplankton in the total phytoplankton community is on the decrease.

3.6.4. *Does Eutrophication Select for Harmful Species?*

It has been suggested (Smayda, 1989, 1990) that eutrophication may actually increase the proportion of harmful species relative to the total phytoplankton biomass. The argument is based on the premise that non-diatom species are most often associated with harmful blooms and the fact that eutrophication does not increase the availability of silicon within the system. Since diatoms have an obligate requirement for silicon, eutrophication should not stimulate diatom abundance. Hence, blooms resulting from eutrophication should be comprised of non-diatom species. Circumstantial support for this argument can be drawn from a number of different regions (see review by Conley et al., 1993) as well as from mesocosm experiments (Egge and Aksnes, 1992). Thus, it appears that there is the potential for a change in the availability of silicon relative to other inorganic macronutrients to alter the biomass of various phytoplankton groups relative to one another. The weakness in the argument that a species shift resulting from a change in the relative abundance of silicon will selectively stimulate the occurrence of harmful phytoplankton is that diatoms have actually been shown to be the second most implicated taxonomic group in harmful blooms. However, they are relatively unimportant in terms of toxic blooms (see Table 1). Thus, it is possible that conditions selecting for non-diatom species may increase the probability of the occurrence of toxic phytoplankton species.

Less attention has been directed towards the role that other nutrient ratios may have played in controlling relative abundance of various phytoplankton groups in relation to each other. However, changes in, for example, $N:P$ ratios have been shown to be related to changes in phytoplankton species succession (Egge and Heimdal, 1994). In addition, certain contaminants in the marine environment are suspected to affect phytoplankton succession (Papathanassiou *et al.*, 1994). Thus, there is the potential for human activities to alter the natural pattern of phytoplankton species succession. However, the role that this influence on species succession may have in changing the relative abundance of exceptional/harmful bloom-forming species is not yet clear.

In addition to influencing species succession *per se*, and thus potentially selecting for exceptional/harmful bloom-forming species, the stimulation of phytoplankton biomass by eutrophication may not affect all phytoplankton groups equally. In general, toxic bloom-forming species are flagellates (see above). The relative proportion of flagellate species in the phytoplankton community is not constant throughout the year. The stimulation of phytoplankton biomass through eutrophication appears to be most pronounced during periods when flagellates dominate the phytoplankton community and thus eutrophication will, in some regions, selectively stimulate flagellate biomass.

A distinct pattern in the size structure of the phytoplankton community has been identified relative to various hydrographic conditions in temperate regions (Cushing, 1989; Kiørboe, 1993). Assuming that adequate light is available for phytoplankton growth, high nutrient conditions tend to promote large cells while low nutrient conditions select for small cells. In many temperate coastal ecosystems, nutrients become mixed into the surface waters as a result of winter storm activity. In the spring, as light conditions increase, the high nutrient/high light conditions give rise to a bloom of phytoplankton dominated by large cells ("spring phytoplankton bloom"). As cells sediment out of surface waters at the end of this bloom, nutrients are transferred from the surface to deeper layers of the water column. Wind conditions are generally calmer during the summer than during the winter. Thus, except under some storm events, nutrients remain in the deep layers and the high light surface waters become nutrient depleted. Because of their greater efficiency relative to large cells in uptake of nutrients under limiting conditions (Cushing, 1989; Kiørboe, 1993), small cells are favoured under these conditions.

There is often a pattern in the size distribution of phytoplankton in surface stratified waters in temperate regions where large cells dominate in the early spring and late autumn (when the onset of winter storms again brings nutrients into the surface waters) while small cells dominate during the summer. In the Kattegat (Figure 9), the small cells observed during summer are mostly flagellates, and dinoflagellates comprise a significant proportion of these (Thomsen, 1992). This dominance during the summer months of flagellates/dinoflagellates is probably typical for surface stratified waters in temperate regions but there are exceptions. Richardson and

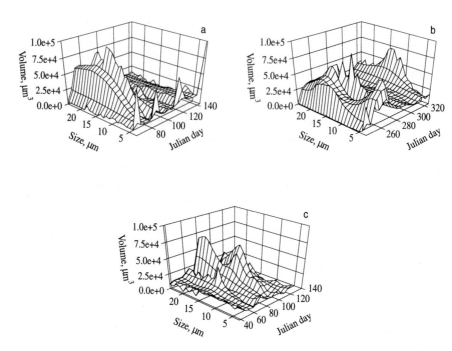

Figure 9 Particle size distribution in surface waters of the Kattegat. Total volume in a given size range (estimated spherical diameter) plotted as a function of a Julian day. (a) Spring/early summer 1994; (b) autumn 1994; (c) spring/early summer 1995. (Richardson, unpublished data.)

Heilmann (1995; see discussion above) have suggested that eutrophication resulting from human activities has led to an increase in phytoplankton production in the Kattegat throughout the annual period in which nutrients are predicted to be limiting for phytoplankton growth—a period which includes the entire summer months when flagellates dominate the phytoplankton community. Although it has been argued above that cultural eutrophication may alter the availability of silicon relative to nitrogen and phosphorus and, thus, select against diatoms, let us, for the sake of the present analysis, assume that all phytoplankton species occurring in this nutrient-limited period will be equally stimulated by the increased availability of nutrients. It can be predicted that the result of this fertilization will be an increase in the magnitude and/or duration of the spring bloom. Several authors (Cushing, 1989; Kiørboe, 1993) have pointed out that the growth rates of the herbivores capable of consuming large phytoplankton cells are slow relative to the growth rates of the algae and that there will be a relatively long lag time between the onset of an increase in growth rate in larger phytoplankton species and a build-up in the biomass of their predators. Thus, algal biomass will accumulate (i.e. a "bloom" will develop) before the grazing community is built up. If this scenario is correct, then eutrophication effects during the spring bloom period ought to be quantifiable in terms of an increase in the magnitude/duration of the spring bloom.

On the other hand, it has been pointed out (see Kiørboe, 1993, and references therein) that there is a much closer coupling between the growth rates of smaller phytoplankton and those of their predators. It is argued that this coupling will discriminate against a build-up in biomass ("bloom") of small phytoplankton cells as grazing pressure will develop at approximately the same rate as phytoplankton biomass. Thus, it could be predicted that the increased nutrient availability associated with eutrophication should not lead to blooms of the small flagellates that dominate during the summer months but, rather, to an increase in the biomass of herbivores or to an increased rate of carbon and nutrient turnover in the pelagic community. An exception here would be in the case of small algae that, for some reason, are not easily grazed.

A number of harmful and toxic species have been identified as having inhibitory effects on potential grazers (Fiedler, 1982; Ives, 1985, 1987; Huntley et al., 1986; Durbin and Durbin, 1989; Gallager et al., 1989; Hansen, 1989, 1995; Hansen et al., 1992; Buskey and Stockwell, 1993). In addition, herbivore distributions (Figure 10) in relation to the distribution of harmful/toxic species suggests, in some cases, avoidance of the phytoplankton by potential grazers (Nielsen et al., 1990). If, as Kiørboe suggests, it is grazing pressure that controls phytoplankton biomass during the periods in which small cells dominate the phytoplankton community, then we may predict that the only blooms likely to occur during summer months in stratified temperate waters are those of "unpalatable" species. Following the same line of reasoning we used above to argue that eutrophication may increase the magnitude and/or the duration of the spring bloom, we can argue that the magnitude and/or duration of blooms of unpalatable (i.e. some toxic/harmful species) may be stimulated during the summer months by eutrophication. Thus, a consequence of eutrophication in some coastal waters may be an increase in the occurrence of biomass accumulations (i.e. "blooms") of toxic or other phytoplankton species that are not easily grazed.

Equally, anthropogenic activities that potentially influence grazing activity (such as the use of pesticides that may be generally harmful to crustaceans) may be predicted to increase the probability of blooms of phytoplankton in periods when algal biomass is controlled by grazing pressure. Following concern that nutrient enrichment occurring in connection with aquaculture activities in Loch Linnhe, Scotland, might lead to an increase in algal blooms, Ross et al. (1993) constructed a model to

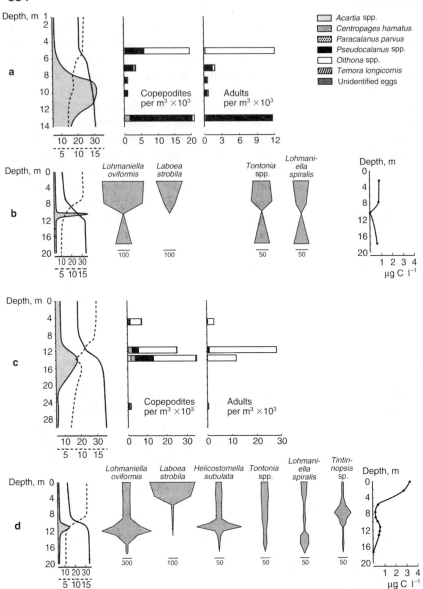

Figure 10 Vertical distribution of: (a) copepods, and (b) ciliates, in the water column at two stations examined during the 1988 *Chrysochromulina polylepis* bloom in the Skagerrak–Kattegat. At both stations, *C. polylepis* dominated the phytoplankton community (relative fluorescence (shaded area), salinity (solid line) and temperature (broken line) shown in left-hand panels). Total ciliate biomass in the water column is illustrated in the far right-hand panels of (b) and (d). When *C. polylepis* was not dominating in the phytoplankton community no avoidance at the pycnocline region was noted in copepods (c) or in ciliates (d). (After Nielsen *et al.*, 1990.)

examine the interaction of aquaculture and the pelagic community. This model suggested that phytoplankton biomass in the loch is determined by grazing pressure. Thus, the authors conclude that activities interfering with zooplankton activity and/or abundance are more likely to give rise to algal blooms than the nutrient enrichment associated with the aquaculture activities. Such activities might include the use of pesticides to combat sea-lice in aquaculture facilities. In such cases, however, it seems likely that the biomass of harmful and non-harmful species alike would be stimulated.

3.6.5. *Spreading of Harmful Species*

In addition to the suggestion that human activities may have increased the number of harmful blooms occurring, several workers (Anderson, 1989; Smayda, 1989; Hallegraeff, 1993) believe there has been a global spreading of harmful species during recent years. Quantifying such a spreading is difficult for several reasons. First, the increased awareness of harmful algae in recent years often makes it impossible to determine whether a "new" siting of a harmful alga is simply the first recorded observation of a species which has always been present in the area or whether the organism has only recently entered the region. Secondly, as pointed out in the introduction, it is known that phytoplankton distributions vary naturally over time.

Wyatt (1995) has considered the types of data series that would be required in order to quantitatively address what he calls the "global spreading hypothesis" relating to harmful algae. He considers first the time and space scales relevant to harmful algal blooms and to the spreading hypothesis (i.e. biogeography) (Figure 11). He then considers the scales of the data sets which would be required to examine these phytoplankton phenomena and relates them to existing monitoring programmes and models (Figure 11b). His analysis suggests that there are data collections underway and models either developed or under development that operate at scales that are relevant to addressing questions relating to the frequency of algal blooms in restricted areas. However, for addressing questions relating to biogeography, few data sets or tools are available. Thus, he concludes that quantitative confirmation of the global spreading hypothesis is not possible at the present time.

Nevertheless, there is strong circumstantial evidence, especially from Australian waters, that several species of toxic dinoflagellates are new introductions (references reviewed by Hallegraeff, 1993). It has been shown that resting cysts of a number of toxic dinoflagellates are able to survive transport over great distances in ballast water tanks (Hallegraeff and Bolch, 1992) and Hallegraeff and Bolch (1991) estimated that one

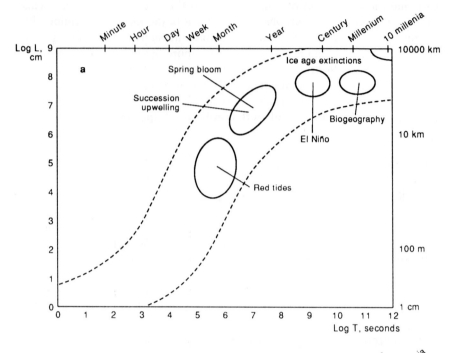

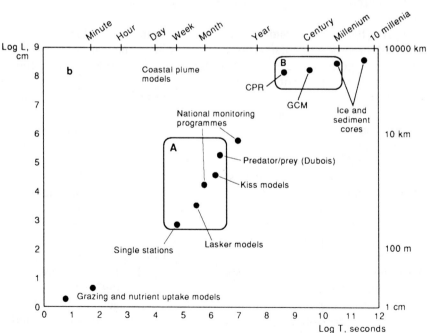

single ballast tank contained over 300 million viable toxic dinoflagellate cysts. Thus, a likely mechanism for new introduction and transfer of harmful species has been identified. The case for this type of transfer is further strengthened by rRNA sequencing studies (Scholin and Anderson, 1991) that have shown strong similarities between *Alexandrium minutum* cultures isolated from the Spanish and Australian coasts.

Other anthropogenic vectors for the potential transport of toxic algae or their cysts have also been identified (i.e. the transfer of shellfish stocks containing phytoplankton in their digestive organs (Hallegraeff, 1993)). Thus, while quantitative demonstration of the global spreading hypothesis may not be possible on the basis of existing data, there does seem to be a potential for anthropogenic activities to increase the global distribution of harmful species.

4. EXCEPTIONAL BLOOMS IN THE CONTEXT OF SEASONAL BLOOM DEVELOPMENT

4.1. Background

It has long been realized that phytoplankton blooms are a natural phenomenon occurring within the pelagic marine ecosystem (Mills, 1989). Cushing (1959) recognized and produced a theoretical model describing the different patterns in the seasonal distribution of blooms as a function of latitude (Figure 12). This generic model is still widely accepted as a basis for understanding of bloom occurrence in different latitudinal regions and appears in most introductory textbooks on marine biology. However, there is increasing appreciation for the fact that the general patterns described are modified by different hydrographic conditions (Cushing, 1989; Richardson, 1985). Blooms are important in channelling carbon (energy) into the marine food web and it is hypothesized that the high productivity (in terms of fisheries yield) of temperate and arctic marine ecosystems compared with tropical systems may result from the more intense blooms that occur in these regions. Following this hypothesis, the occurrence of intense algal blooms may actually be a prerequisite for highly productive marine ecosystems. Intense blooms occur at temperate and polar latitudes because the

Figure 11 Stommel diagram showing spatial and temporal scales of some planktonic phenomena. (a) Space–time scales relevant to population dynamics and various modelling/monitoring programmes. (b) Time-scales relevant to the global spreading hypothesis (see text). Only the continuous plankton recorder (CPR) programme and global climate models (GCM) are identified as operating as space–time scales relevant to addressing the global spreading hypothesis. (After Wyatt, 1995.)

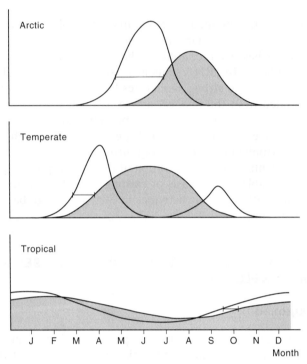

Figure 12 Seasonal variations in phytoplankton and zooplankton grazers (hatched) in different latitudinal regions. Horizontal bar indicates delay period between increases in zooplankton and phytoplankton biomasses. (After Cushing, 1959.)

hydrographic processes occurring in these regions allow phytoplankton, nutrients and light to be present simultaneously in the surface waters prior to the development of grazing populations (Gran and Braarud, 1935; Sverdrup, 1953) while the permanent stratification in tropical seas spatially separates light and nutrients.

The general mechanisms leading to the occurrence of the spring (vernal) phytoplankton blooms in temperate and arctic waters are predictable and well studied and it is not the purpose of the present chapter to consider these blooms in detail. Some consideration of vernal blooms is, however, relevant in the context of exceptional/harmful blooms both because some of the species which can occur in or immediately following vernal blooms can be harmful (i.e. *Phaeocystis*) and because sedimentation of vernal blooms can, under some conditions, lead to hypoxia or anoxia.

Following the lead of Cushing (1959), the seasonal development of phytoplankton blooms as it relates to exceptional/harmful blooms will here be considered for each latitudinal region.

4.2. Temperate Regions

4.2.1. *Spring Bloom*

As indicated above, the most striking non-exceptional bloom that occurs annually in temperate waters is the spring bloom. In this bloom, chlorophyll concentrations often reach levels of about two orders of magnitude greater than those observed during the dark winter months. The spring bloom in temperate regions is recognized to be important for the secondary production occurring in these regions and this bloom produces a substantial proportion of the organic material annually entering the food web (Figure 13) (for other examples, see Parsons *et al.*, 1984).

The growth rates of herbivores that are supported by relatively large phytoplankton that usually dominate during the spring bloom (Kiørboe, 1993) are temperature-dependent (Huntley and Lopez, 1992). Because of the low temperatures at the time of the spring bloom, grazing pressure is low at this time. Thus, much of the organic material produced during the spring bloom may sediment directly to the bottom rather than being degraded in the pelagic zone (Figure 14). In some cases, this sedimentation and subsequent degradation can give rise to hypoxia or anoxia which can cause mortality of benthic marine organisms (Morrison *et al.*, 1991).

While the occurrence of the spring bloom is predictable, it has long been

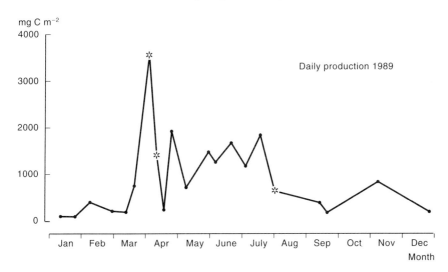

Figure 13 Annual cycle of water column primary production at a station in the southern Kattegat in 1989. Stars indicate weeks in which more than one sample was taken. For these weeks, an average of all primary production measurements is plotted. (From Richardson and Christoffersen, 1991.)

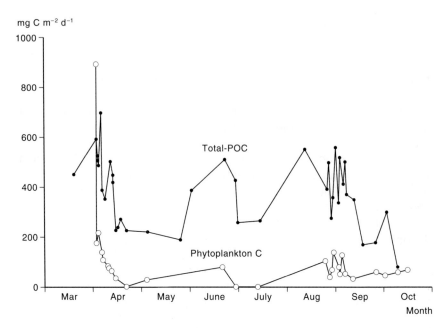

Figure 14 Sedimentation of particulate organic carbon (POC) and phytoplankton carbon at a station in the southern Kattegat during the period March–October 1989. Note that it is only during the spring bloom that substantial quantities of intact phytoplankton reach the bottom of the water column (~ 26 m). (After Olesen and Lundsgaard, 1992.)

recognized that its timing varies from year to year (Bigelow *et al.*, 1940). Until recently, however, there has been little interest shown in how these differences in timing of the spring bloom might affect the role of these blooms in the marine ecosystem. Townsend *et al.* (1994) hypothesized that the timing of the bloom is important for the ultimate fate of its products and, thus, is potentially critical for the survival of the organisms dependent upon the carbon flux initiated by the bloom. These workers developed a numerical simulation model to examine the effect of the temperature at the time of the bloom on the fate of the bloom products. Their results (Figure 15) indicate that differences in the timing of the bloom between years can have a very significant influence on the fate of the phytoplankton comprising the spring bloom. Using input data from a "cold" year (1974) and a "warm" year (1978) for the shallow coastal waters of the Gulf of Maine on the North American east coast, these workers compared model predictions concerning the timing and magnitude of the spring phytoplankton and its fate. During the cold year (water temperatures during the bloom 0–2°C), the phytoplankton bloom occurred about one week later than in the warm

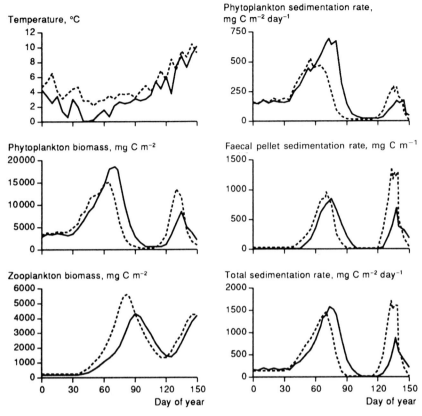

Figure 15 Results from Townsend *et al.*'s (1994) model demonstrating differences in "export" of phytoplankton carbon in "cold" (solid line) and "warm" (broken line) water spring blooms in shallow inshore waters of the Gulf of Maine. The double phytoplankton and zooplankton peaks result from the collapse of the first phytoplankton pulse due to light limitation by self-shading, followed by recovery and a second peak.

(water temperatures 2–4°C) year. In addition, the zooplankton bloom which occurs following the phytoplankton bloom was delayed by about two weeks in the cold year as opposed to the warm. Net phytoplankton production was greater during the cold year (7%) and zooplankton grazing was 11% less. As a result, Townsend *et al.*'s model indicates that about 30% more of the phytoplankton associated with the spring bloom settled to the bottom during the cold year than in the warm year.

Conventional wisdom indicates that the spring bloom in temperate offshore waters occurs as a single peak following the onset of stratification of the water column. However, another interesting result of Townsend *et al.*'s model was the suggestion that the spring bloom may occur as a series of "pulses" rather than a single peak during some years and that stratification is not a necessary prerequisite for bloom formation. Support for the model conclusions can be found in a number of reports of field data (Townsend *et al.*, 1992; Garside and Garside, 1993). Townsend *et al.* (1994) suggest that spring bloom pulses occur in the period prior to the establishment of stratification when the wind is below a certain threshold level (and phytoplankton in surface waters are not mixed deeply in the water column). It is hypothesized that these pulses end either because of nutrient exhaustion or self-shading but that subsequent pulses may occur when the necessary conditions are again met. The "chain" of pulses that may occur in any given year will end with the pulse associated with the onset of stratification as nutrients will no longer be delivered to the high light surface waters above the pycnocline until stratification is broken down.

If the spring bloom occurs as a series of pulses in offshore temperate waters, as suggested by these workers, then the magnitude of the production occurring in association with the spring bloom(s) in some regions may be even greater than previously believed. In addition, interannual variability in weather conditions (especially in relation to wind and insolation) will be critical in determining both the magnitude of the spring bloom(s) and the fate of the bloom(s) in any given year.

This example concerning newer studies on the conditions surrounding the spring bloom emphasizes how limited our understanding of bloom dynamics actually is—even when the bloom in question is a predictable and well-known phenomenon. This is an important point to consider in light of the major national and international efforts currently underway to understand the occurrence and dynamics of *harmful* algal blooms. Perhaps it would be wiser for these programmes to focus on a better understanding of the dynamics of algal blooms, in general, rather than to focus on the distinct but very heterogeneous (see Sections 1–3) subgroup of harmful algal blooms.

4.2.2. *Summer Blooms in Offshore Surface Waters*

The traditionally accepted distribution of algal biomass in offshore surface waters of temperate regions (Figure 12) indicates that the period following the spring bloom until the autumnal breakdown of stratification is characterized by low phytoplankton biomasses. Despite these low biomasses it has become clear in recent years, with the help of satellite remote sensing (Figure 16), that huge blooms of coccolithophorids occur regularly during the summer and early autumn months in temperate and subarctic regions and it is possible that these blooms can account for a

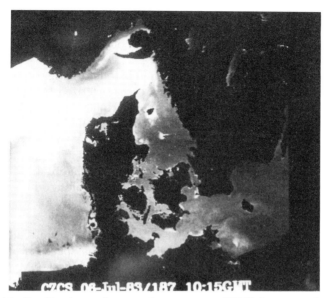

Figure 16 Satellite (coastal zone colour scanner, CZCS) image on 6 July 1983 showing the eastern North Sea, Skagerrak, Kattegat and western Baltic Seas. The light area in the North Sea extending into the Skagerrak shows the area affected by a bloom of the coccolithophorid, *Emiliania huxleyi*. Satellite image provided by T. Aarup.

substantial proportion of the primary production occurring in some years in these regions (Holligan *et al.*, 1983, 1989; Brown and Yoder, 1994).

Coccolithophorids comprise a group within the Prymnesiophyceae which is characterized by the fact that the organisms form calcite particles intracellularly which are then extruded. Together, these extruded "coccoliths" form a "coccosphere" around the cell and it is light reflected from these coccoliths (both those in the coccospheres and those released into the water column) that produces the strong signal observed in the satellite pictures. Brown and Yoder (1994) examined satellite pictures and estimated that a spectral signal similar to those recorded in association with well-studied coccolithophorid blooms annually covered an average of $1.4 \times 10^6 \, km^{-2}$ of the world's ocean area during the period 1979–85. Of these signal recordings, 71% were from subarctic latitudes. The most affected area was the subpolar North Atlantic. However, a similar spectral signal was recorded in the North Pacific and Southern Ocean, as well as off the Chilean and New Zealand coasts. There did not appear to be any indication in the spectral signal data of coccolithophorid blooms in the open equatorial waters. These workers emphasize that there is good biogeographical data to support their conclusion that the "coccolithophorid-like" spectral signal observed in temperate and subarctic latitudes actually represents coccolithophorid blooms. On the other hand, it is not known whether or not the signals observed at the lower latitudes represent coccolithophorid blooms.

Emiliania huxleyi is the most common bloom forming coccolithophorid in the North Sea where its blooms are characterized by only moderate concentrations of chlorophyll ($1-2 \mu g^{-1}$) (Holligan *et al.*, 1993b). This phytoplankter also forms blooms throughout the rest of the North Atlantic. Interestingly, although these blooms are "non-exceptional" in the sense that they are not noticeable to the general public because of their effects, they may well be considered to be "harmful" under some circumstances. This is because coccolithophorids (along with a number of other phytoplankton species including *Phaeocystis*) are known to produce dimethylsulphide (DMS).

DMS is a sulphur-containing gas that may be involved in climate regulation. A number of different phytoplankton species are implicated in DMS production. It has been suggested that most DMS producers belong to the Prymnesiophyceae or Dinophyceae (Keller *et al.*, 1989) and there is some indication that DMS production may be greatest at the late stages of a bloom (Matrai and Keller, 1993). The phytoplankton produce a precursor to DMS, dimethylsulphonium proprionate (DMSP) which may be enzymatically cleaved to form DMS which is volatile (Andreae, 1990). Thus, the production of DMS will affect the sea-to-air flux of sulphur and, potentially, the geochemical sulphur cycle. As DMS is an important precursor for cloud condensation nucleii in the maritime atmosphere (Brown and Yoder, 1994), it is argued that DMS release may influence regional albedo (Charlson *et al.*, 1987). In addition, as a source of sulphur to the atmosphere, DMS may be implicated as a potential cause of "acid rain". Brown and Yoder (1994) concluded that DMS sulphur production by coccolithophorid blooms produces only a minor portion (0.03–0.07%) of the total amount of DMS sulphur (0.5–1.1 Tmol DMS-S (Tmol = 32×10^{12} mols)) produced annually on a global scale. However, there is evidence that DMS production by coccolithophorid and other phytoplankton blooms may be an important factor in determining sulphur cycling at the regional level (Holligan *et al.*, 1993a; Malin *et al.*, 1993; Brown and Yoder, 1994).

Because coccolithophorids produce calcite plates or "coccoliths", considerable interest has also been directed towards this group of phytoplankton in recent years with respect to the potential effect that they may have on oceanic carbon cycling (Fernández *et al.*, 1993; Holligan *et al.*, 1993a; Robertson *et al.*, 1994; van der Wal *et al.*, 1995). It is clear that the sedimentation of coccoliths is potentially an important mechanism for removal of carbon from the water column. However, on the basis of theoretical chemical considerations (see discussion by Robertson *et al.*, 1994), the calcification process involved in the production of coccoliths should cause a drop in alkalinity, thus shifting the inorganic carbon equilibria in the sea water medium where a coccolithophorid bloom is underway in the direction of the dissolved gas. By affecting the partial pressure difference with respect to CO_2 in the atmosphere and in sea water, a coccolithophorid bloom may actually decrease CO_2 uptake in sea water from the atmosphere relative to when non-calcifying organisms are present.

Robertson *et al.* (1994) have, indeed, demonstrated that a coccolithophorid bloom occurring in the north-east Atlantic in 1991 apparently reduced the air–sea gradient in dissolved CO_2 by a mean of 15 μatm. These workers further estimated, by comparing measurements taken in 1990 (when few coccolithophorids were present) and those made during the 1991 bloom, that the presence of a bloom such as that observed in 1991 could reduce the uptake of atmospheric CO_2 over the spring–summer period by about 17%. Given these observations, it seems clear that knowledge of the magnitude of the occurrence of coccolithophorids in the total phytoplankton community is important in developing models to describe global carbon cycling.

These examples concerning the potential influence of coccolithophorid blooms on the geochemical cycling of sulphur and carbon illustrate that, while it is the local effects of algal blooms where and when they come in contact with human activities that attract public attention, changes in the magnitude or frequency of phytoplankton blooms may have more than local implications. Such changes may affect global geochemical cycling and, thus, climate. A better understanding of the role that phytoplankton blooms may play in the geochemical cycling processes occurring in the sea is, thus, necessary in order to predict the potential impact of global warming on climate processes and *vice versa*.

4.2.3. *Subsurface Blooms in Summer Months*

A further modification of the general picture of phytoplankton biomass distribution and blooms in temperate waters that has become obvious in recent years is the fact that large phytoplankton biomasses (often with peak chlorophyll concentrations similar to or greater than those recorded during the spring bloom) can occur during the late spring and summer months in association with the pycnocline (Holligan and Harbour, 1977; Pingree *et al.*, 1982; Holligan *et al.*, 1984; Richardson, 1985; Cushing, 1989; Riegman *et al.*, 1990; Kaas *et al.*, 1991; Bjørnsen *et al.*, 1993; Nielsen *et al.*, 1993b). In some cases, these large biomasses associated

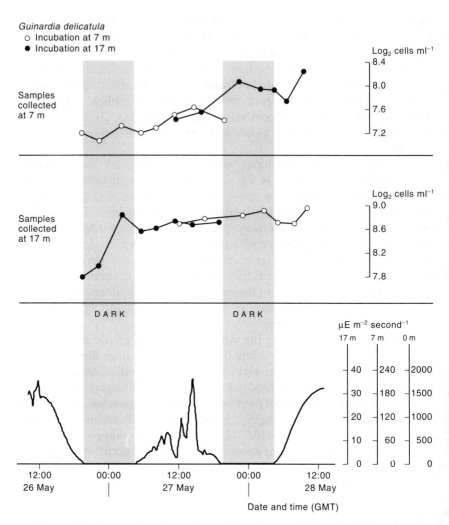

Figure 17 Changes in cell number of the diatom, *Guinardia delicatula*, when naturally occurring phytoplankton populations from 7 and 17 m at a station in the Irish Sea were enclosed in sterile plastic bags and returned to either 7 or 17 m for incubation. Sampling was carried out with the incubation bags remaining at depth. During the first 24 h sampling period, samples were returned to their collection depth. *G. delicatula* was the dominant species at 17 m and was common at 7 m. At 17 m, the *G. delicatula* population underwent a synchronized doubling during the dark period. Although an increase in cell numbers was also recorded at 7 m, it did not amount to a doubling of the population. After 24 h, new samples were taken from the water column. At both depths, the numbers of *G. delicatula* in the incubation bags after 24 h were similar to those in the water column suggesting that the changes recorded in the bags reflected those occurring in nature. The sample taken from the water column at 7 m after 24 h was enclosed in a plastic

with the pycnocline are essentially unialgal as was the case for the *Chrysochromulina polylepis* bloom recorded in the open Kattegat in 1988 (Kaas *et al.*, 1991).

There has been considerable scientific discussion as to whether these subsurface chlorophyll peaks represent accumulations of senescent and moribund phytoplankton sinking out of the water column (Bienfang *et al.*, 1983; le Fèvre, 1986) or represent active populations. However, the presence of actively photosynthesizing phytoplankton and/or high concentrations of grazers in association with these pycnocline peaks has been demonstrated on a number of occasions (Holligan *et al.*, 1984; Kiørboe *et al.*, 1990a; Richardson and Christoffersen, 1991; Nielsen *et al.*, 1993a). Active photosynthesis is, of course, not necessarily proof of *in situ* growth of phytoplankton along the pycnocline. Although fewer studies have examined growth than photosynthesis in subsurface chlorophyll peaks, there is also evidence that some species grow better at depth within the water column.

Heath (unpublished data) took samples from a subsurface phytoplankton peak at 17 m dominated by *Guinardia delicatula*. After placing them in sterile plastic incubation bags, he returned one sample to the sampling depth and one to 7 m. He also took samples from 7 m where *G. delicatula* was also present but at lower concentrations and treated them in the same manner as those taken from 17 m. All incubation bags were fitted with plastic tubes which extended up to the ship's deck, thus allowing sampling from the incubation bags without removal of the incubation bags from the incubation depths. The results (Figure 17) showed better growth in terms of increase in cell numbers in the 17 m samples both for those samples taken from 17 m and those taken from 7 m.

While this growth may not, strictly speaking, be "rapid" because of the limited *in situ* photon flux densities recorded at the pycnocline, there appears to be the potential for many of the phytoplankton biomass accumulations observed at the pycnocline to be the result of *in situ* growth. Thus, we may consider them as phytoplankton "blooms". Quantification of the extent of these pycnocline "blooms" and their importance in terms of the overall carbon fixation by phytoplankton is difficult as this pycnocline layer is often very narrow and difficult to sample. However, Richardson and Christoffersen (1991) estimated that pycnocline blooms

---

bag and transferred to 17 m. Similarly, the 17 m sample was incubated at 7 m. During the second incubation period, there was again a greater increase in *G. delicatula* cell numbers at 17 m (i.e. sample taken from 7 m) than at 7 m (sample taken from 17 m). These results indicate faster *in situ* growth for *G. delicatula* at 17 m than at 7 m. Light measurements at the various depths and at the surface were made using a cosine collector. (M. R. Heath, unpublished data.)

in the Kattegat were responsible for approximately 33% of the primary production occurring in this region. On a cruise in the North Sea in May 1992, Richardson et al. (unpublished data), recorded chlorophyll concentrations in association with the pycnocline of $>10 \mu g \, l^{-1}$. At a number of stations, these workers estimated that the photosynthesis occurring within the pycnocline peak accounted for approximately 75% of total water column primary production (Figure 18).

As indicated above, these subsurface "blooms" can be dominated by single species. It is not only harmful species that can come to dominate in these subsurface blooms. However, it is interesting to note that there are a number of harmful phytoplankton that often appear in association with the pycnocline. Examples of such phytoplankton include *Alexandrium* sp. (Anderson and Stolzenbach, 1985; Carreto et al., 1986; Yentsch et al., 1986), *Gyrodinium aureolum* (Holligan, 1979; Lindahl, 1983, 1986; Richardson and Kullenberg, 1987; Dahl and Tangen, 1993) and *Chrysochromulina polylepis* (Dahl et al., 1989; Kaas et al., 1991). In some cases (see, e.g., Lindahl, 1986; Richardson and Kullenberg, 1987; Dahl and Tangen, 1993) it appears that these pycnocline blooms may provide the "seed" population that gives rise to the exceptional or harmful coastal blooms. A mechanism for this seeding appears to be that the subsurface phytoplankton population becomes mixed into surface waters in frontal zones where stratified and mixed waters meet (Holligan, 1979). Given the proper combination of wind and current conditions, these populations can then be transported inshore (Dahl and Tangen, 1993). Thus, an important point to note with respect to the exceptional blooms that attract so much attention in surface waters at coastal sites is that these blooms are not always initiated at the geographic location where they are observed (see Section 5.1).

It is not known why phytoplankton sometimes appear to "bloom" under the low light conditions associated with the pycnocline. However, Richardson et al. (1983) argued, on the basis of a literature survey of data relating to the light requirements for growth of various marine phytoplankton species, that dinoflagellates, in general, have a relatively low light compensation point for growth (mean 6.6 μmol photons m^{-2} s^{-1}) and that, on average, growth saturation for this class occurs at about 47 μmol photons m^{-2} s^{-1} in laboratory studies (Figure 19). Thus, the phytoplankton class most often associated with harmful and toxic blooms (see Table 1) may have a light requirement that is best met when these phytoplankton are in an environment with a relatively low light climate and with limited extremes in the photon flux densities occurring during the day. Such a light climate can often be found near the pycnocline in temperate waters. Thus, although conventional wisdom argues that surface waters with high light intensities should be the most conducive to

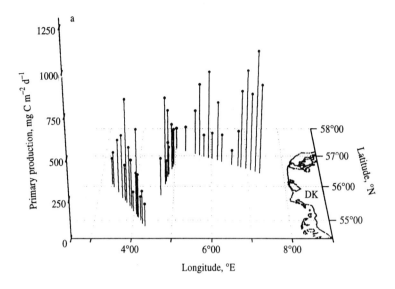

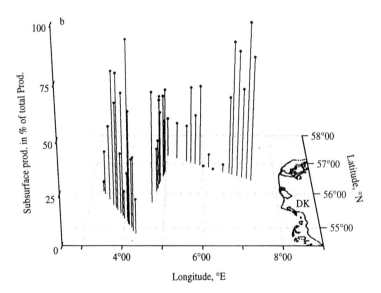

Figure 18 Water column primary production (a) and the percentage of total water column production estimated to occur in the subsurface chlorophyll peak (b) at various stations in the North Sea in May 1992 (K. Richardson *et al.*, unpublished data.)

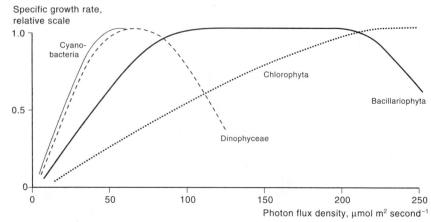

Figure 19 Relative specific growth rates plotted as a function of photon flux density in four major taxonomic groups of marine phytoplankton. (After Raven and Richardson, 1986.)

phytoplankton growth, many dinoflagellates (harmful and non-harmful alike) may actually "prefer" the conditions found further down in the water column.

In laboratory studies, it has been shown (Rasmussen and Richardson, 1989) that the toxic dinoflagellate *Alexandrium tamarense* (formerly *Gonyaulax tamarensis*) actively accumulates at the pycnocline in an artificial water column when it is introduced in the water column below the pycnocline. When introduced above or in the absence of a pycnocline, the algae accumulated at the surface of the water column where photon flux densities were highest. By adding dye to the surface waters to reduce the incident photon flux density at the pycnocline, these workers showed that the accumulation at the pycnocline was not caused by an inability of the organisms to cross the density gradient. When the dye was present, the dinoflagellates penetrated the pycnocline and concentrated in the high-light surface waters (Figure 20). Thus, it would appear that this organism actively concentrates at a pycnocline if certain conditions concerning light availability are met.

There could be some advantage to such a strategy in nature for an obligate autotroph in a stratified water column as nutrients will normally be available from waters below the pycnocline. Thus, accumulation at a pycnocline where a minimal light requirement is met will optimize the chances of being in the part of the water column where there is the greatest chance of having access to adequate nutrient and light supplies. It is not known how widespread the response observed in *A. tamarense* is within the phytoplankton flora but there is the suggestion of a similar response

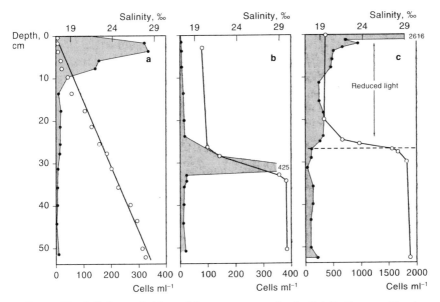

Figure 20 Salinity and *Alexandrium tamarense* depth distribution profiles in: (a) an artificial water column with gradually increasing salinity from 19‰ near the surface to 29‰ at the bottom. (b) Heterogeneous water column with a salinity difference across the halocline of 10‰. (c) The same water column as in (b) but with the addition of food colour above the pycnocline. Areas shown above the stippled line represent the region where food colour reduced the incident light. ●, Concentration of cells; ○, salinity. (After Rasmussen and Richardson, 1989.)

by the dinoflagellate *Gyrodinium aureolum* (Nielsen *et al.*, 1993a), an organism which has been associated with numerous fish kills (Tangen, 1977; Jones *et al.*, 1982; Dahl and Tangen, 1993).

4.3. Tropical and Subtropical Regions

Cushing's (1959) model of phytoplankton bloom development in open tropical waters suggests the occurrence of only low biomass and very little seasonal fluctuation in these biomasses. However, many of the considerations relating to phytoplankton bloom development at the pycnocline in temperate regions are probably also applicable to tropical regions where it is common that the largest phytoplankton concentrations observed in the water column are located at depth and in association with a pycnocline or nutricline (Jochem and Zeitzschel, 1993; Michaels *et al.*, 1994).

While numerous studies in open tropical waters have indicated the presence of subsurface phytoplankton peaks, little effort has been devoted to quantification of the potential role that these subsurface populations might play in the overall carbon dynamics in tropical regions. However, Goldman *et al.* (1992) isolated three large diatoms (*Stephanopyxis palmeriana*, *Pseudoguinardia recta* and *Navicula* sp.) from the Sargasso Sea and, on the basis of results concerning the growth characteristics of these phytoplankton in the laboratory, suggested that a single 21 day bloom of one of these species would supply the entire new production estimated to occur annually in a region such as the Sargasso Sea. For *S. palmeriana*, about 50% of the particulate organic carbon production occurred after the cultures of the organism had entered the stationary phase. While this theoretical exercise does not demonstrate that blooms of these organisms actually occur in the Sargasso Sea, it does indicate the potential importance of bloom events on carbon dynamics at these latitudes.

Recent studies suggest also that bloom formation in surface waters of the tropical oceans may be much more predictable than previously thought. Longhurst (1993) used coastal zone colour scanner (CZCS) satellite images to examine the timing of tropical surface water blooms occurring in the tropical oceans. He identifies different patterns for the tropical Pacific, Atlantic and Indian Oceans and relates these patterns to the different hydrographic regimes that dominate in the three regions. Generally, offshore tropical waters have not been considered as sites for exceptional/harmful phytoplankton blooms. However, the recent demonstration by Hawser and Codd (1992) of toxicity associated with a bloom of *Trichodesmium thiebautii* in tropical waters suggests that these regions also experience blooms that would be considered as exceptional or harmful if they occurred in an area where they attracted human attention.

4.4. Arctic and Antarctic Regions

The Arctic and Antarctic regions, also, have not really been considered as sites for the occurrence of exceptional/harmful phytoplankton blooms. Again, however, the lack of reports of such blooms probably reflects the fact that their occurrence would be unlikely to interfere with any human activities. Support for this assertion comes from the fact that the dominant phytoplankter in the spring bloom in regions of both the Arctic and Antarctic is often reported to be *Phaeocystis* sp. (Wassmann *et al.*, 1990; Smith, 1993)—a genus considered to be an exceptional/harmful bloom former in temperate waters. While there are differences in the timing and characteristics of the spring bloom between regions (see, e.g., Wassmann *et al.*, 1991), the general pattern of a single bloom has been observed in open waters in a number of Arctic (e.g. the Greenland Sea: Smith, 1993; arctic Canada: Hsiao, 1988; Barents Sea: Wassmann *et al.*, 1990) and antarctic regions (i.e. western Ross Sea: Nelson and Smith, 1986; Bransfield Strait: Smetacek *et al.*, 1990). Nevertheless, it is increasingly apparent that such single peaked vernal blooms do not occur at all polar localities and that the open polar waters may not be responsible for as much primary production as previously assumed (Smetacek *et al.*, 1990; Codispoti *et al.*, 1991).

It is now accepted that phytoplankton blooms occur sporadically in association with hydrographic events or features throughout much of the year in many polar regions. In particular, there is now increased awareness concerning the role of the ice-edge and the occurrence of blooms. Niebauer (1991) has estimated that approximately 7% of the world's oceans are subject to the advance and retreat of sea ice. The melting of sea ice can stimulate phytoplankton blooms by promoting stability in the nearby water column through the delivery of a layer of relatively fresh water to the region. Thus, blooms can occur locally in the open waters bordering the ice-edge (Conover and Huntley, 1991; Niebauer, 1991).

Even more dramatic than the blooms occurring in association with melting ice, is the occurrence of phytoplankton blooms associated with ice formation (Smetacek *et al.*, 1992). Sakshaug (1989) termed such blooms "superblooms" as the concentrations of chlorophyll recorded in them can exceed $100\,\mu g\,l^{-1}$. Such blooms have been recorded in the Weddell Sea on a number of occasions (Figure 21). If blooms of such magnitude were occurring in coastal regions of the more densely populated temperate lands, they would almost certainly be considered as being exceptional.

Smetacek *et al.* (1992) examined a "superbloom" of centric diatoms (*Thalassiosira antarctica*, *Porosira pseudodenticulata* and *Stellarima microtrias*) associated with ice platelets underlying pack ice. The maximum chlorophyll concentration recorded in this bloom was $220\,\mu g\,l^{-1}$ and the diatom cells appeared not to be attached to the ice platelets. The bloom covered an area of approximately $20\,000\,km^2$. These workers argue that drifting platelet ice offers an ideal habitat for localized algal blooms and that phytoplankton blooms occurring in association with platelet ice under pack ice provide a potentially important food source for grazers (krill) during periods when the rest of the Weddell Sea is at its most barren.

The relationship between ice algae and the open water phytoplankton blooms described by Cushing's (1959) theoretical model is not clear. However, there is at least some evidence that ice algae may "seed" the water column with phytoplankton prior to the spring bloom (Michel *et al.*, 1993). Thus, the importance of algal blooms occurring in association with

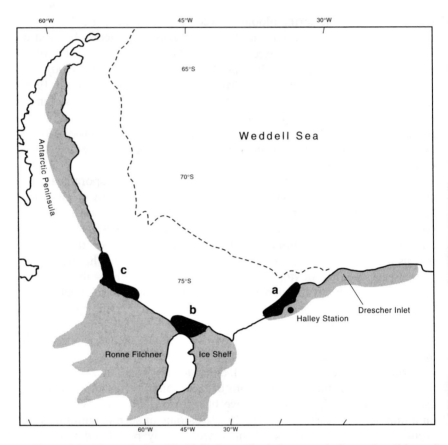

Figure 21 Map of the Weddell Sea. Shaded areas indicate localities of superblooms observed in: (a) October/November 1986 by Smetacek *et al.* (1992); (b) February/March 1983 by Sakshaug (1989); and (c) February/March 1968 by El-Sayed (1971). Dotted areas indicate floating ice shelves, the likely production sites of platelet ice. The stippled line represents the 1000 m isobath. (After Smetacek *et al.*, 1992.)

the edges of the ice pack may be much more important to total productivity and the structure of polar ecosystems than previously realized.

The data relating to the distribution of algal blooms in time and space discussed above suggest that phytoplankton blooms can and do occur whenever and wherever adequate light and nutrient conditions for the bloom species are coincident. In addition to occurring in surface waters, the examples cited above indicate that appropriate conditions for some phytoplankton blooms can also be found, at least at times, at such unlikely sites as under ice cover and in association with hydrographic features deep in the water column. Exceptional or harmful phytoplankton blooms have most often been recorded in relatively densely populated coastal areas. However, there is at least circumstantial evidence that such blooms are occurring—but go unnoticed—in more remote areas such as the open waters of the tropics and in polar regions.

5. BLOOM DYNAMICS WITH SPECIAL REFERENCE TO EXCEPTIONAL/HARMFUL SPECIES

5.1. Bloom Initiation

In the post-mortem following many exceptional or harmful bloom events, unusual wind/weather/hydrographic conditions are identified as having preceded or accompanied the bloom in question (Cosper *et al.*, 1989; Maestrini and Granéli, 1991; Rhodes *et al.*, 1993). Often, it is not possible to identify a direct (cause and effect) link between the unusual event and the bloom. However, regions exhibiting specific hydrographic conditions (such as coastal upwellings and river discharge areas) are often identified as regions susceptible to phytoplankton blooms (Fraga, 1993). The association of harmful bloom species and specific hydrographic phenomena such as pycnoclines and coastal upwelling features has led to the suggestion that some of the predicted hydrographic changes which would be associated with climate change (global warming) may increase the occurrence of the toxic dinoflagellate, *Gymnodinium catenatum*, along the Galician coast (Fraga and Bakun, 1993).

In some cases, it appears that the wind, weather, tidal or other hydrographic conditions have been related to initiation of nearshore blooms by delivery of the organism into the bloom region (Balch, 1981; Lindahl, 1986, 1993; Richardson and Kullenberg, 1987; Dahl and Tangen, 1993; Delmas *et al.*, 1993; Keafer and Anderson, 1993; Taylor *et al.*, 1994). In some situations, large concentrations of phytoplankton may be transported inshore so that one can say that the bloom, itself, is advected to the coastal site. In other cases, it may be more appropriate to speak of a "seed" population being transported to the coastal site. In addition to transport of phytoplankton blooms from offshore to inshore regions, long-shore transport of blooms from one coastal region to another has also been recorded (Franks and Anderson, 1992).

Coastal blooms of *Gyrodinium aureolum* (see Section 3.3.7) may often result from advection of an offshore population to the coast (Lindahl, 1986, 1993; Richardson and Kullenberg, 1987; Dahl and Tangen, 1993).

This species is known to be associated with shelf sea fronts where it first occurs in dense concentrations at the pycnocline (see Section 4) and later blooms in surface waters associated with the front. It is the blooms associated with surface waters at or near the front that can be advected under the proper wind conditions towards shore. That some coastal blooms of this organism have an offshore origin can often be seen in the salinity characteristics of the associated water (Figure 22).

A number of other potentially harmful dinoflagellates are also known to form pycnocline and surface blooms in association with offshore frontal regions. For a number of these species, there are specific coastal bloom incidents for which the source population can be traced to an offshore bloom. Rasmussen and Richardson (1989) suggested that some dinoflagellates may be adapted to particularly exploit subsurface (pycnocline) conditions (see Section 4). Dinoflagellates represent the phytoplankton group most often associated with harmful and toxic blooms (see Table 1). Thus, this mechanism of bloom initiation in coastal waters whereby blooms or seed populations of phytoplankton are advected onshore from offshore sources may be even more important for the initiation of exceptional/harmful blooms in coastal waters than commonly realized.

There may also be a link between pycnocline populations (blooms) of phytoplankton in offshore stratified waters and coastal blooms of harmful phytoplankton belonging to taxonomic groups other than the Dinophyceae. The observation of dense pycnocline concentrations of *Chrysochromulina polylepis* in the Kattegat and Skagerrak combined with the sudden appearance of the organism in Gulmar Fjord along the Swedish west coast during the 1988 *C. polylepis* bloom in this region has caused some workers (Kaas *et al.*, 1991) to speculate as to whether the coastal bloom may have been initiated by upwelling and advection of a pycnocline population into the coastal region. In any event, it seems clear that the *C. polylepis* bloom in coastal waters was transported along the Scandinavian coast by the Norwegian Coastal Current (Figure 23). Hydrographic influence on the location of this bloom is also evident from the fact that the bloom suddenly appeared in surface waters of the open Skagerrak in mid-May. This was presumably caused by westerly winds in this period which caused a reduction in the outflow of water from the Skagerrak in the Norwegian Coastal Current. Thus, waters in the Norwegian Coastal Current were redirected from the coast forming a gyre in the open Skagerrak.

A number of toxic phytoplankton species (for example, *Alexandrium* spp.; *Pyrodinium* and *Chattonella*) form resting cysts that settle out of the water column. These cysts can be resuspended in the water column and/or transported from one region to another through hydrographic processes. Thus, it is not only blooms or seed populations comprised of active

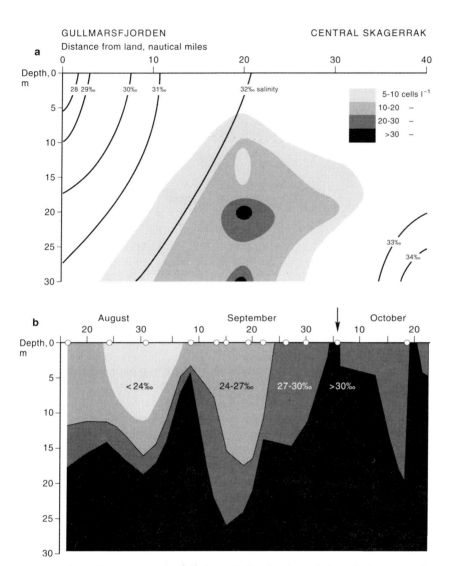

Figure 22 (a) Vertical and horizontal distribution of *Gyrodinium aureolum* (cells l^{-1}) and salinity (‰) along a transect running westward from Gullmarfjorden (Swedish west coast) on 28 and 29 August 1983. (b) Salinity in Gullmarfjorden in 1983. Dots on time axis indicate sampling dates. The arrow shows when *G. aureolum* was found. (After Lindahl, 1986.)

Figure 23 Dates of appearance of *Chrysochromulina polylepis* along the Swedish and Norwegian coasts in 1988. The organism was first observed in Gullmar Fjord on 9 May. (After Dahl, 1989.)

phytoplankton cells that can be advected to a new region via hydrographic processes.

Quantifying the relationship(s) between offshore hydrographic events and the initiation of coastal blooms is hampered by the fact that the focus of studies on harmful blooms is often geographically limited to the bloom site itself, and temporally limited to the period after the onset of the bloom. These limitations obviously do not allow for a description of the mechanisms by which offshore and inshore bloom events may be related. In order to examine these mechanisms more closely, long-term studies are needed in which the advection of phytoplankton (both harmful and non-harmful) from specific hydrographic features is examined in relation to local wind and weather conditions. Identification of the forcing conditions required to transport offshore phytoplankton blooms inshore is necessary in order to be able to predict which offshore events can be of potential coastal significance.

5.2. Bloom Composition

The real enigma with respect to phytoplankton blooms relates not to the fact that they occur but rather to the factors that control which species will bloom at a given time and place. During recent years there has been an increased appreciation of the fact that individual phytoplankton species have different requirements and responses to the physical environment. These differences have, perhaps, been best studied with respect to requirements for light and different strategies for light adaptation in phytoplankton (Richardson *et al.*, 1983) and both "high light" and "shade" species have been identified. In addition, there is some indication of a pattern with respect to the responses to light availability of the most common taxonomic groups of marine phytoplankton (see Figure 19).

There is also a general appreciation of the fact that different species have different requirements for macronutrients (see discussion and references in Section 3). Kiørboe (1993) points out that size of phytoplankton cells is vitally important to the uptake of nutrients (diffusion-limited nutrient uptake is proportional to 1/cell radius^2). Thus, on the basis of size alone, differences in nutrient requirements for different phytoplankton species can be predicted. Superimposed on the limitations/opportunities that size places on the cell with respect to the nutrient environment, are the various taxonomic requirements (diatoms have, e.g., an obligate silicon requirement) and individual or taxonomic requirements for various micronutrients. Our overall understanding of the various phytoplankton species' requirements and tolerances with respect to micronutrients is at a very primitive stage. However, it is clear that there are differences between species in terms of, for example, vitamin (Guillard *et al.*, 1991; Granéli *et al.*, 1993; Honjo, 1993) and trace metal (Sunda, 1989; Price and Morél, 1991; Honjo, 1993) requirements.

At present, it is sometimes possible in the aftermath of a harmful phytoplankton bloom to argue or show that the environmental conditions at the time of the bloom development were especially well suited to the individual requirements of the bloom organism (Maestrini and Granéli, 1991). However, it will not be possible to develop accurate predictive models for the occurrence of harmful phytoplankton blooms until a much better understanding of all of the different requirements that individual species or groups of species have with respect to their environment and the interaction of these requirements with one another. In theory, it should, ultimately, be possible to identify a "fingerprint" for each species or group of species which describes the total set of requirements and tolerances for each species and group of species. Matching the environmental conditions at any given time with the various "fingerprints" in our phytoplankton catalogue would form the basis of a predictive model concerning blooms of individual species but much work is still required to describe the more subtle requirements of individual species.

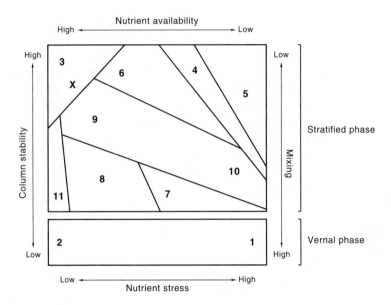

Figure 24 Matrix showing the most likely phytoplankton assemblages as a function of nutrient availability and turbulence during the vernal (spring) bloom and the summer stratified period in a lake system. Assemblages are as follows: 1, *Asterionella, Melosira italica*; 2, *Asterionella, Stephanodicus astraea*; 3, *Eudorina, Volvox*; 4, *Sphaerocystis*; 5, Chrysophytes; 6, *Anabaena, Aphanizomenon*; 7, *Tabellaria, Fragilaria, Staurastrum*; 8, *Melosira granulata, Fragilaria, Closterium*; 9, *Microcystis*; 10, *Ceratium*; 11, *Pediastrum, Coelastrum*; 12, *Oscillatoria agardhii*. (After Reynolds, 1980.)

5.3. Bloom Species Succession

Given the diversity of the marine phytoplankton flora (see Table 1), the description of "fingerprints" describing the various macro- and micronutritional and light requirements for each species is a formidable task indeed. In the short term, valuable information may be gained by elucidating the environmental conditions that lead to the development of different assemblages of phytoplankton species. The seasonal succession of phytoplankton assemblages has been rather better studied in limnological than in marine systems (e.g. Reynolds, 1980, 1984), probably because the easy accessibility of many lake systems has allowed accumulation of more detailed and longer time series of data relating to phytoplankton speciation than has been possible for most marine systems. Already in 1980, Reynolds was able to present a probability matrix identifying the various phytoplankton assemblages and relating their occurrence to nutrient and mixing conditions in the spring bloom period and when the water column was permanently stratified in (temperate) limnological systems (Figure 24).

While our understanding of phytoplankton assemblage succession is not as advanced for marine systems as is the case for fresh waters, some general patterns are now obvious. Figure 9 shows the distribution of different sizes of phytoplankton in surface waters of the Kattegat in different time periods extending over the periods of the establishment and breakdown of strong summer stratification in these waters. The largest cells are associated with the spring bloom period. While small cells dominate during the period of summer stratification, there are also events during this period when larger cells briefly reappear.

The seasonal succession of large (mostly diatom) cells during the spring bloom followed by smaller cells has been identified and related to hydrographic features (i.e. nutrient and light availability induced by different hydrographic features) by a number of workers (Cushing, 1989; Kiørboe, 1993; Légendre and Rassoulzadegan, 1995). Kiørboe (1993) has also presented evidence that the occurrence of large cells during the period of summer stratification may be related to hydrographic events which, for short periods, create mixing and nutrient conditions that mimic those found at the time of the spring bloom. Thus, a general pattern of phytoplankton succession, at least for seasonally stratified waters, has emerged but not yet at the species level.

For some particularly well-studied marine systems, it is also possible to describe succession in terms of dominating phytoplankton assemblages. One such system is Narragansett Bay, where the relatively long database on species distributions (see Table 4) makes generalizations concerning species succession patterns possible. Smayda and Villareal (1989) have examined the 1985 "brown tide" bloom of *Aureococcus anophagefferens* from the long-term data available. They argue that a characteristic feature of phytoplankton succession in this region is the existence of an "open" phytoplankton niche during the summer in which flagellates or other (mostly) non-diatomaceous phytoplankton often form blooms. They point out that there is poor predictive capability with respect to which species will come to dominate within this open niche and that "predictability is also compromised by significant inter-annual variations in key deter- minants of niche availability and occupancy, such as factor interactions and hysteresis effects from winter–spring diatom bloom dynamics". With respect to the *Aureococcus* bloom, they point out that a number of other species bloomed concurrently with this species. In all, they conclude that there was a succession of bloom events in Narragansett Bay during the summer of 1985 that involved 15 different species (Figure 25). For two of these species (*Minutocellus polymorphus* and *Fibrocapsa japonica*), they argue that the bloom events were just as "novel" as in the case of

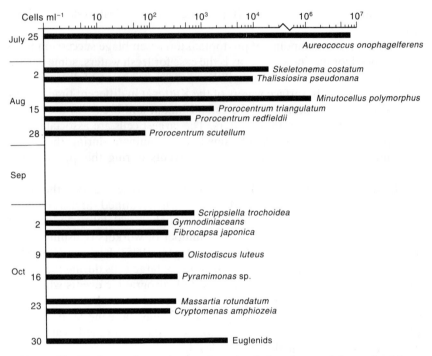

Figure 25 Timing and magnitude of maximal abundance of the major bloom species occurring at a station in Narragansett Bay during late summer and autumn 1985 (during a "brown tide" event—see text. (After Smayda and Villareal, 1989b.)

Aureococcus anophagefferens. In order to understand the dynamics of a harmful algal bloom such as the "brown tide" which occurred in Narragansett Bay, it is essential to understand the interactions between the different species and the factors which lead to their occurrence.

Often, reports of harmful algal blooms focus only on the organism in question and not on the other phytoplankton species present during the bloom. Thus, valuable information concerning the assemblages of phytoplankton which occur together and the conditions which lead to their occurrence may be lost. In times of limited economic resources for scientific research, studies involving the routine collection of taxonomic data concerning phytoplankton speciation often come under severe pressure. It is important to remember, however, that such data are crucial to a better understanding of phytoplankton assemblage succession and, ultimately, to the development of predictive models relating to the occurrence of "harmful" phytoplankton species.

5.4. Bloom Maintenance

In order to be maintained, the growth rate of the phytoplankton in a bloom must exceed the loss rate of phytoplankton in the population. Loss can occur by grazing or through cell lysis sinking or advection away from the bloom site. Various exceptional/harmful bloom-forming species have evolved mechanisms for reducing these loss terms and this may, at least in part, explain why these organisms form blooms.

It has already been pointed out that grazing impact can be important in the development of harmful algal blooms because some phytoplankton species appear to be unpalatable to herbivores (see Section 3). The limited grazing pressure on the unpalatable (often toxic) phytoplankters may be important in allowing large biomasses of these phytoplankton to develop.

Another mechanism by which grazing pressure may affect the development of phytoplankton blooms is by acting as a selection factor in phytoplankton species succession (Smayda and Villareal, 1989b; Olsson et al., 1992). This can be demonstrated by taking the example from Narragansett Bay discussed above where Smayda and Villareal (1989) argued that grazing structure influences the species which occur in the "open" phytoplankton niche during summer months. In particular, they argued that copepod predation determines whether *Skeletonema*, dinoflagellates or other phytoplankton groups will dominate. Thus, factors affecting the abundance of copepods will also impact algal speciation. As harmful algae are not equally distributed among the various phytoplankton groups (see Table 1), grazing structure, by affecting the relative abundance of various phytoplankton groups, will potentially also affect the probability of harmful blooms occurring. Detracting grazers and, thus, reducing grazing pressure is an obvious mechanism by which bloom loss factors may be reduced.

A more subtle strategy (Crawford and Purdie, 1992) for reducing loss factors which may be playing a role in the maintenance of some blooms may be a reduction in loss through advection. This reduction in the loss term may be brought about by vertical migration of the phytoplankter in the water column. It has been suggested (Garcon et al., 1986) that *Alexandrium tamarense* blooms occurring in a tidally flushed embayment near Woods Hole were only able to bloom there because the phytoplankton were essentially confined to regions of the water column below the outflowing surface waters. Their argument is based on the observation that the growth rate of this organism is similar to the flushing rate of the embayment. Thus, if the phytoplankton were equally distributed in the water column, the loss rate through advection would be similar to the increase rate (through growth) of the bloom and increases in biomass should not be observed.

Yet another mechanism by which the maintenance of a particular phytoplankton species may be encouraged is by reduction of the competition for limiting resources experienced by the blooming organism. If nutrients are, for example, limiting, then inhibition of other phytoplankters that will be competing for the same nutrients will serve to optimize the growth conditions for the bloom organism and, thus, maximize the production rate or length of production period of the bloom-forming phytoplankter. Several exceptional/harmful bloom-forming species have been shown to have inhibitory or repressing effects on other phytoplankton species (i.e. *Gynodinium aureolum*: Arzul *et al.*, 1993; *Heterosigma akashiwo*: Honjo, 1993). Thus, this mechanism of optimizing the production conditions for the bloom organism may be important in the maintenance of some phytoplankton blooms where single species are dominant.

5.5. Bloom Termination and Fate of Bloom Products

As discussed above, a phytoplankton bloom will decline when the loss factors together exceed the increase factors (i.e. growth and/or advective accumulation of the phytoplankton bloom species). Usually, the immediate cause of bloom decline is a decrease in the growth rate of the bloom-forming phytoplankton species brought about by light and/or nutrient limitation of the bloom species—in other words, a change in the environmental conditions which matched the environmental "fingerprint" of the bloom species (see Section 4.2). Often, when physiological stress of the bloom-forming organism has begun and the bloom begins to decline, the process is hastened by bacterial, fungal or viral infestation of the phytoplankton cells (van Donk, 1989; Bruning *et al.*, 1992; Nagasaki *et al.*, 1994).

Generally speaking, the phytoplankton cells from declining phytoplankton blooms can meet two fates (assuming that they are not grazed): they can either sink out of the water column or lyse and release their contents to the surrounding water column. Flagellate blooms appear often to be degraded in the water column (Boekel *et al.*, 1992) while diatom blooms more often sink out of the water column (Smetacek, 1985; Cushing, 1992; Waite *et al.*, 1992a). Different diatoms have different sinking rates and it has been proposed (Waite *et al.*, 1992b) that sinking rate, in addition to being influenced by cell volume, may be affected by physiological state (as expressed by respiration rates).

Often, a sudden collapse or sinking of a diatom bloom is recorded (see, e.g., Cushing, 1992). Recent developments in the application of coagulation theory to phytoplankton cells occurring under bloom conditions suggests that aggregation may be an important mechanism in terms of the sudden disappearance of diatom blooms (Jackson, 1990; Jackson and Lochmann, 1992). Essentially, it is argued that phytoplankton cells have an inherent natural "stickiness" that will allow them, under some conditions, to stick together (aggregate) upon contact with one another. The rate of the aggregation will be a function of the size, concentration and "stickiness" of the phytoplankton cells as well as the turbulent shear rate that makes the phytoplankton collide with one another. Kiørboe *et al.* (1990b) demonstrated that, under laboratory conditions, different phytoplankton species exhibit different degrees of "stickiness". In addition, age (physiological state) can influence "stickiness" within an individual species.

Jackson's (1990) model implies that a given phytoplankton population will, initially, grow exponentially. Ultimately, it will approach an equilibrium concentration ("critical concentration") where growth and coagulation (leading to sedimentation) will balance. Thus, according to this model, it ought to be possible to predict the maximum concentration that a given algal bloom will attain under various turbulent conditions. Kiørboe *et al.* (1994) compared the maximum concentrations of various algal species recorded during a spring bloom in a Danish fjord with the calculated "critical concentrations" for these species under the given conditions and found good agreement between the predicted and observed maxima in cell numbers for these species (Figure 26). Thus, they argue that aggregate formation controls the vertical flux (sedimentation) of phytoplankton blooms and sets an upper limit for the concentration of the phytoplankton cells.

Much work remains to elucidate the influence of, for example, bacterial and viral invasions on the "stickiness" of individual species before it will be possible to model the sedimentation of specific blooms. However, the application of coagulation theory to phytoplankton populations appears to provide the opportunity to describe the dynamics surrounding phytoplankton bloom culmination and sedimentation in much greater detail than previously possible.

6. SUMMARY AND CONCLUSIONS

Phytoplankton blooms are natural phenomena that play an important role in relation to carbon and energy flow as well as geochemical cycling in marine ecosystems. Phytoplankton growth will occur wherever and whenever adequate light and nutrient (both macro- and micronutrient) conditions occur. Blooms (accumulations of phytoplankton biomass) will

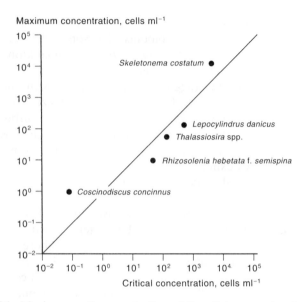

Figure 26 Maximum cell concentration of five diatom species during a study of the dynamics of the spring bloom in a Danish fjord plotted as a function of "critical concentrations" (Jackson, 1990; and see text). The line is $x = y$. (After Kiørboe *et al.*, 1994.)

result when the combined effect of population loss factors (grazing, advection and lysis or sedimentation) is less than the biomass accumulation due to growth of the phytoplankton and via advective processes. Until recently, it was assumed that adequate conditions for bloom development would only be found in surface waters where photon flux densities are greatest. It is now clear, however, that phytoplankton blooms also can and do occur under ice cover and at the base of the euphotic zone in regions where surface waters are nutrient depleted.

It is estimated (see Table 1) that the world's phytoplankton flora consists of approximately 4000 different species. Of these, about 200 species (i.e. approximately 6%) have been identified as causing "harmful" or "exceptional" (i.e. noticeable particularly to the general public through their effects) phytoplankton blooms. About 2% of the world's phytoplankton flora have been implicated as toxin producers. The toxins produced vary in their chemical structures and their effects but their public health implications are important as some are neurotoxins that can cause death at low concentrations and others are, apparently, carcinogenic. Much scientific attention is directed specifically towards elucidating the dynamics of harmful blooms—especially those in which toxic phytoplankton species are implicated. However, there are no phylogenetic, physiological or structural features that are common to "harmful" phytoplankton and there is no obvious scientific basis to treat "harmful" bloom species as a distinct subset of the total phytoplankton flora.

A number of workers have suggested that there has been an increase in harmful phytoplankton blooms due to anthropogenic activities in recent decades. Owing to the subjectivity in identifying harmful blooms and the lack of relevant long-term comprehensive data sets, it is, for most areas, difficult to address quantitatively the question of whether or not a real increase in such blooms has occurred.

Harmful and toxic species are a small subset of the total phytoplankton flora and we do not know what the naturally occurring interannual variability is with respect to the relative frequency of occurrence of different phytoplankton species. Thus, it is often difficult—even when changes in the relative occurrence of harmful species can be identified—to ascertain what role anthropogenic influences may have had in producing the change. It is clear that harmful phytoplankton blooms occurred in prehistoric times so anthropogenic influence is certainly not a prerequisite for the occurrence of such blooms. Nevertheless, a number of mechanisms by which anthropogenic activities can, theoretically, affect (directly or indirectly) the occurrence of harmful blooms are discussed in this review. These include eutrophication, the use of pesticides which may influence herbivore abundance, and the distribution and transport of harmful species from one geographic region to another via ballast water or in connection with aquaculture activities.

The most interesting question with respect to phytoplankton blooms is not why they occur but, rather, what mechanisms control the species which occur at a given time and place. There is increasing appreciation of the fact that the marine environment provides many different niches that can be exploited by different phytoplankton species and that each species has its own specific combination of requirements to the external environment (light, macro- and micronutrients). Thus, it ought to, at least in theory, be possible to identify a "fingerprint" for each species that describes these external requirements. Matching the environmental conditions at any given time to the "fingerprints" of phytoplankton species potentially occurring in an area would provide a basis for predictive models concerning the potential development of harmful blooms. In practice, however, a total description of the environmental requirements of individual phytoplankton species lies in the far distant future. Nevertheless, understanding the factors that control phytoplankton species succession is crucial to the understanding of why and where harmful phytoplankton blooms occur and research into these factors ought to be an urgent priority.

ACKNOWLEDGEMENTS

All of my good intentions with regard to this review article would never have resulted in a concrete product without the technical assistance of Grethe Hedeager and Charlotte Jørgensen or the detective skills of the librarians at the Danish Institute for Fisheries Research, Carina Anderberg and Bent Gaardestrup, and to them I am grateful. In addition, I would like to thank a number of colleagues who have allowed me to present previously unpublished material and/or read and commented on a draft manuscript. Finally, I would like to thank Professor J. H. S. Blaxter for stimulating me to write this review.

REFERENCES

Aikman, K. E., Tindall, D. R. and Morton, S. L. (1993). Physiology and potency of the dinoflagellate *Prorocentrum hoffmannianum* (Faust) during one complete growth cycle. *In* "Toxic Phytoplankton Blooms in the Sea" (T. J. Smayda and Y. Shimuzu, eds), vol. 3, pp. 463–468. Elsevier, Amsterdam, The Netherlands.

Anderson, D. M. (1989). Toxic algae blooms and red tides: a global perspective. *In* "Red Tides: Biology. Environmental Science and Toxicology" (T. Okaichi, D. M. Anderson and T. Nemoto, eds), pp. 11–21. Elsevier, New York, USA.

Anderson, D. M. and Stolzenbach, K. D. (1985). Selective retention of two dinoflagellates in a well mixed estuarine embayment: the importance of diel vertical migration and surface avoidance. *Marine Ecology Progress Series* **25**, 39–50.

Anderson, D. M. and White, A. W. (1992). Marine biotoxins at the top of the food chain. *Oceanus* **35**, 55–61.

Anderson, D. M., Kulis, D. M., Sullivan, J. J., Hall, S. and Lee, C. (1990). Dynamics and physiology of saxitoxin producton by the dinoflagellates *Alexandrium* spp. *Marine Biology* **104**, 511–524.

Andreae, M. O. (1990). Ocean–atmospheric interactions in the global biogeochemical sulfur cycle. *Marine Chemistry* **30**, 1–29.

Arzul, G., Erard-Le Denn, Videau, C., Jegou, A. M. and Gentien, P. (1993). Diatom growth repressing factors during an offshore bloom of *Gyrodinium* cf. *aureolum*. *In* "Toxic Phytoplankton Blooms in the Sea" (T. J. Smayda and Y. Shimuzu, eds), vol. 3, pp. 719–730. Elsevier, Amsterdam, The Netherlands.

Aune, T., Skulberg, O. M. and Underdal, B. (1991). A toxic phytoflagellate bloom of *Chrysochromulina* cf. *leadbeateri* in coastal waters in the north of Norway May–June 1991. *AMBIO* **21**, 471–474.

Aure, J. and Rey, F. (1992). Oceanographic conditions in the Sandsfjord system, western Norway, after a bloom of the toxic prymnesiophyte *Prymnesium parvum* Carter in August 1990. *Sarsia* **76**, 247–254.

Baddyr, M. (1992). PSP in Morocco. *In* "Harmful Algae News", No. 2 (T. Wyatt and Y. Pazos, eds), p. 5. UNESCO, Paris, France.

Baden, D. G. (1989). Brevetoxin analysis. Toxic dinoflagellates and marine mammal mortalities. *Technical Reports. Woods Hole Oceanographic Institution* 47–52.

Baden, D. G. (1995). Structure/function relationships of the brevetoxins: inter-ferences from molecular modeling, organic chemistry, and specific receptor binding protocols. In "Harmful Marine Algal Blooms" (P. Lassus, G. Arzul, E. Erard, P. Gentien and C. Marcaillou, eds), pp. 257–266. Lavoisier/Intercept, Paris, France.

Balch, W. M. (1981). An apparent lunar tidal cycle of phytoplankton blooming and community succession in the Gulf of Maine. Journal of Experimental Marine Biology and Ecology 55, 65–77.

Bates, S. S., Bird, C. J., de Freitas, A. S. W., Foxall, R., Gilgan, M., Hanic, L. A., Johnson, G. R., McCulloch, A. W., Odense, P., Pocklington, R., Quilliam, M. A., Sim, P. G., Smith, J. C., Subba Rao, D. V., Todd, E. C. D., Walter, J. A. and Wright, J. L. C. (1989). Pennate diatom Nitzschia pungens as the primary source of domoic acid, a toxin in shellfish from eastern Prince Edward Island, Canada. Canadian Journal of Fisheries and Aquatic Sciences 46, 1203–1215.

Bates, S. S., de Freitas, A. S. W., Milley, J. E., Pocklington, R., Quilliam, M. A., Smith, J. C. and Worms, J. (1991). Controls on domoic acid production by the diatom Nitzschia pungens f. multiseries in culture: nutrients and irradiance. Canadian Journal of Fisheries and Aquatic Sciences 48, 1136–1144.

Bates, S. S., Douglas, D. J., Doucette, G. J. and Léger, C. (1995). Effects of reintroducing bacteria on domoic acid production by axenic cultures of the diatom Pseudonitzschia pungens f. multiseries. In "Harmful Marine Algal Blooms" (P. Lassus, G. Arzul, E. Erard, P. Gentien and C. Marcaillou, eds), pp. 401–406. Lavoisier/Intercept, Paris, France.

Bell, G. R. (1961). Penetration of spines from a marine diatom into the gill tissue of lingcod Ophiodon elongatus. Nature, London 192, 279–280.

Berg, J. and Radach, G. (1985). Trends in nutrient and phytoplankton concentra-tions at Helgoland Reede (German Bight) since 1962. International Council Meeting Paper 1985/L, 2 (mimeo).

Bienfang, P., Szyper, J. and Laws, E. (1983). Sinking rate and pigment responses to light limitation of a marine diatom: implications to dynamics of chlorophyll maximum layers. Oceanologica Acta 6, 52–62.

Bigelow, H. B., Lillick, L. C. and Sears, M. (1940). Phytoplankton and planktonic protozoa in the offshore waters of the Gulf of Maine. Part I. Numerical distribution. Transactions of the American Philosophical Society 21, 149–191.

Bjørnsen, P. K. and Nielsen, T. G. (1991). Decimeter scale heterogeneity in the plankton during a pycnocline bloom of Gyrodinium aureolum. Marine Ecology Progress Series 73, 263–267.

Bjørnsen, P. K. and Nielsen, T. G. (1992). Planktondynamik i skillefladen. In "Plankton Dynamics and Nutrient Turnover in the Kattegat" (T. Fenchel, ed.). Havforskning fra Miljøstyrelsen, No. 10, pp. 151–166. The Danish Ministry of Environment and Energy, The Danish Environmental Protection Agency, Copenhagen, Denmark. (In Danish.)

Bjørnsen, P. K., Kaas, H., Nielsen, T. G., Olesen, M. and Richardson, K. (1993). Dynamics of a subsurface phytoplankton maximum in the Skagerrak. Marine Ecology Progress Series 95, 279–294.

Black, E. A., Whyte, J. N. C., Bagshaw, J. W. and Ginther, N. G. (1991). The effects of Heterosigma akashiwo on juvenile Oncorhynchus tshawytscha and its implications for fish culture. Journal of Applied Ichthyology 7, 168–175.

Blackburn, S. I. and Jones, G. J. (1995). Toxic *Nodularia spumigena* Mertens blooms in Australian waters—a case study from Orielton Lagoon, Tasmania. *In* "Harmful Marine Algal Blooms" (P. Lassus, G. Arzul, E. Erard, P. Gentien and C. Marcaillou, eds), pp. 121–126. Lavoisier/Intercept, Paris, France.

Boalch, G. T. (1983). Recent dinoflagellate blooms in the Plymouth area. *British Phycological Journal*, **18**, 200–291 (abstract).

Boekel, W. H. M. van, Hansen, F. C., Rigman, R. and Bak, R. P. M. (1992). Lysis-induced decline of a Phaeocystis spring bloom and coupling with the microbial food web. *Marine Ecology Progress Series* **81**, 269–276.

Boyer, G. L., Sullivan, J. J., Anderson, R. J., Harrison, P. J. and Taylor, F. J. R. (1987). Effects of nutrient limitation on toxin production and composition in the marine dinoflagellate *Protogonyaulax tamarenses*. *Marine Biology* **96**, 123–128.

Braarud, T. and Heimdal, B. R. (1970). Brown water on the Norwegian coast in autumn 1966. *Nytt Magasin for Botanik* **17**, 91–97.

Brown, C. W. and Yoder, J. A. (1994). Coccolithophorid blooms in the global ocean. *Journal of Geophysical Research* **99**, 7467–7482.

Bruning, K., Lingeman, R. and Ringelberg, J. (1992). Estimating the impact of fungal parasites on phytoplankton populations. *Limnology and Oceanography* **37**, 252–260.

Bullock, A. M., Turner, M. F. and Gowen, R. J. (1985). The toxicity of *Gyrodinium aureolum*. *Bulletin of Marine Science* **37**, 763–764.

Burkholder, J. M., Noga, E. J., Hobbs, C. H. and Glasgow, H. B. Jr (1992). New "phantom" dinoflagellate is the causative agent of major estuarine fish kills. *Nature, London* **358**, 407–410.

Burkholder, J. M., Glasgow, H. B. Jr and Hobbs, C. N. (1995). Distribution and environmental conditions for fish kills linked to a toxic ambush-predator dinoflagellate. *Marine Ecology Progress Series* **124**, 43–61.

Buskey, E. J. and Stockwell, D. A. (1993). Effects of a persistent "brown tide" on zooplankton populations in the Laguna Madre of southern Texas. *In* "Toxic Phytoplankton Blooms in the Sea" (T. J. Smayda and Y. Shimuzu, eds), pp. 659–666. Elsevier, Amsterdam, Netherlands.

Cadée, G. C. (1986). Increased phytoplankton primary production in the Marsdiep area (western Dutch Wadden Sea). *Netherlands Journal of Sea Research* **20**, 285–290.

Cadée, G. C. and Hegeman, J. (1986). Seasonal and annual variation in *Phaeocystis pouchetti* (Haptophyceae) in the westernmost inlet of the Wadden Sea during the 1973 to 1985 period. *Netherlands Journal of Sea Research* **20**, 29–36.

Carmichael, W. W. (1992). "Status report on Planktonic Cyanobacteria (Blue-green Algae) and their Toxins." Ecological Research Service US Environmental Protection Agency, 149 pp.

Carreto, J. I., Benavides, H. R., Negri, R. M. and Glorioso, P. D. (1986). Toxic red-tide in the Argentina Sea. Phytoplankton distribution and survival of the toxic dinoflagellate *Gonyaulax tamarenses* in a frontal area. *Journal of Plankton Research* **8**, 15–28.

Change, F. H., Anderson, C. and Boustead, N. C. (1990). First record of a Heterosigma (Raphidophyceae) bloom with associated mortality of cage-reared salmon in Big Glory Bay, New Zealand. *New Zealand Journal of Marine and Freshwater Research* **24**, 461–469.

Charlson, R. J., Lovelock, J. E., Andreae, M. O. and Warren, S. G. (1987).

Oceanic phytoplankton, atmospheric sulphur, cloud albedo and climate. *Nature, London* **326**, 655–661.

Codispoti, L. A., Friederich, G. E., Sakamoto, C. M. and Gordon, L. I. (1991). Nutrient cycling and primary production in the marine systems of the Arctic and Antarctic. *Journal of Marine Systems* **2**, 359–384.

Conley, D. J., Schelske, C. L. and Stoermer, E. F. (1993). Modification of the biogeochemical cycle of silica with eutrophication. *Marine Ecology Progress Series* **101**, 179–192.

Conover, R. J. and Huntley, M. (1991). Copepods in ice-covered seas—distribution, adaptations to seasonally limited food, metabolism, growth patterns and life cycle strategies in polar seas. *Journal of Marine Systems* **2**, 1–41.

Cosper, E. M., Dennison, W., Carpenter, E., Bricelj, V. M., Mitchell, J., Kuenstner, S., Colflesh, D. and Dewey, M. (1987). Recurrent and persistent brown tide blooms perturb coastal marine ecosystems. *Estuaries* **10**, 284–290.

Cosper, E. M., Dennison, W., Milligan, A., Carpenter, E. J., Lee, C., Holzapfel, J. and Milanese, L. (1989). An examination of the environmental factors important to initiating and sustaining "brown tide" blooms. *In* "Novel Phytoplankton Blooms" (E. M. Cosper, V. M. Bricelj and E. J. Carpenter, eds), pp. 317–340. Springer, Berlin, Germany.

Cosper, E. M., Garry, R. T., Milligan, A. J. and Doall, M. H. (1993). Iron, selenium and citric acid are critical to the growth of the brown tide microalga, *Aurecoccus anphagefferens*. *In* "Toxic Phytoplankton Blooms in the Sea" (T. J. Smayda and Y. Shimuzu, eds), vol. 3, pp. 667–673. Elsevier, Amsterdam, The Netherlands.

Crawford, D. W. and Purdie, D. A. (1992). Evidence for avoidance of flushing from an estuary by a planktonic, phototrophic ciliate. *Marine Ecology Progress Series* **79**, 259–265.

Cushing, D. H. (1959). The seasonal variation in oceanic production as a problem in population dynamics. *Journal du Conseil Permanent International pour l'Exploration de la Mer* **24**, 455–464.

Cushing, D. H. (1989). A difference in structure between ecosystems in strongly stratified waters and those that are only weakly stratified. *Journal of Plankton Research* **11**, 1–13.

Cushing, D. H. (1992). The loss of diatoms in the spring bloom. *Philosophical Transactions of the Royal Society B* **335**, 237–246.

Dahl, E. (1989). Algal blooms along the coast of Norway in 1988. *In* "Problems of Algal Blooms in Aquaculture", 2nd Sherkin International Workshop and Conference, Sherkin Island Marine Station, Co. Cork, Ireland.

Dahl, E. and Tangen, K. (1993). 25 years' experience with *Gyrodinium aureolum* in Norwegian waters. *In* "Toxic Phytoplankton Blooms in the Sea" (T. J. Smayda and Y. Shimuzu, eds.), vol. 3, pp. 15–21. Elsevier, Amsterdam, The Netherlands.

Dahl, E., Lindahl, O., Paasche, E. and Throndsen, J. (1989). The *Chrysochromulina polylepis* bloom in Scandinavian waters during spring 1988. *In* "Novel Phytoplankton Blooms" (E. M. Cosper, V. M. Bricelj and E. J. Carpenter, eds), pp. 383–405. Springer, Berlin, Germany.

Dale, B. and Nordberg, K. (1993). Possible environmental factors regulating prehistoric and historic blooms of the toxic dinoflagellate *Gymnodinium catenatum* in the Kattegat–Skagerrak region of Scandinavia. *In* "Toxic Phytoplankton Blooms in the Sea" (T. J. Smayda and Y. Smimuzu, eds), pp. 53–57. Elsevier, Amsterdam, The Netherlands.

Dale, B. and Yentsch, C. M. (1978). Red tide and paralytic shellfish poisoning. *Oceanus* **21**, 41–49.

Dale, B., Madsen, A., Nordberg, K. and Thorsen, T. A. (1993). Evidence for prehistoric and historic "blooms" of the toxic dinoflagellate *Gymnodinium catenatum* in the Kattegat–Skagerrak region of Scandinavia. *In* "Toxic Phytoplankton Blooms in the Sea" (T. J. Smayda and Y. Shimuzu, eds), pp. 53–57. Elsevier, Amsterdam, The Netherlands.

Delmas, D., Herbland, A. and Maestrini, S. Y. (1993). Do *Dinophysis* spp. come from the "open sea" along the French Atlantic coast? *In* "Toxic Phytoplankton Blooms in the Sea" (T. J. Smayda and Y. Shimuzu, eds), pp. 489–494. Elsevier, Amsterdam, The Netherlands.

Desbiens, M. and Cembella, A. D. (1995). Occurrence and elimination kinetics of PSP toxins in the American lobster (*Homarus americanus*). *In* "Harmful Marine Algal Blooms" (P. Lassus, G. Arzul, E. Erard, P. Gentien and C. Marcaillou, eds), pp. 433–438. Lavoisier/Intercept, Paris, France.

van Donk, E. (1989). The role of fungal parasites in phytoplankton succession. *In* "Plankton Ecology Succession in Plankton Communities" (U. Sommer, ed.), pp. 171–194. Springer Verlag, Berlin.

Doucette, G. J. (1995). Assessment of the interaction of prokaryotic cells with harmful algal species. *In* "Harmful Marine Algal Blooms" (P. Lassus, G. Arzul, E. Erard, P. Gentien and C. Marcaillou, eds), pp. 385–394. Lavoisier/Intercept, Paris, France.

Doucette, G. J. and Trick, C. G. (1995). Characterization of bacteria associated with different isolates of *Alexandrium tamarenses*. *In* "Harmful Marine Algal Blooms" (P. Lassus, G. Arzul, E. Erard, P. Gentien and C. Marcaillou, eds), pp. 33–38. Lavoisier/Intercept, Paris, France.

Douglas, D. J., Bates, S. S., Bourgue, L. A. and Selvia, R. C. (1993). Domoic acid production by axenic and non-axenic cultures of the pennate diatom *Nitzschia pungens* f. *multiseries*. *In* "Toxic Phytoplankton Blooms in the Sea" (T. J. Smayda and Y. Shimuzu, eds), pp. 595–600. Elsevier, Amsterdam, The Netherlands.

Drum, A. S., Siebens, T. L., Crecelius, E. A. and Elston, R. A. (1993). Domoic acid in the Pacific razor clam *Siliqua patula* (Dixon, 1789). *Journal of Shellfish Research* **12**, 443–450.

Dugdale, R. C. and Goering, J. J. (1967). Uptake of new and regenerated forms of nitrogen in primary productivity. *Limnology and Oceanography* **12**, 196–206.

Durbin, A. G. and Durbin, E. G. (1989). Effect of the "brown tide" on feeding, size and egg laying rate of adult female *Acartia tonsa*. *In* "Novel Phytoplankton Blooms" (E. M. Cosper, V. M. Bricelj and E. J. Carpenter, eds), pp. 626–646, Springer, Berlin, Germany.

Edvardsen, P., Moy, F. and Paasche, E. (1990). Hemolytic activity in extracts of *Chrysochromulina polylepis* grown at different levels of selenite and phosphate. *In* "Toxic Marine Phytoplankton" (E. Graneli, B. Sundstrøm, L. Edler and D. M. Anderson, eds), pp. 284–289. Elsevier, New York, USA.

Egge, J. K. and Aksnes, D. L. (1992). Selicate as regulating nutrient in phytoplankton competition. *Marine Ecology Progress Series* **83**, 281–289.

Egge, J. K. and Heimdal, B. R. (1994). Blooms of phytoplankton including *Emiliania huxleyi* (haptophyta). Effects of nutrient supply in different N : P ratios. *Sarsia* **79**, 333–348.

Eilertsen, H. C. and Raa, J. (eds) (1994). Phytoplankton toxins in sea-water. *In*

"3rd Marine Biotechnology Conference, Tromsø International University, Norway", p. 69.

El-Sayed, S. Z. (1971). Observation on a phytoplankton-bloom in the Weddel Sea. *In* "Biology of the Antarctic Seas", vol. IV (G. A. Llano and J. E. Wallen, eds). Antarctic Research Series, vol. 17, pp. 301–312. American Geophysical Union, Washington, DC, USA.

Falconer, I. R. (1991). Tumour promotion and liver injury caused by oral consumption of cyanobacteria. *Environmental Toxicology and Water Quality* **6**, 177–184.

Falconer, I. R. (ed.) (1993). "Algal Toxins in Seafood and Drinking Water." Academic Press, London, UK.

Farrington, C. W. (1988). Mortality and pathology of juvenile chinook salmon (*Oncorhynchus tshawytscha*) and chum salmon (*Oncorhynchus keta*) exposed to cultures of the marine diatom *Chaetoceros convolutus*. *Alaska Sea Grant Report*, 80 pp.

Fernández, E., Boyd, P., Holligan, P. M. and Harbour, D. S. (1993). Production of organic and inorganic carbon within a large-scale coccolithophore bloom in the northeast Atlantic Ocean. *Marine Ecology Progress Series* **97**, 271–285.

Fiedler, P. (1982). Zooplankton avoidance and reduced grazing responses to *Gymnodinium splendens* (Dinophyceae). *Limnology and Oceanography* **27**, 961–965.

Flynn, K. J. and Flynn, K. (1995). Dinoflagellate physiology: nutrient stress and toxicity. *In* "Harmful Marine Algal Blooms" (P. Lassus, G. Arzul, E. Erard, P. Gentien and C. Marcaillou, eds), pp. 541–550. Lavoisier/Intercept, Paris, France.

Fraga, S. (1993). Harmful algal blooms in relation to wind induced coastal upwelling and river plumes. *International Council Meeting* Paper 1993/L:38 (mimeo).

Fraga, S. and Bakun, A. (1993). Global climate change and harmful algal blooms: the example of *Gymnodinium catenatum* on the Galician coast. *In* "Toxic Phytoplankton Blooms in the Sea" (T. J. Smayda and Y. Shimuzu, eds), vol. 3, pp. 59–65. Elsevier, Amsterdam, The Netherlands.

Franks, P. J. S. and Anderson, D. M. (1992). Alongshore transport of a toxic phytoplankton bloom in a buoyancy current: *Alexandrium tamarense* in the Gulf of Maine. *Marine Biology* **112**, 153.

Frémy, J. M. (1991). Marine biotoxins: a short overview. *In* "Proceedings of Symposium on Marine Biotoxins", Centre National d'Études Vétèrinaires et Alimentaires, pp. 3–6. Maisons Alfort, Paris, France.

Gainey, L. F. Jr and Shumway, S. E. (1991). The physiological effect of *Aureococcus anophagefferens* ("brown tide") in the lateral cilia of bivalve mollusks. *Biological Bulletin, Marine Biological Laboratory, Woods Hole* **181**, 298–306.

Gallager, S. M., Stoecker, D. K. and Bricelj, V. M. (1989). Effects of the brown tide alga on growth, feeding physiology and locomotive behavior of scallop larvae (*Argopecten irradens*). *In* "Novel Phytoplankton Blooms" (E. M. Cosper, V. M. Bricelji and E. J. Carpenter, eds), pp. 511–541. Springer, Berlin, Germany.

Garcon, V. C., Stolzenbach, K. D. and Anderson, D. M. (1986). Tidal flushing of an estuarine embayment subject to recurrent dinoflagellate blooms. *Estuaries* **9**, 179–187.

374 K. RICHARDSON

Garside, C. and Garside, J. C. (1993). The "*f*-ratio" on 20°W during the North Atlantic Bloom Experiment. *Deep-Sea Research II* **40**, 75–90.

Goldman, J. C., Hansell, D. A. and Dennett, M. R. (1992). Chemical characterization of three large oceanic diatoms: potential impact on water column chemistry. *Marine Ecology Progress Series* **88**, 257–270.

Gosselin, S., Fortier, L. and Gagne, J. A. (1989). Vulnerability of marine fish larvae to the toxic dinoflagellate *Protogonyaulax tamarenses*. *Marine Ecology Progress Series* **57**, 1–10.

Gran, H. H. and Braarud, T. (1935). A quantitative study of the phytoplankton of the Bay of Fundy and Gulf of Maine (including observations on hydrography, chemistry and turbidity). *Journal of the Biological Board of Canada* **1**, 279–467.

Granéli, E., Paasche, E. and Maestrini, S. Y. (1993). Three years after the *Chrysochromulina polylepis* bloom in Scandinavian waters in 1988. Some conclusions of recent research and monitoring. *In* "Toxic Phytoplankton Blooms in the Sea" (T. J. Smayda and Y. Shimuzu, eds), vol. 3, pp. 23–32. Elsevier, Amsterdam, The Netherlands.

Granmo, A., Havenhand, J., Magnusson, K. and Svane, I. (1988). Effects of the planktonic flagellate *Chrysochromulina polylepis* Manton et Park on fertilization and early development of the ascidian *Ciona intestinalis* (L.) and the blue mussel *Mytilus edulis* L. *Journal of Experimental Marine Biology and Ecology* **124**, 65–71.

Guillard, R. R. L., Keller, M. D., O'Kelly, C. J. and Floyd, G. L. (1991). *Pycnococcus provasolii* gen. et sp. nov., a coccoid prasinoxanthin-containing phytoplankter from the western North Atlantic and Gulf of Mexico. *Journal of Phycology* **27**, 39–47.

Guo, C. and Tester, P. A. (1994). Toxic effect of the bloom-forming *Trichodesmium* sp. (Cyanophyta) to the copepod *Acartia tonsa*. *Natural Toxins* **2**, 222–227.

Haigh, R. and Taylor, F. J. R. (1990). The distribution of potentially harmful phytoplankon species in the northern Strait of Georgia, British Columbia. *Canadian Journal of Fisheries and Aquatic Sciences* **47**, 2339–2350.

Hallegraeff, G. M. (1993). A review of harmful algae blooms and the apparent global increase. *Phycologia* **32**, 79–99.

Hallegraeff, G. M. and Bolch, C. J. (1991). Transport of toxic dinoflagellate cysts via ships' ballast water. *Marine Pollution Bulletin* **22**, 27–30.

Hallegraeff, G. M. and Bolch, C. J. (1992). Transport of diatom and dinoflagellate resting spores in ships' ballast water: implications for plankton biogeography and aquaculture. *Journal of Plankton Research* **14**, 1067–1084.

Hansen, P. J. (1989). The red tide dinoflagellate *Alexandrium tamarense*: effects on behaviour and growth of a tintinnid ciliate. *Marine Ecology Progress Series* **53**, 105–116.

Hansen, P. J. (1995). Growth and grazing response of a ciliate feeding on the red tide dinoflagellate *Gyrodinium aureolum* in monoculture and in mixture with a non-toxic alga. *Marine Ecology Progress Series* **121**, 65–72.

Hansen, P. J., Cembella, A. D. and Moestrup, Ø. (1992). The marine dinoflagellate *Alexandrium ostenfeldii*: paralytic shellfish toxin concentration, composition, and toxicity to a tintinnid ciliate. *Journal of Phycology* **28**, 873.

Hawser, S. P. and Codd, G. A. (1992). The toxicity of *Trichodesmium* blooms from Caribbean waters. *In* "Marine Pelagic Cyanobacteria: *Trichodesmium* and other Diazotrophs" (E. J. Carpenter, D. G. Capone and J. G. Reuter, eds), pp. 319–329. Kluwer, Dordrecht, The Netherlands.

Haya, K., Martin, J. L., Burridge, L. E., Wainwood, B. A. and Wildish, D. J. (1991). Domoic acid in shellfish and plankton from the Bay of Fundy, New Brunswick, Canada. *Journal of Shellfish Research* **10**, 113–118.

Heilmann, J. P., Richardson, K. and Ærtebjerg, G. (1994). Annual distribution and activity of phytoplankton in the Skagerrak–Kattegat frontal region. *Marine Ecology Progress Series* **112**, 213–223.

Heinig, C. S. and Campbell, D. E. (1992). The environmental context of a *Gyrodinium aureolum* bloom and shellfish kill in Maquoit Bay, Maine, September 1988. *Journal of Shellfish Research* **11**, 111–122.

Hofman, R. J. (1989). Actions necessary to assess the possible impacts of marine biotoxins on marine mammals. *In* "Toxic Dinoflagellates and Marine Mammal Mortalities" Technical Report, Woods Hole Oceanographic Institute, MA, USA, pp. 53–55.

Holligan, P. M. (1979). Dinoflagellate blooms associated with tidal fronts around the British Isles. *In* "Toxic Dinoflagellate Blooms" (D. L. Taylor and H. H. Seliger, eds), pp. 249–256. Elsevier/North-Holland, New York, USA.

Holligan, P. M. and Harbour, D. S. (1977). The vertical distribution and succession of phytoplankton in the western English Channel in 1975 and 1976. *Journal of the Marine Biological Association of the UK* **57**, 1075–1093.

Holligan, P. M., Viollier, M., Harbour, D. S., Camus, P. and Champagne-Philippe, M. (1983). Satellite and ship studies of coccolithophore production along a continental shelf edge. *Nature, London* **304**, 339–342.

Holligan, P. M., Harris, R. P., Newell, R. C., Harbour, D. S., Head, R. W., Linley, E. A. S., Lucas, M. I., Trantor, P. R. G. and Weekley, C. M. (1984). The vertical distribution and partitioning of organic carbon in mixed, frontal and stratified waters in the English Channel. *Marine Ecology Progress Series* **14**, 111–127.

Holligan, P. M., Aarup, T. and Groom, S. B. (1989). The North Sea: satellite colour atlas. *Continental Shelf Research* **9**, 665–766.

Holligan, P. M., Fernández, E., Aiken, J., Balch, W. M., Boyd, P., Burkhill, P. H., Finch, M., Groom, S. B., Malin, G., Muller, K., Purdie, D., Robinson, C., Trees, C. C., Turner, S. M. and van der Wal, P. (1993a). A biogeochemical study of the coccolithophore, *Emiliania huxleyi*, in the North Atlantic. *Global Biogeochemical Cycles* **7**, 879–900.

Holligan, P. M., Groom, S. B. and Harbour, D. S. (1993b). What controls the distribution of the coccolithophore, *Emiliania huxleyi*, in the North Sea? *Fisheries Oceanography* **2**, 175–183.

Honjo, T. (1993). Overview on bloom dynamics and physiological ecology of *Heterosigma akashiwo*. *In* "Toxic Phytoplankton Blooms in the Sea" (T. J. Smayda and I. Shimuzu, eds), vol. 3, pp. 33–42. Elsevier, Amsterdam, The Netherlands.

Honjo, T. (1994). The biology and prediction of representative red tides associated with fish kills in Japan. *Review in Fisheries Science* **2**, 225–253.

Hsiao, S. I. C. (1988). Spatial and seasonal variations in primary production of sea ice microalgae and phytoplankton in Frobisher Bay, Arctic Canada. *Marine Ecology Progress Series* **44**, 275–285.

Huber, A. L. and Hamel, K. S. (1985). Phosphatase activities in relation to phosphorus nutrition in *Nodularia spumigena* (Cyanobacteriaccac). 1. Field studies. *Hydrobiologia* **123**, 145–152.

Huntley, M. E. and Lopez, M. D. G. (1992). Temperature dependent production of marine copepods: a global synthesis. *American Naturalist* **140**, 201–242.

Huntley, M. E., Stykes, P., Rohan, S. and Marin, V. (1986). Chemically mediated rejection of dinoflagellate prey by the copepods *Calanus pacificus* and *Paracalanus parvus*: mechanism, occurrence and significance. *Marine Ecology Progress Series* **28**, 105–120.

Hurley, D. E. (1982). The "Nelson Slime". Observations on past occurrences. *New Zealand Oceanographic Institute Oceanographic Summary* **20**, 1–11.

ICES (1984). Report of the ICES Special Meeting on the Causes, Dynamics and Effects of Exceptional Marine Blooms and related events. *International Council Meeting* Paper 1984/E, 42 (mimeo).

ICES (1991). Report of the Working Group on Phytoplankton and their Effects. *International Council Meeting* Paper 1991/Poll, 3 (mimeo).

Ives, J. D. (1985). The relationship between *Gonyaulax tamarenses* cell toxin levels and copepod ingestion levels. *In* "Toxic Dinoflagellates" (D. M. Anderson, A. W. White and D. G. Baden, eds), pp. 413–418. Elsevier, New York, USA.

Ives, J. D. (1987). Possible mechanisms underlying copepod grazing responses to levels of toxicity in red tide dinoflagellates. *Journal of Experimental Marine Biology and Ecology* **112**, 131–145.

Jackson, G. A. (1990). A model of the formation of marine algal flocs by physical coagulation processes. *Deep-Sea Research* **37**, 1197–1211.

Jackson, G. A. and Lochmann, S. E. (1992). Effect of coagulation on nutrient and light limitation of an algal bloom. *Limnology and Oceanography* **37**, 77–89.

Jenkinson, I. R. (1989). Increases in viscosity may kill fish in some blooms. *In* "Red Tides: Biology, Environmental Science and Toxicology" (T. Okaichi, D. M. Anderson and T. Nemoto, eds), pp. 435–438. Elsevier, New York, Amsterdam.

Jenkinson, I. R. (1993). Viscosity and elasticity of *Gyrodinium* cf. *aureolum* and *Noctiluca scientillans* exudates, in relation to mortality of fish and damping of turbulence. *In* "Toxic Phytoplankton Blooms in the Sea" (T. J. Smayda and Y. Shimuzu, eds), vol. 3, pp. 757–762. Elsevier, Amsterdam, The Netherlands.

Jenkinson, I. R. and Biddanda, B. A. (1995). Bulk-phase viscoelastic properties of seawater: relationship with plankton components. *Journal of Plankton Research* **17**, 2251–2274.

Jochem, F. J. and Zeitzschel, B. (1993). Productivity regime and phytoplankton size structure in the tropical and subtropical North Atlantic in spring 1989. *Deep-Sea Research II* **40**, 495–519.

Jones, J. B. and Rhodes, L. L. (1994). Suffocation of pilchards (*Sardinops sagax*) by a green microalgal bloom in Wellington Harbour, New Zealand. *New Zealand Journal of Marine and Freshwater Research* **28**, 379–383.

Jones, K. J., Ayres, P., Bullock, A. M., Roberts, R. J. and Tett, T. (1982). A red tide of *Gyrodinium aureolum* in sea lochs of the Firth of Clyde and associate mortality of pond-reared salmon. *Journal of the Marine Biological Association of the UK* **62**, 771–782.

Justic, D., Legovic, T. and Rottini-Sandrini, L. (1987). Trends in oxygen content 1911–1984 and occurrence of benthic mortality in the northern Adriatic Sea. *Estuarine and Coastal Shelf Science* **25**, 435–445.

Kaas, H., Larsen, J., Møhlenberg, F. and Richardson, K. (1991). The

Chrysochromulina polylepis Manton and Parke bloom in the Kattegat (Scandinavia) May–June, 1988: distribution, primary production and nutrient dynamics in the late stage of the bloom. *Marine Ecology Progress Series* **79**, 151–161.

Keafer, B. A. and Anderson, D. M. (1993). Use of remotely sensed sea surface temperature in studies of *Alexandrium tamarense* bloom dynamics. *In* "Toxic Phytoplankton Blooms in the Sea" (T. J. Smayda and Y. Shimizu, eds), vol. 3, pp. 763–768. Elsevier, Amsterdam, The Netherlands.

Keller, M. D. (1989). Dimethyl sulfide production and marine phytoplankton: the importance of species composition and cell size. *Biological Oceanography* **6**, 375–382.

Keller, M. D., Bellows, W. K. and Guillard, R. R. L. (1989). Dimethylsulfide production and marine phytoplankton: an additional impact of unusual blooms. *In* "Novel Phytoplankton Blooms" (E. M. Cosper, V. M. Bricelj and E. J. Carpenter, eds), pp. 101–115. Springer, Berlin, Germany.

Kent, M. L., Whyte, J. N. C. and La Trace, C. (1995). Gill lesions and mortality in seawater pen-reared Atlantic salmon *Salmo salar* associated with a dense bloom of *Skeletonema costatum* and *Thalassiosira* species. *Diseases of Aquatic Organisms* **22**, 77–81.

Kiørboe, T. (1993). Turbulence, phytoplankton cell size, and the structure of pelagic food webs. *Advances in Marine Biology* **29**, 1–72.

Kiørboe, T., Kaas, H., Kruse, B., Møhlenberg, F., Tiselius, and Ærtebjerg, G. (1990a). The structure of the pelagic food web in relation to water column structure in the Skagerrak. *Marine Ecology Progress Series* **59**, 19–32.

Kiørboe, T., Andersen, K. P. and Dam, H. (1990b). Coagulation efficiency and aggregate formation in marine phytoplankton. *Marine Biology* **107**, 235–245.

Kiørboe, T., Lundsgaard, C., Olesen, M. and Hansen, J. L. S. (1994). Aggregation and sedimentation processes during a spring phytoplankton bloom: a field experiment to test coagulation theory. *Journal of Marine Research* **52**, 297–323.

Kirschbaum, J., Hummert, C. and Bernd, L. (1995). Determination of paralytic shellfish poisoning (PSP) toxins by application of ion-exchange HPLC, electrochemical oxidation, and mass detection. *In* "Harmful Marine Algal Blooms" (P. Lassus, G. Arzul, E. Erard, P. Gentien and C. Marcaillou, eds), pp. 309–314. Lavoisier/Intercept, Paris, France.

Kodama, M. (1990). Possible links between bacteria and toxin production in algal blooms. *In* "Toxic Marine Phytoplankton" (E. Granéli, B. Sundstrøm, L. Edler and D. M. Anderson, eds), pp. 52–61. Elsevier, New York, USA.

Kononen, K. (1992). "Dynamics of the Toxic Cyanobacterial Blooms in the Baltic Sea." Finnish Marine Research, No. 261. Helsinki, Finland.

Lam, C. W. Y. and Ho, K. C. (1989). Red tides in Tolo Harbour, Hong Kong. *In* "Red Tides: Biology, Environmental Science and Toxicology" (T. Okaichi, D. M. Anderson and T. Nemoto, eds), pp. 49–52. Elsevier, New York, USA.

Langlois, G. W., Kizer, K. W., Hansen, K. H., Howell, R. and Loscutoff, S. M. (1993). A note on domoic acid in California coastal molluscs and crabs. *Journal of Shellfish Research* **12**, 467–468.

le Fèvre, J. (1986). Aspects of the biology of frontal systems. *Advances in Marine Biology* **23**, 163–299.

Légendre, L. and Rassoulzadegan, F. (1995). Plankton and nutrient dynamics in marine waters. *Ophelia* **41**, 153–172.

Lenanton, R. C. J., Loneragan, N. R. and Potter, I. C. (1985). Blue-green algal blooms and the commercial fishery of a large Australian estuary. *Marine Pollution Bulletin* **16**, 477–482.

Lindahl, O. (1983). On the development of a *Gyrodinium aureolum* occurrence on the Swedish west coast in 1982. *Marine Biology* **77**, 143–150.

Lindahl, O. (1986). Offshore growth of *Gyrodinium aureolum* (Dinophyceae). The cause of coastal blooms in the Skagerrak area? *Sarsia* **71**, 27–33.

Lindahl, O. (1993). Hydrodynamical processes: a trigger and source for flagellate blooms along the Skagerrak coasts? *In* "Toxic Phytoplankton Blooms in the Sea" (T. J. Smayda and Y. Shimuzu, eds), vol. 3, pp. 775–782. Elsevier, Amsterdam, The Netherlands.

Longhurst, A. (1993). Seasonal cooling and warming in tropical oceans. *Deep-Sea Research I* **40**, 2145–2165.

Lundholm, N., Skov, J., Pocklington, R. and Moestrup, Ø. (1994). Domoic acid, the toxic amino acid responsible for amnesic shellfish poisoning, now in *Pseudonitzschia seriata* (Baciaellariophyceae) in Europe. *Phycologia* **33**, 475–478.

MacKenzie, L. (1991). Toxic and noxious phytoplankton in Big Glory Bay, Stewart Island, New Zealand. *Journal of Applied Phycology* **3**, 19–34.

Maestrini, S. Y. and Granéli, E. (1991). Environmental conditions and ecophysiological mechanisms which led to the 1988 *Chrysochromulina polylepis* bloom: an hypothesis. *Oceanologica Acta* **14**, 397–413.

Malin, G., Turner, S., Liss, P., Holligan, P. and Harbour, D. (1993). Dimethylsulphide and dimethylsulphoniopropionate in the northeast Atlantic during the summer coccolithophore bloom. *Deep Sea Research I* **40**, 1487–1508.

Mallin, M. A., Burkholder, J. M., Larsen, L. M. and Glasgow, H. B. Jr (1995). Response of two zooplankton grazers to an ichthyotoxic estuarine dinoflagellate. *Journal of Plankton Research* **17**, 351–363.

Maranda, L., Wang, R., Masuda, K. and Shimizu, Y. (1990). Investigation of the source of domoic acid in mussels. *In* "Toxic Marine Phytoplankton" (E. Granéli, B. Sundström, L. Edler and D. M Anderson, eds), pp. 300–304. Elsevier, New York, USA.

Martin, J. L., Haya, K. and Wildish, D. J. (1993). Distribution and domoic acid content of *Nitszchia pseudodelicatissima* in the Bay of Fundy. *In* "Toxic Phytoplankton Blooms in the Sea" (T. J. Smayda and Y. Shimuzu, eds), vol. 3, pp. 613–618. Elsevier, Amsterdam, The Netherlands.

Matrai, P. A. and Keller, M. D. (1993). Dimethylsulfide in a large-scale coccolithophore bloom in the Gulf of Maine. *Continental Shelf Research* **13**, 831–843.

Mead, A. D. (1898). *Peridinium* and the red water in Narragansett Bay. *Science, NY* **8**, 707–709.

Meldahl, A.-S., Ivenstuen, J., Grasbakken, G. J., Edvardsen, B. and Fonnum, F. (1995). Toxic activity of *Prymnesium* spp. and *Chrysochromulina* spp. tested by different test methods. *In* "Harmful Marine Algal Blooms" (P. Lassus, G. Arzul, E. Erard, P. Gentien and C. Marcaillou, eds), pp. 315–320. Lavoisier/Intercept, Paris, France.

Mendez, S. (1992). Update from Uruguay. *In* "Harmful Algae News" (T. Wyatt and Y. Pazos, eds), pp. 2, 5. UNESCO, Paris, France.

Michaels, A. F., Knap, A. H., Dow, R. L., Gunderson, K., Johnson, F. J., Sørensen, J., Close, A., Knauer, G. A., Lohrenz, S. E., Asper, V. A., Tuci, M. and Bidigare, R. (1994). Seasonal patterns of the ocean biogeochemistry at the US JGOFS Bermuda Atlantic Time-series Study site. *Deep-Sea Research I* **41**, 1013–1038.

Michel, C., Légendre, L., Therriault, J. C., Demers, S. and Vandevelde, T. (1993). Springtime coupling between ice algal and phytoplankton assemblages in southeastern Hudson Bay, Canadian Arctic. *Polar Biology* **13**, 441–449.

Mills, E. L. (1989). "Biological Oceanography. An Early History, 1870–1960." Cornell University Press, Ithaca, NY, USA.

Moestrup, Ø. (1994). Economic aspects: "blooms", nuisance species, and toxins. *In* "The Haptophyte Algae" (J. C. Green and B. S. C. Leadbeater, eds), Systematics Association Special Volume 51, pp. 265–285. Clarendon Press, Oxford, UK.

Morrison, J. A., Napier, I. R. and Gamble, J. C. (1991). Mass mortality of herring eggs associated with a sedimenting diatom bloom. *ICES Journal of Marine Science* **48**, 237–245.

Moshiri, G. A., Crompton, W. G. and Blaylock, D. A. (1978). Algal metabolites and fish kills in a bayou estuary, an alternative explanation to the low dissolved oxygen controversy. *Journal of Water Pollution Control Federation* **50**, 2043–2046.

Nagasaki, K., Ando, M., Itakura, S., Imai, I. and Ishida, Y. (1994). Viral mortality in the final stage of *Heterosigma akashiwo* (Raphidophyceae) red tide. *Journal of Plankton Research* **16**, 1595–1599.

Nelson, D. M. and Smith, W. O. (1986). Phytoplankton bloom dynamics of the western Ross Sea ice-edge. II. Mesoscale cycling of nitrogen and silicon. *Deep-Sea Research* **33**, 1389–1412.

Niebauer, H. J. (1991). Bio-physical oceanographic interactions at the edge of the Arctic ice pack. *Journal of Marine Systems* **2**, 209–232.

Nielsen, T. G., Kiørboe, T. and Bjørnsen, P. K. (1990). Effects of a *Chrysochromulina polylepis* subsurface bloom on the plankton community. *Marine Ecology Progress Series* **62**, 21–35.

Nielsen, T. G., Kaas, H. and Ravn, H. (1993a). Vertical migration of *Gyrodinium aureolum* in an artificial water column. *In* "Toxic Phytoplankton Blooms in the Sea" (T. J. Smayda and Y. Shimuzu, eds), vol. 3, pp. 789–794. Elsevier, Amsterdam, The Netherlands.

Nielsen, T. G., Løkkegaard, B., Richardson, K., Pedersen, F. B. and Hansen, L. (1993b). The structure of the plankton communities in the Dogger Bank area (North Sea) during a stratified situation. *Marine Ecology Progress Series* **95**, 115–131.

Nixon, S. W. (1989). An extraordinary red tide and fish kill in Narragansett Bay. *In* "Novel Phytoplankton Blooms" (E. M. Cosper, W. M. Bricelj and E. J. Carpenter, eds), pp. 429–447. Springer, Berlin, Germany.

Nixon, S. W. (1995). Coastal marine eutrophication: a definition, social causes, and future concerns. *Ophelia*, **41**, 199–220.

Noe-Nygaard, F., Surlyk, K. F. and Piasecki, S. (1987). Bivalve mass mortality caused by toxic dinoflagellate blooms in a Berrisian–Valanginian lagoon, Bornholm, Denmark. *Palaios* **2**, 263–273.

Nuzzi, R. and Waters, R. M. (1989). The spatial and temporal distribution of "brown tide" in eastern Long Island. *In* "Novel Phytoplankton Blooms" (E. M. Cosper, V. M. Bricelj and E. J. Carpenter, eds), pp. 117–137. Springer, Berlin, Germany.

380 K. RICHARDSON

Ogata, T., Ishimarua, T. and Kodoma, M. (1987). Effect of water temperature and light intensity on growth rate and toxicity change in *Protogonyaulax tamarenses*. *Marine Biology* **95**, 217–220.

Olesen, M. and Lundsgaard, C. (1992). Sedimentation of organic material from the photic zone in the Southern Kattegat. *Marine Research from the Danish Environmental Protection Agency* **10**, 167–183. (In Danish.)

Olesen, M. and Lundsgaard, C. (1995). Seasonal sedimentation of autochthonous material from the euphotic zone of a coastal system. *Estuarine and Coastal Shelf Science* **41**, 475–490.

Olsgard, F. (1993). Do toxic algal blooms affect subtidal soft-bottom communities? *Marine Ecology Progress Series* **102**, 269–286.

Olsson, P., Granéli, E., Carlsson, P. and Abreu, P. (1992). Structuring of a postspring phytoplankton community by manipulation of trophic interactions. *Journal of Experimental Marine Biology and Ecology* **158**, 249–266.

O'Shea, T. J., Rathbun, G. B., Bonde, R. K., Buergelt, C. D. and Odell, D. K. (1991). An epizootic of Florida manatees associated with a dinoflagellate bloom. *Marine Manual Science* **7**, 165–179.

Papathanassiou, E., Christaki, U., Christou, E. and Milona, A. (1994). *In situ* toxicity of dispersants. Controlled environmental pollution experiments. *Mediterranean Action Plan, Technical Report Series* **79**, 91–112.

Parsons, T. R., Takahashi, M. and Hargrave, B. (1984). "Biological Oceanographic Processes", 3rd edn. Pergamon Press, Oxford, UK, 330 pp.

Pauley, K. E., Seguel, M. R., Smith, J. C., McLachlan, J. L. and Worms, J. (1993). Occurrences of phycotoxins and related phytoplankton at winter temperatures in the southeastern Gulf of St Lawrence, Canada. *In* "Toxic Phytoplankton Blooms in the Sea" (T. J. Smayda and Y. Shimuzu, eds), vol. 3, pp. 311–316. Elsevier, Amsterdam, The Netherlands.

Pingree, R. D., Holligan, P. M., Mardell, G. T. and Harris, R. P. (1982). Vertical distribution of plankton in the Skagerrak in relation to doming of the seasonal thermocline. *Continental Shelf Research* **1**, 209–219.

Poulet, S. A., Ianora, A., Miralto, A. and Meijer, L. (1994). Do diatoms arrest embryonic development in copepods? *Marine Ecology Progress Series* **11**, 79–86.

Prakash, A. (1987). Coastal organic pollution as a contributing factor to red-tide development. *Rapports et Procés-verbaux des Réunions du Conseil Permanent International pour l'Exploration de la Mer* **187**, 61–65.

Premazzi, G. and Volterra, L. (1993). "Microphyte Toxins. A Manual for Toxin Detection, Environmental Monitoring and Therapies to Counteract Intoxications." Joint Research Centre, Commission of the European Communities.

Price, N. M. and Morél, F. M. (1991). Colimitation of phytoplankton growth by nickel and nitrogen. *Limnology and Oceanography* **36**, 1071–1077.

Qi, Y., Zhang, Z., Hong, Y., Lu, S., Zhu, C. and Li, Y. (1993). Occurrence of red tides on the coasts of China. *In* "Toxic Phytoplankton Blooms in the Sea" (T. J. Smayda and Y. Shimuzu, eds), vol. 3, pp. 43–46. Elsevier, Amsterdam, The Netherlands.

Rao, D. V. S. and Pan, Y. (1995). Apart from toxin production, are the toxigenic algae physiologically different? *In* "Harmful Marine Algal Blooms" (P. Lassus, G. Arzul, E. Erard, P. Gentien and C. Marcaillou, eds), pp. 675–680. Lavoisier/Intercept, Paris, France.

Rasmussen, J. and Richardson, K. (1989). Response of *Gonyaulax tamarenses* to the presence of a pycnocline in an artificial water column. *Journal of Plankton Research* **11**, 747–762.

Raven, J. A. and Richardson, K. (1986). Marine environments. *In* "Photosynthesis in Contrasting Environments" (N. R. Baker and S. P. Long, eds), pp. 337–398. Elsevier, Amsterdam, The Netherlands.

Reguera, B. and Oshimo, Y. (1990). Response of *Gymnodinium catenatum* to increasing levels of nitrate: growth patterns and toxicity. *In* "Toxic Marine Phytoplankton" (E. Graneli, B. Sundström, B. Edler and D. M. Anderson, eds), pp. 316–319. Elsevier, New York, USA.

Rensel, J. E. (1993). Severe blood hypoxia of Atlantic salmon (*Salmo salar*) exposed to the marine diatom *Chaetoceros concavicornis*. *In* "Toxic Phytoplankton Blooms in the Sea" (T. J. Smayda and Y. Shimuzu, eds), vol. 3, pp. 625–630. Elsevier, Amsterdam, The Netherlands.

Reynolds, C. S. (1980). Phytoplankton assemblages and their periodicity in stratifying lake systems. *Holarctic Ecology* **3**, 141–159.

Reynolds, C. S. (1984). Phytoplankton periodicity: the interactions of form, function and environmental variability. *Freshwater Biology* **14**, 111–142.

Rhodes, L. L., Haywood, A. J., Ballantine, W. J. and Mackenzie, A. L. (1993). Algal blooms and climate anomalies in north-east New Zealand, August–December 1992. *New Zealand Journal of Marine and Freshwater Research* **27**, 419–430.

Richardson, K. (1985). Plankton distribution and activity in the North Sea/Skagerak–Kattegat frontal area in April 1984. *Marine Ecology Progress Series* **26**, 233–244.

Richardson, K. (1991). Comparison of ^{14}C primary production determinations made by different laboratories. *Marine Ecology Progress Series* **72**, 189–201.

Richardson, K. (1996). Conclusion, research and eutrophication control. *In* "Eutrophication in a Coastal Marine Ecosystem" (B. Barker Jørgensen and K. Richardson, eds). American Geophysical Union, Washington, USA, in press.

Richardson, K. and Christoffersen, A. (1991). Seasonal distribution and production of phytoplankton in the southern Kattegat. *Marine Ecology Progress Series* **78**, 217–227.

Richardson, K. and Heilmann, J. P. (1995). Primary production in the Kattegat: past and present. *Ophelia* **41**, 317–328.

Richardson, K. and Kullenberg, G. (1987). Physical and biological interactions leading to plankton blooms. *Rapports et Procés-verbaux des Réunions du Conseil Permanent International pour l'Exploration de la Mer* **187**, 19–26.

Richardson, K., Beardall, J. and Raven, J. (1983). Adaptation of unicellular algae to irradiance: an analysis of strategies. *New Phytologist* **93**, 157–191.

Richardson, K., Lavín-Peregrina, M. F., Mitchelson, E. G. and Simpson, J. H. (1995). Seasonal distribution of chlorophyll *a* in relation to physical structure in the western Irish Sea. *Oceanologica Acta* **8**, 77–86.

Riegman, R., Malschaert, H. and Colijn, F. (1990). Primary production at a frontal zone located at the northern slope of the Dogger Bank (North Sea). *Marine Biology* **105**, 329–336.

Riley, C. M., Holt, S. A., Holt, G. J., Buskey, E. J. and Arnold, C. R. (1989). Mortality of larval red drum (*Sciaenops ocellatus*) associated with a *Ptychodiscus brevis* red tide. *Contributions in Marine Science, University of Texas* **31**, 137–146.

Robertson, J. E., Robinson, C., Turner, D. R., Holligan, P., Watson, A. J., Boyd, P., Fernandez, E. and Finch, M. (1994). The impact of a coccolithophore bloom on oceanic carbon uptake in the northeast Atlantic during summer 1991. *Deep-Sea Research I* **42**, 297–314.

Robineau, B., Gagné, J. A., Fortier, L. and Cembella, A. D. (1991). Potential impact of a toxic dinoflagellate (*Alexandrium excavatum*) bloom on survival of fish and crustacean larvae. *Marine Biology* **108**, 293–301.

Ross, A. H., Gurney, W. S. C. and Heath, M. R. (1993). Ecosystem models of Scottish sea lochs for assessing the impact of nutrient enrichment. *ICES Journal of Marine Science*, **50**, 359–367.

Sakshaug, E. (1989). The physiological ecology of polar phytoplankton. *In* "Marine Living Systems of the Far North" (L. Rey and V. Alexander, eds), pp. 61–89. E. J. Brill, Leiden, The Netherlands.

Scholin, C. A. and Anderson, D. M. (1991). Population analysis of toxic and nontoxic *Alexandrium* species using ribosomal RNA signature sequences. *In* "Toxic Phytoplankton Blooms in the Sea" (T. J. Smayda and Y. Shimuzu, eds), vol. 3, pp. 95–102. Elsevier, Amsterdam, The Netherlands.

Shimuzu, Y., Watanabe, N. and Wrensford, G. (1995). Biosynthesis of brevetoxins and heterotrophic metabolism in *Gymnodinium breve*. *In* "Harmful Marine Algal Blooms" (P. Lassus, G. Arzul, E. Erard, P. Gentien and C. Marcaillou, eds), pp. 351–357. Lavoisier/Intercept, Paris, France.

Shumway, S. E. (1992). A review of the effects of algal blooms on shellfish and aquaculture. *Journal of Shellfish Research* **11**, 556.

Smayda, T. J. (1989). Primary production and the global epidemic of phytoplankton blooms in the sea. *In* "Novel Phytoplankton Blooms" (E. M. Cosper, V. M. Bricelj and E. J. Carpenter, eds), pp. 449–483. Springer, Berlin, Germany.

Smayda, T. (1990). Novel and nuisance phytoplankton blooms in the sea: evidence for a global epidemic. *In* "Toxic Marine Phytoplankton" (E. Granéli, B. Sundstrøm, L. Edler and D. M. Anderson, eds), pp. 29–40. Elsevier, New York, USA.

Smayda, T. J. and Fofonoff, P. (1989). An extraordinary noxious "brown tide" in Narragansett Bay. II. Inimical effects. *In* "Red Tides: Biology, Environmental Science and Toxicology" (T. Okaichi, D. M. Anderson and T. Nemoto, eds), pp. 131–134. Elsevier, New York, USA.

Smayda, T. J. and Villareal, T. A. (1989a). An extraordinary, noxious "brown tide" in Narragansett Bay. I. The organism and its dynamics. *In* "Red Tides: Biology, Environmental Science and Toxicology" (T. Okaichi, D. M. Anderson and T. Nemoto, eds), pp. 127–130. Elsevier, New York, USA.

Smayda, T. J. and Villareal, T. A. (1989b). The 1985 "brown tide" and the open phytoplankton niche in Narragansett Bay during summer. *In* "Novel Phytoplankton Blooms" (E. M. Cosper, V. M. Bricelj and E. J. Carpenter, eds), pp. 159–187. Springer, Berlin, Germany.

Smetacek, V. S. (1985). Role of sinking in diatom life-history cycles: ecological, evolutionary and geological significance. *Marine Biology* **84**, 239–251.

Smetacek, V. S., Scharek, R. and Nöthig, E. M. (1990). Seasonal and regional variation in the pelagial and its relationship to the life history cycle of conservation. *In* "Antarctic Ecosystems" (K. R. Kerry and G. Hempel, eds), pp. 103–114. Springer, Berlin, Germany.

Smetacek, V. S., Scharek, R., Gordon, L. I., Eicken, H., Fahrbach, E., Rohardt, G. and Moore, S. (1992). Early spring phytoplankton blooms in ice platelet layers of the southern Weddell Sea, Antarctica. *Deep-Sea Research* **39**, 153–168.

Smith, P., Chang, F. Hoe and MacKenzie, L. (1993). Toxic phytoplankton and algal blooms, summer 1992/93. *In* "Marine Toxins and New Zealand Shellfish" (J. A. Jasperse, ed.), vol. 24, pp. 11–17. Royal Society of New Zealand, Wellington, New Zealand.

Smith, W. O. Jr (1993). Nitrogen uptake and new production in the Greenland Sea: the spring *Phaeocystis* bloom. *Journal of Geophysical Research* **98**, 4681–4688.

Sournia, A. (1995). Red-tide and toxic marine phytoplankton of the world ocean: an enquiry into biodiversity. *In* "Harmful Marine Algal Blooms" (P. Lassus, G. Arzul, E. Erard, P. Gentien and C. Marcaillou, eds), pp. 103–112. Lavoisier/Intercept, Paris, France.

Sousa-Silva, E. (1990). Intracellular bacteria: the origin of dinoflagellate toxicity. *Journal of Environmental Pathology, Toxicology and Oncology* **10**, 124–128.

Steemann Nielsen, E. (1952). The use of radio-active carbon (C^{14}) for measuring organic production in the sea. *Journal du Conseil Permanent International pour l'Exploration de la Mer* **18**, 117–140.

Steemann Nielsen, E. (1964). Investigations of the rate of primary production at two Danish lightships in the transition area between the North Sea and the Baltic. *Meddelelser fra Danmarks Fiskeri- og Havundersøgelser* **4**, 31–77.

Steidinger, K. A., Burkholder, J. M., Glasgow, H. B. Jr, Truby, E. W., Garrett, J. K., Noga, E. K. and Smith, S. A. (1996). *Pfiesteria piscicida*, a new toxic dinoflagellate genus and species of the order Dinamoebales. *Journal of Phycology* **32**, 157–164.

Steimle, F. W. and Sindermann, C. J. (1978). Review of oxygen depletion and associated mass mortalities of shellfish in the Middle Atlantic Bight in 1976. *Marine Fisheries Review* **40**, 17–26.

Suganuma, M., Fujiki, H. and Suguri, H. (1988). Okadaic acid: an additional non-phorboltetradecanoate-13-acetata-type tumor promoter. *Proceedings of the National Academy of Sciences USA* **85**, 1768–1771.

Sunda, W. G. (1989). Trace metal interactions with marine phytoplankton. *Biological Oceanography* **6**, 411–442.

Sverdrup, H. U. (1953). On conditions for the vernal blooming of phytoplankton. *Journal du Conseil Permanent International pour l'Exploration de la Mer* **18**, 287–295.

Tangen, K. (1977). Blooms of *Gyrodinium aureolum* (Dinophyceae) in north European waters, accompanied by mortality in marine organisms. *Sarsia* **63**, 123–133.

Taylor, F. J. R., Taylor, N. J. and Walsby, J. R. (1985). A bloom of the planktonic diatom, *Cerataulina pelagica*, off the coast of northwestern New Zealand in 1983, and its contribution to an associated mortality of fish and benthic fauna. *International Revue der Gesamten Hydrobiologie* **70**, 773–795.

Taylor, F. J. R., Haigh, R. and Sutherland, T. F. (1994). Phytoplankton ecology of Sechelt Inlet, a fjord system on the British Columbia coast. 2. Potentially harmful species. *Marine Ecology Progress Series* **103**, 151–164.

Tester, P. A. and Mahoney, B. (1995). Implication of the diatom, *Chaetoceros convolutus*, in the death of Red King Crabs, *Paralithodes camtschatica*, Captains Bay, Unalaska Island, Alaska. *In* "Harmful Marine Algal Blooms" (P. Lassus, G. Arzul, E. Erard, P. Gentien and C. Marcaillou, eds), pp. 95–100. Lavoisier/Intercept, Paris, France.

Thompson, A. M., Esaias, W. E. and Iverson, R. L. (1990). Two approaches to determining the sea-to-air flux of dimethyl sulfide: satellite ocean color and a photochemical model with atmospheric measurements. *Journal of Geophysical Research* **95**, 20 551–20 558.

Thomsen, H. A. (ed.) (1992). Plankton in the inner Danish waters. *Havforskning fra Miljøstyrelsen*, No. 11, 331 pp. (In Danish.)

Todd, E. C. D., Kuiper-Goodman, T., Watson-Wright, W., Gilgan, M. W., Stephen, S., Marr, J., Pleasance, S., Quilliam, M. A., Klix, H., Luu, H. A. and Holmes, C. F. B. (1993). Recent illnesses from seafood toxins in Canada: paralytic, amnesic and diarrhetic shellfish poisoning. *In* "Toxic Phytoplankton Blooms in the Sea" (T. J. Smayda and Y. Shimuzu, eds), vol. 3, pp. 335–340. Elsevier, Amsterdam, The Netherlands.

Townsend, D. W., Keller, M. D., Sieracki, M. E. and Ackleson, S. G. (1992). Spring phytoplankton blooms in the absence of vertical water column stratification. *Nature, London* **360**, 59–62.

Townsend, D. W., Cammen, L. M., Holligan, P. M., Campbell, D. E. and Pettigrew, N. R. (1994). Causes and consequences of variability in the timing of spring phytoplankton blooms. *Deep-Sea Research I* **41**, 747–766.

Tracey, G. A. (1988). Feeding reduction, reproductive failure and mass mortality of mussels (*Mytilus edulis*) during the 1985 "brown tide" in Narragansett Bay, Rhode Island. *Marine Ecology Progress Series* **50**, 73–81.

Trainer, V. L., Edwards, R. A., Szmant, A. M., Stuart, A. M., Mende, T. J. and Baden, D. G. (1990). Brevetoxins: unique activators of voltage-sensitive sodium channels. *In* "Marine Toxins, Origins, Structure and Molecular Pharmacology" (S. Hall and G. Strichartz, eds), vol. 418, pp. 168–175. Woods Hole, MA, USA.

Turriff, N., Runge, J. A. and Cembella, A. D. (1995). Toxin accumulation and feeding behaviour of the planktonic copepod *Calanus finmarchicus* exposed to the red-tide dinoflagellate *Alexandrium excavatum*. *Marine Biology* **123**, 55–64.

van der Wal, P., Kempers, R. S. and Weldhuis, M. J. W. (1995). Production and downward flux of organic matter and calcite in a North Sea bloom of the coccolithophore *Emiliania huxleyi*. *Marine Ecology Progress Series* **95** (D12), 20, 551, 558.

Villac, M. C., Roelke, D. L., Chavez, F. P., Cifuentes, L. A. and Fryxell, G. A. (1993). *Pseudonitzschia australis* Frenguelli and related species from the west coast of the USA: occurrence and domoic acid production. *Journal of Shellfish Research* **12**, 457–465.

Waite, A. M., Thompson, P. A. and Harrison, P. J. (1992a). Does energy control the sinking rates of marine diatoms? *Limnology and Oceanography*, **37**, 468–477.

Waite, A., Bienfang, P. K. and Harrison, P. J. (1992b). Spring bloom sedimentation in a subarctic ecosystem. 2. *Journal of Marine Biology* **114**, 131–138.

Wassmann, P., Vernet, M., Mitchell, G. and Rey, F. (1990). Mass sedimentation of *Phaeocystis pouchetii* in the Barents Sea. *Marine Ecology Progress Series* **66**, 183–195.

Wassmann, P., Peinert, R. and Smetacek, V. (1991). Patterns of production and sedimentation in the boreal and polar northeast Atlantic. *Polar Research* **10**, 209–228.

White, A. W. (1977). Dinoflagellate toxins as probable cause of an Atlantic herring (*Clupea harengus harengus*) kill, and pteropods as apparent vector. *Journal of the Fisheries Research Board of Canada* **34**, 2421–2424.

WHO (1984). "Aquatic (Marine and Freshwater) Biotoxins." Environmental Health Criteria, No. 37. World Health Organisation, Geneva, Switzerland, 95 pp.

Work, T. M., Beale, A. M., Fritz, L., Quilliam, M. A., Silver, M., Buck, K. and Wright, J. L. C. (1993). Domoic acid in toxication of brown pelicans and cormorants in Santa Cruz, California. *In* "Toxic Phytoplankton Blooms in the Sea" (T. J. Smayda and Y. Shimuzu, eds), vol. 3, pp. 643–649. Elsevier, Amsterdam, The Netherlands.

Wyatt, T. (1980). Morrell's seals. *Journal du Conseil Permanent International pour l'Exploration de la Mer* **39**, 1–6.

Wyatt, T. (1995). Global spreading, time series models and monitoring. *In* "Harmful Marine Algal Blooms" (P. Lassus, G. Arzul, E. Erard, P. Gentien and C. Marcaillou, eds), pp. 755–764. Lavoisier/Intercept, Paris, France.

Yang, C. Z. and Albright, L. J. (1994). Oxygen radical mediated effects of the toxic phytoplankter, *Heterosigma akashiwo* on juvenile salmonids. *In* "International Symposium on Aquatic Animal Health", University of California, Davis, CA, USA, p. W15.3.

Yasumoto, T., Oshima, Y. and Sugaware, W. (1978). Identification of *Dinophysis fortii* as the causative organism of diarrhetic shellfish poisoning. *Bulletin of the Japanese Society for Scientific Fisheries* **46**, 1405–1411.

Yasumoto, T., Underdal, B., Aune, T., Hormazabal, V., Skulberg, O. M. and Oshima, Y. (1990). Screening for hemolytic and ichthyotoxic components of *Chrysochromulina polylepis* and *Gyrodinium aureolum* from Norwegian waters. *In* "Toxic Marine Phytoplankton" (E. Granéli, B. Sundstrøm, L. Edler and D. M. Anderson, eds), pp. 436–440. Elsevier, New York, USA.

Yentsch, C. M., Holligan, P. M., Balch, W. M. and Tvirbutas, A. (1986). Tidal stirring vs stratification: microalgal dynamics with special reference to cyst-forming, toxin producing dinoflagellates. *In* "Lecture Notes on Coastal and Estuarine Studies", vol. 17: "Tidal Mixing and Plankton Dynamics" (M. Bowman, C. M. Yentsch and W. T. Peterson, eds), pp. 224–252. Springer, Berlin, Germany.

Taxonomic Index

Note: **Page references in** *italics* refer to Figures; those in **bold** refer to Tables

Subject Index

Note: **Page references in** *italics* refer to Figures; those in **bold** refer to Tables

Cumulative Index of Titles

Cumulative Index of Authors